The X-ray background

This book presents a review of the current observational knowledge and understanding of the cosmic X-ray background, discovered 30 years ago. The most relevant observational features of the cosmic X-ray background, its spectrum, high galactic latitude isotropy on all angular scales and its source content, are reviewed in detail. The contribution of the Ginga, Rosat and BBXRT space missions, play a major role in the discussions of the most recent estimates of the contribution to the X-ray intensity of different classes of sources, mostly Active Galactic Nuclei and Clusters of Galaxies. The fraction of resolved background intensity approaches 50 per cent at soft energies, where QSOs dominate. At higher photon energies, where most of the energy density of the background resides, the fraction directly resolved is smaller. Models for the residual intensity are discussed in the book, with particular emphasis on spectral as well as isotropy constraints. The necessity of future space missions where the X-ray sky can be mapped with good angular resolution at energies from 0.3 to 10 keV is highlighted. The material is presented principally in the form of review chapters by experts, and the book is therefore an ideal introduction to the cosmic X-ray background.

The X-ray Background

Edited by

X. Barcons
Departmento de Física Moderna, Universidad de Cantabria, Spain

A. C. Fabian
Institute of Astronomy, Cambridge, UK

CAMBRIDGE
UNIVERSITY PRESS

Published by the Press Syndicate of the University of Cambridge
The Pitt Building, Trumpington Street, Cambridge CB2 1RP
40 West 20th Street, New York, NY 10011–4211, USA
10 Stamford Road, Oakleigh, Victoria 3166, Australia

First published 1992

Printed in Great Britain at the University Press, Cambridge

British Library cataloguing in publication data available

Library of Congress cataloguing in publication data available

ISBN 0 521 41651 5 hardback

Contents

Invited reviews are listed in **bold**

CAMB 8.13.92

CHAPTER IV: THE CONTRIBUTION OF KNOWN CLASSES OF SOURCES

CHAPTER V: MODELS FOR SOURCES OF THE X–RAY BACKGROUND

CHAPTER VI: RECENT OBSERVATIONAL RESULTS

CHAPTER VII: FUTURE MISSIONS

CHAPTER VIII: SUMMARY

Preface

Nearly thirty years have passed since the discovery of cosmic X-ray sources and the first extragalactic background radiation, the X-ray Background (XRB), by Riccardo Giacconi and his colleagues. In that time, X-ray Astronomy has matured into a fully-developed area of astronomy in which all classes of cosmic objects have been detected and in which studies of both galactic and extragalactic objects have become (almost) routine. The XRB, although it is present in every X-ray image and scan, remains a puzzle. Its observed properties distinguish it from every known class of source, except in restricted energy bands such as those below a few keV, or above 100 keV. The origin of the bulk of the XRB is still unknown.

Observationally there has been much progress in our knowledge of the XRB. Its spectrum has been accurately measured, especially in the 3 – 60 keV band, and it has been shown to be isotropic at high galactic latitudes to limits of a few per cent or less. A diffuse origin in a hot intergalactic medium, one of the first suggested sources, has been ruled out by the observed lack of any Compton distortion in the spectrum of the microwave background.

Many extragalactic sources have been discovered and their contributions to the XRB estimated. In the 3 – 60 keV band, where most of the energy density in the XRB resides, these contributions are small and of order ten per cent or so. The spectra of the sources also leads to what has been dubbed by Elihu Boldt as the 'spectral paradox'; the spectra of the observed sources is steep whereas the spectrum of the XRB is flat, so that when the source contribution is subtracted the residual XRB is yet flatter. No known class of source has the correct flat spectral shape. There is a suspicion that many absorbed (and/or 'reflected') sources might sum to give the observed shape but some form of spectral evolution is definitely required. It is the first time that a class of object has been predicted on the basis of the integrated background spectrum rather than the reverse. Studies of the XRB should define the X-ray evolution of these and all X-ray active objects, and describe the smoothness or otherwise of their distribution through the Universe.

The observational understanding of the XRB has benefitted from the technological advances that have caused the enormous progress of X-ray Astronomy. The use of orbiting X-ray telescopes, from the *Einstein Observatory* that was the reference point for the last decade to *Rosat* which was launched in 1990, has meant that the soft X-ray Sky (below photon energies of about 3 keV) has been mapped in much detail. In particular it has meant that a significant excess soft XRB has been resolved into quasars. At the present time the spectra of individual quasars suggest that the quasar contribution is small at higher photon energies and that quasar contribute little to the 3 – 60 keV XRB, or even to its extrapolation to lower energies. At the higher photon energies above 3 keV, traditional collimated proportional counters and scintillator detectors have been responsible for the results, with *HEAO-1* and more recently *Ginga* having played important rôles in XRB studies. The observational situation is now changing with the launch of imaging X-ray telescopes sensitive in this band, with the BBXRT mission providing a taste of what should emerge from *ASTRO-D* next year.

Studies and discussion of the XRB have become an increasingly important part of the work of X-ray Astronomy. This is easily seen by the contributions to most meetings on the subject. Given the breadth of interest and the recent and forthcoming advances in the field, we organized and held the first Workshop solely devoted to the X-ray Background in Laredo, Spain, in September 1990.

The invited reviews and contributed papers from the Laredo Workshop are collected in this Volume. The papers have been organized into Chapters covering different aspects of the XRB. The introductory talk given by R. Burg on behalf of R. Giacconi deserves a Chapter of its own. Chapter II deals with the spectrum of the XRB in different energy bands and contains the reviews by D. McCammon (soft energies), S. Holt (intermediate energies) and D. Gruber (hard energies).

Chapter III is devoted to the isotropy of the XRB. Here R. Mushotzky reviews the large-scale features and L. Danese concentrates on the small-scale results. The contributions of known classes of source, surveys and source counts are presented in Chapter IV. E. Boldt discusses the local and high redshift volume emissivity, T. Hamilton presents a general view of the deepest X-ray surveys, and D. Schwartz reviews the AGN contribution to the XRB. Several models for the residual XRB and the physics of its sources are discussed in Chapter V under the introduction by G. Setti. An account of the most relevant recent results is presented in Chapter VI: G. Stewart reviews the *Ginga* results, G. Hasinger the ones from *Rosat* and K. Jahoda the preliminary information from *BBXRT*. Chapter VII is a compendium of contributions on several space projects which will certainly have an impact in the study of the XRB and we end with the Summary of the Workshop by A.C. Fabian (Chapter VIII).

We thank all the participants for their enthusiasm in coming to Laredo, and delivering their talks. Special thanks are due to the invited speakers who made every effort to present comprehensive reviews of the subjects they were asked and for writing such excellent papers in this book. To them, contributors and participants many thanks for keeping the sessions so alive and enjoyable.

The organization of this Workshop would not have been possible without the help from our colleagues in the Organising Commitee Francisco J. Carrera and Luigi Toffolatti to whom we are especially grateful. The inclusion of this Workshop in the Summer University scheme of the Universidad de Cantabria in the marvellous villa of Laredo was most appropriate, since the Workshop was provided with many facilities otherwise unavailable. We are grateful to the Director of the Summer University of Laredo, Prof. J. Martínez Maurica, for his help, encouragement and support and we thank our sponsors: the Research Vice–director office of the Universidad de Cantabria, the Department of Modern Physics of the Universidad de Cantabria, the "Consejería de Transportes, Turismo, Comunicaciones e Industria" of the Governement of Cantabria, The British Council and Cambridge University Press.

X. Barcons and A.C. Fabian
February 1992

List of participants

X. Barcons   Universidad de Cantabria, Spain

A. Blanchard   Observatoire de Meudon, France

E. Boldt   Goddard Space Flight Center, USA

R. Burg   Space Telescope Science Institute, USA

D.N. Burrows   Penn. State University, USA

J.A. Butcher   Leicester University, UK

F.J. Carrera   Universidad de Cantabria, Spain

L.W. Chen   Institute of Astronomy, Cambridge, UK

M.T. Ceballos   Universidad de Cantabria, Spain

R.A. Daly   Princeton University, USA

L. Danese   Osservatorio Astronomico di Padova, Italy

A.C. Fabian   Institute of Astronomy, Cambridge, UK

G.B. Field   Smithsonian Astrophysical Observatory, USA

A. Giménez   Laboratorio de Astronomía Espacial y Física Fundamental, Spain

R.E. Griffiths   Space Telescope Science Institute, USA

D.E. Gruber   University of California San Diego, USA

T. Hamilton   Columbia University, USA

G. Hasinger   Max–Planck–Inst. für Extrat. Physik, Germany

K. Hayashida   Osaka University, Japan

S.S. Holt   Goddard Space Flight Center, USA

H. Inoue   Institute of Space and Astronautical Science, Japan

K. Jahoda   Goddard Space Flight Center, USA

O. Lahav   Institute of Astronomy, Cambridge, UK

D. MacCammon   University of Wisconsin, USA

J.M. Martín-Mirones   Universidad de Cantabria, Spain

M. Mouchet   Observatoire de Meudon, France

R.F. Mushotzky   Goddard Space Flight Center, USA

R.D. Rogers   Smithsonian Astrophysical Observatory, USA

R.E. Rothschild   University of California San Diego, USA

I.E. de Salamanca   Observatoire de Meudon, France

W.T. Sanders   University of Wisconsin, USA

R. Schaeffer   Cen–Saclay, France

M. Schmidt   California Institute of Technology, USA

D.A. Schwartz   Smithsonian Astrophysical Observatory, USA

G. Setti   Istituto di Radioastronomia, Bologna, Italy

S.L. Snowden   Max–Planck–Inst. für Extrat. Physik, Germany

G.C. Stewart   Leicester University, UK

L. Toffolatti   Universidad de Cantabria, Spain

M.A. Treyer   Observatoire de Meudon, France

R.S. Warwick   Leicester University, UK

D.M. Worrall   Smithsonian Astrophysical Observatory, USA

Chapter I

Introduction

The X-ray Background: Past, Present and Future

Riccardo Giacconi and Richard Burg

Space Telescope Science Institute and The Johns Hopkins University

1. INTRODUCTION

The existence of an isotropic X-ray background has been established since the discovery flight of 1962 (Giacconi et al. 1962), Figure 1. The intensity of 1.7 photons cm^{-2} s^{-1} sr^{-1} (in the 2 to 10 keV range) is consistent with current measurements. Due to the high degree of isotropy it soon became evident that the radiation is mainly extragalactic in origin and that its study could furnish data of cosmological interest. Immediately after this discovery, Hoyle (1963) pointed out that the observed X-ray flux was inconsistent by two orders of magnitude from that predicted by the hot steady state cosmological model of Hoyle and Gold.

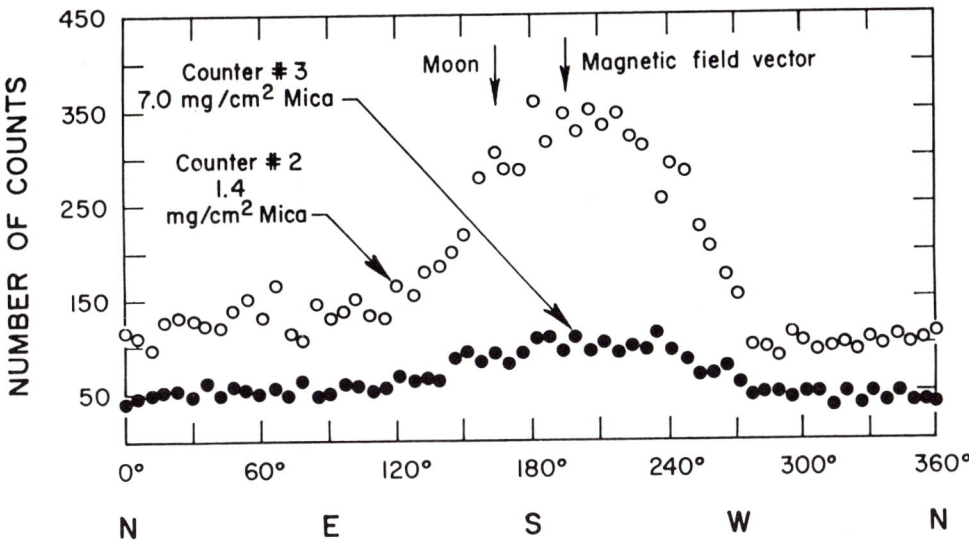

Figure 1: Data from the rocket flight in which the XRB was discovered. (Giacconi and Gursky, 1974)

Over the nearly thirty years that have passed, a great deal of effort has been expended, both experimentally and theoretically, to explain the nature of this background. Two main points of view have guided this work: the point source hypothesis and the diffuse hypothesis. On the one hand, the paradigm that the background is due to the superposition of discrete unresolved sources can be adopted. The apparent uniformity of the background would result from the superposition of very large numbers of them. The study of the X-ray background in this view would lead naturally to the following:

a)   Higher angular resolution deep surveys of blank fields to detect the individual discrete sources and optical follow ups to identify them.

b)   Study of the luminosity, density and spectral evolution of the detected sources.

On the other hand, if one adopted the working hypothesis that the background was truly diffuse then one would be led naturally to:

c)   Attempt experimentally, as accurate a measurement of the spectrum as possible in order to reveal the physical production process.

d)   Theoretically, one would endeavor to explain the source of energy and its cosmological evolution. Most diffuse theories adopted a thermal bremmstrahlung origin for the background. The existence of an X-ray background would then be directly related to the baryonic mass in the universe and the original spectrum of the density fluctuations.

This working hypothesis never had, in our opinion, a very strong observational basis, but as we will discuss later, was always extremely attractive from a theoretical point of view. By analogy to the explanation of the microwave background, if one could devise a unique process which could give rise to the X-ray background, then this model would have profound cosmological implications.

The two working hypothesis led various groups of scientists to design very different sets of experiments with their own internal logic and progression.

2. EVIDENCE FOR THE POINT SOURCE HYPOTHESIS

After the initial discovery, the most interesting new information on the X-ray background was provided by the Uhuru mission launched in December 1970. In that mission the background was measured in the energy range between 1 and 10 keV. It was found to be isotropic to better then 3% on angular scales of 10 degrees (Figure 2). This provided strong confirmation of the extragalactic nature of the background and it was shown by Schwartz and Gursky (1974) that under reasonable assumptions, at least 20% of it should originate at distances $z > 1$.

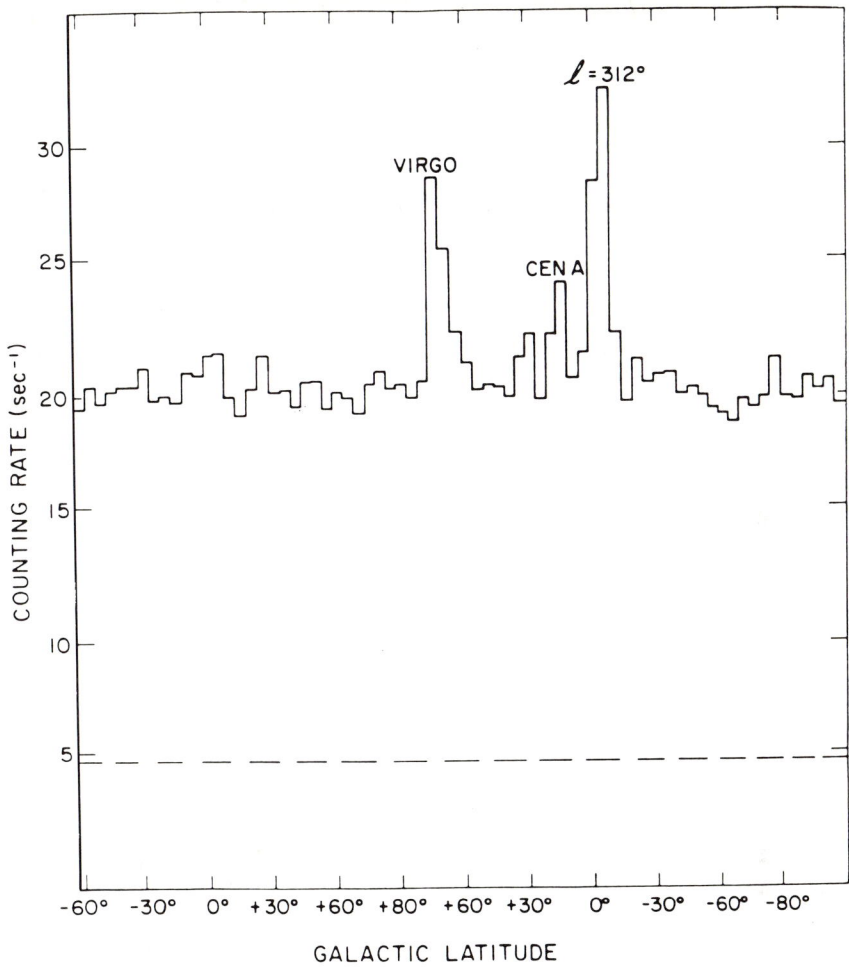

Figure 2: UHURU scan. (Giacconi and Gursky 1974)

A very important finding in the Uhuru mission, which did not attract sufficient notice at the time, is shown in Figure 3. Using the Uhuru catalogue Matilsky (Matilsky et al. 1973) attempted the first plots of number of sources versus limiting flux. He plotted separately the data for high latitude ($|b| > 20°$) and low latitude sources ($|b| < 20°$) and showed that the number count of high latitude sources was consistent with a 3/2 power law. This was interpreted as the result of a luminosity function independent of distance, in which case sources much nearer than $z = 0.1$ would dominate the counts and appear to follow an Euclidian space distribution. We now know that this is not the case and that the apparent Euclidian slope is due to the fortuitous combination of cosmological and evolutionary effects.

This finding together with the direct detection of several classes of extragalactic sources including Normal Galaxies, Giant Radio Galaxies, Seyferts, Rich Clusters of Galaxies and QSO's made it clear

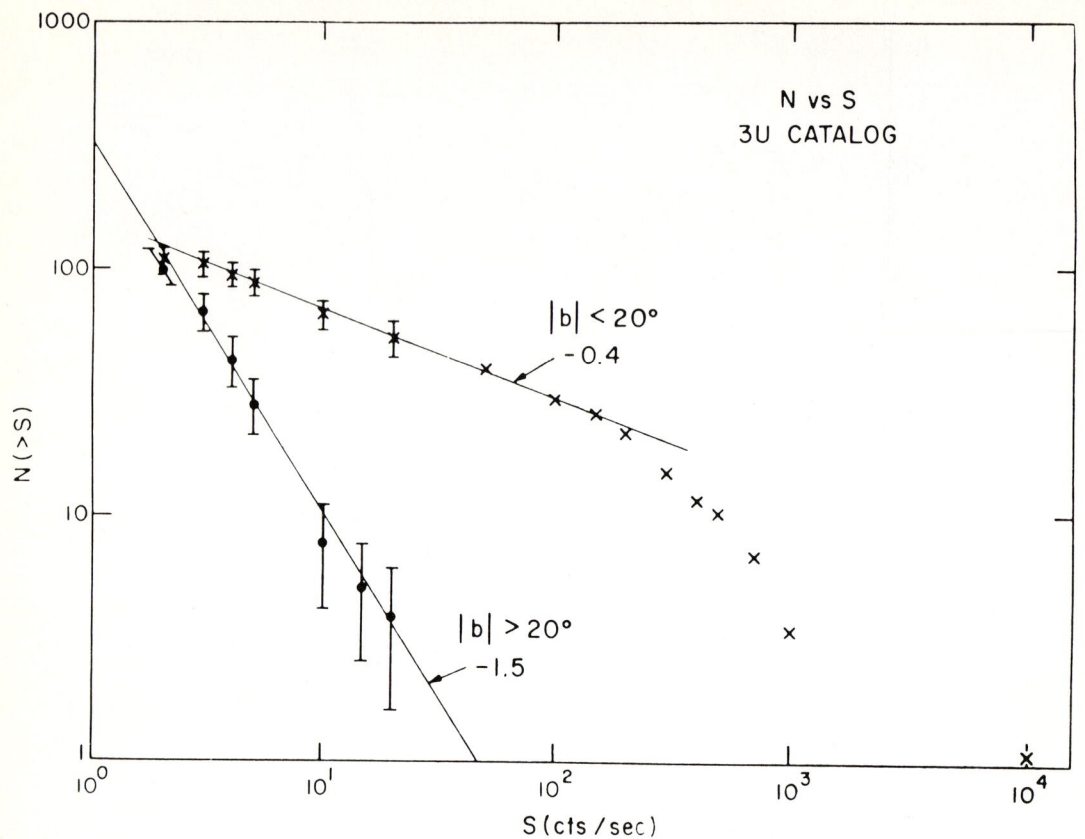

Figure 3: UHURU Log N-Log S. (Giacconi and Gursky, 1974)

that at least a fraction (then estimated at >22%) and possibly all of the background could be due to individual sources. Setti and Woltjer (1970) had in fact pointed out as early as 1970 that if the X-ray luminosity of 3C273 (the only QSO then known) was considered typical of all QSO's then the integrated contribution of all QSO's by itself could explain the entire X-ray background.

Following these initial findings it became clear that to further study the individual sources and their evolution one needed observations at higher sensitivity and angular resolution. This last point is particularly important to avoid source confusion and to permit optical identifications at the faint end. The ultimate sensitivity of the Einstein mission was originally set so that 100% of the background would be resolved if Log N - Log S could be extrapolated with slope -3/2 to fluxes as faint as 3×10^{-15} erg cm^{-2} s^{-1} in the (.3 to 3.5 keV range) (Figure 4). The only available technology for reaching such faint fluxes was the use of grazing incidence optics X-ray telescopes. These telescopes could provide the required angular resolution and field of view, at a sensitivity some three orders of magnitudes below "Uhuru". However, they operated in a range of energy (.3 to 3.5 keV) at the lower end, and below that, of conventional proportional counter experiments as flown on Uhuru, Ariel and HEAO I.

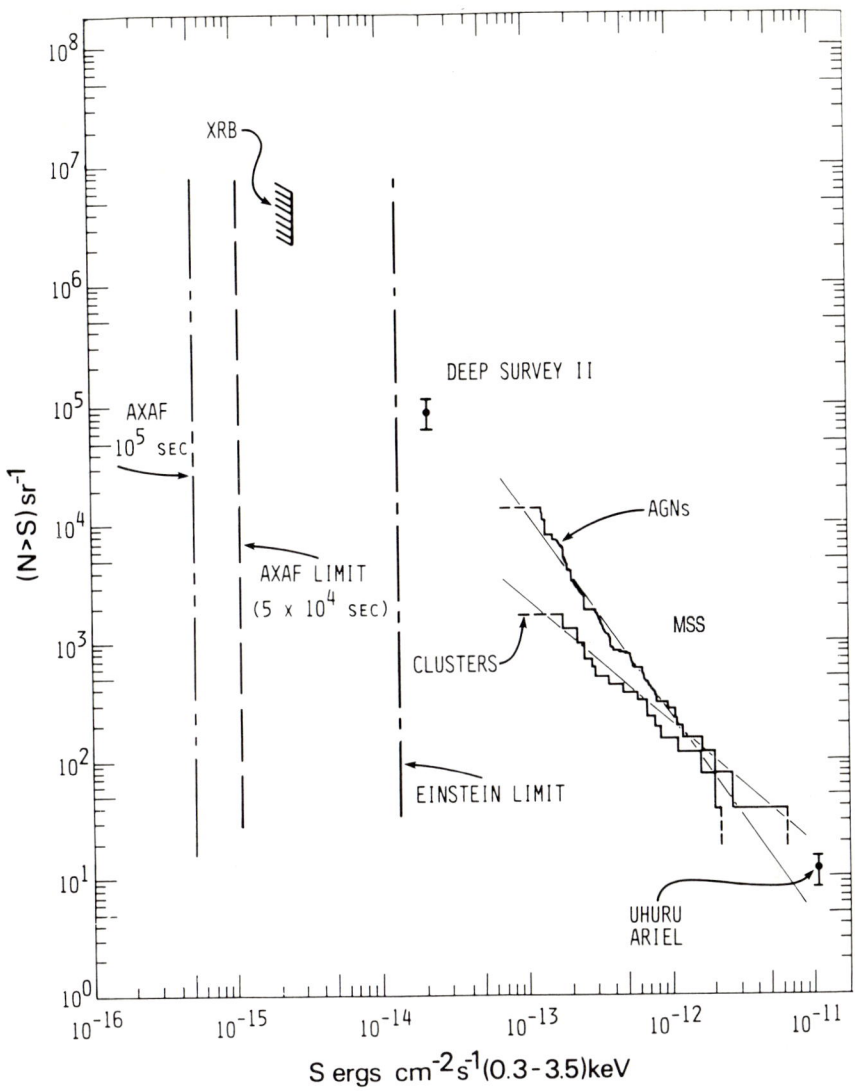

Figure 4: Einstein Log N-Log S.

If one adopts the point of view that the background is made of point sources and that their optical identification and study are the most important objectives, then it makes little difference in which range of X-ray energy one observes them provided one can achieve the required sensitivity. In fact, it is useful to keep in mind that the universe is essentially transparent to X-rays of about 1 keV energy (optical depth is one at z = 7 for $\Omega = 1$). Most spectra of high energy sources show a rapid decrease in number of photons as a function of energy. Highest sensitivity and statistical significance for detection is therefore obtained for most sources at the lowest energy for which the source spectrum shows no self absorption. Most extragalactic sources including clusters of galaxies and AGN, show either flat (thermal bremmstrahlung) or rising (power laws) spectra at low energies. However the issue of a different range of energy could be raised by proponents of the diffused hypothesis to suggest that lower energy detections were irrelevant to the study of the "hard" background.

Notwithstanding this concern, the Einstein medium and deep surveys (Giacconi et al 1979, Maccacaro et al 1982) succeeded in demonstrating that at least 25% of the background is due to discrete sources to a flux limit of 3×10^{-14} erg s^{-1} cm^{-2}, that the bulk of the sources at these flux limits are AGN's (Figure. 4) and that their properties are similar to those found in brighter surveys (Griffiths et al. 1983, Primini et al. 1990). In the Pavo survey with the high resolution detector, Griffiths detected 16 sources in .5 square degree of which 12 are identified as AGN's of z = .8 and $(m_R) = 20.3$

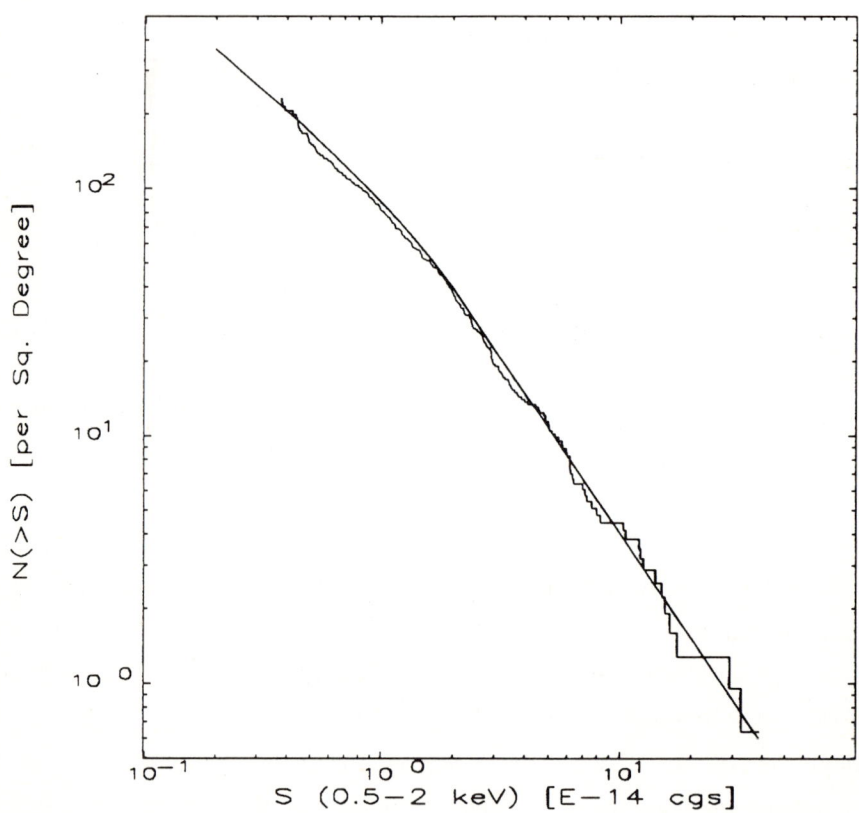

Figure 5: ROSAT Deep Survey Lockman field plus RMSS (Hasinger, this proceedings)

The Deep Survey conducted with ROSAT, for which we are now obtaining the first results, has confirmed, much extended, and strengthened the Einstein findings. The deepest survey is that which has been conducted in the Lockman Hole by Hasinger, Trumper, Hartner, Schmidt, Zamorani, Giacconi, MacKenty and Burg. This survey has reached a limiting sensitivity o

5×10^{-15} erg s^{-1} cm^{-2} in the .1 to 2.5 keV range in 85,000 seconds of exposure. The Lockman Hole (Lockman et al. 1986) was chosen because it has the lowest neutral hydrogen line of sight density in the galaxy, 5×10^{-19} cm^{-2}. Given the low energy response characteristics of the ROSAT system this increases the sensitivity per exposure by a factor of two with respect to that obtained in the average galactic line of sight. Figure 5 shows the integral Log N - Log S obtained from preliminary data (with 16 shallower fields added). Caution should be used in utilizing these result to extract exact parameters, due to the relatively crude current stage of data reduction and calibration. Notwithstanding the above, it is immediately evident that the source counts continue to rise beyond the deepest Einstein Surveys, albeit with a significant change in slope. This is consistent with the fluctuation analysis of Hamilton (1987) and Barcons (1989).

A very interesting result is shown in Figure 6 where we show the directly resolved fraction of the X-ray background as a function of energy. From these data one can derive the integrated fraction of the XRB resolved at 1 keV, which results to be equal to 45%. As described in detail by Hasinger in this proceedings the first estimate of the source spectrum yields (for a power law model) an index of -1.2.

Full understanding of the nature of the sources which emit this radiation will have to await for optical spectroscopy of the candidate objects. However, preliminary investigation of the ratio of X-ray to optical emission has been carried out as a function of the magnitude m_R of the brightest optical candidate within 15" of the source position (Figure 7). These results come from comparison of the preliminary PSPC positions with the CCD surveys of the field obtained at Mauna Kea and Palomar mountains. The mosaic of CCD exposure in R (Figure 8) obtained with the 2.2 meter University of Hawaii telescope is shown superimposed on a Palomar Schmidt plate of the region containing the Lockman hole. It is interesting to note that most of the candidates yield $f_x/f_R > 10^{-1}$, which represents the dividing line between stars and extragalactic objects including AGNs and clusters of galaxies. The surface density of these objects is of order of 200 sources/sq. degree, close to that observed for AGN at corresponding limiting magnitudes in the optical domain.

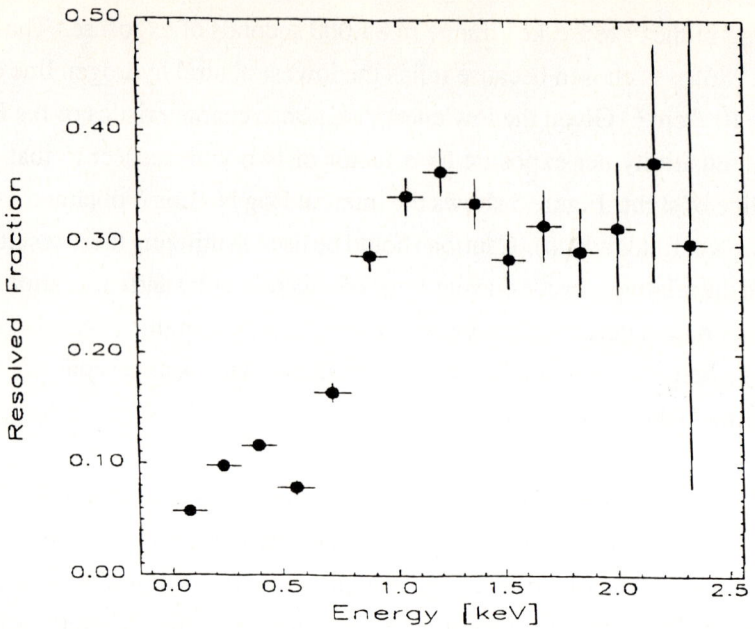

Figure 6: Resolved fraction of the XRB observed in the ROSAT Deep Survey of the Lockman field.

(Hasinger, this proceedings)

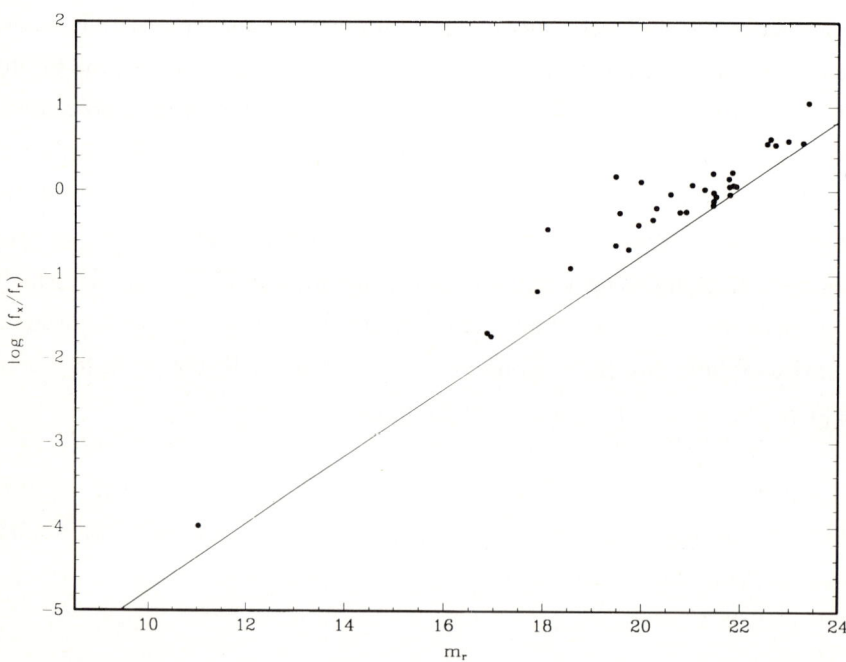

Figure 7: Ratio of X-ray to optical fluxes for candidate objects in the Lockman field.

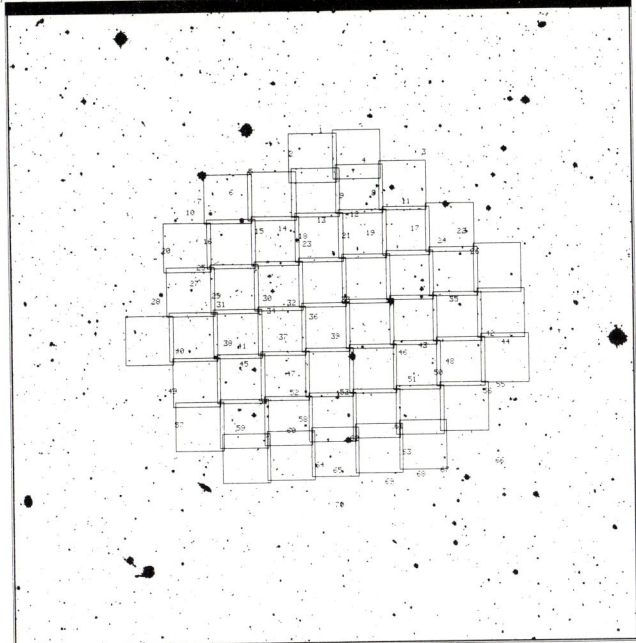

Figure 8: Optical CCD mosaic of observation in the Lockman field; numbers indicate sources

It is important to note that although rising more slowly than with a -1.5 power law the slope of the Log N - Log S is still consistent with a slope of ~ -1. Thus the intensity contribution to the XRB by fainter sources may increase at least logarithmically at fainter fluxes, if this slope is maintained. A factor of 10 decrease in limiting fluxes should therefore provide almost 100% of the XRB. Whether this will in fact occur within the next decade of sensitivity or not we are not yet in a position to assert. It seems inevitable however to conclude from these data that a very large fraction or possibly the whole of the background is made up of discrete sources.

3. THE SPECTRUM OF THE XRB

The spectrum of the XRB has been the subject of considerable study since the early rocket experiments prior to 1970. Although the data obtained from a number of different groups using a variety of techniques were difficult to compare and were subject to substantial uncertainty, it was attractive to attempt fits of simple spectral laws in the hope they would reveal a single underlying physical process. It is clear that in the hypothesis of discrete source superpositions the integral spectrum is much more difficult to interpret. The actual shape depends on the relative contribution of different classes of objects, on their evolutionary properties and space density as a function of cosmic time, on the specific intrinsic spectrum of the individual sources and the detailed integration of their redshifted contribution with appropriate K corrections. Attempts at computing the results of this integration, although of some pedagogical value had little relevance in advancing our knowledge of its origin.

Inverse compton emission by cosmic electrons on the 3°K radiation had been proposed as a possible mechanism for emission by Felten and Morrison (1966). Single power law fits to the spectrum, as would be expected from a single power law spectrum of cosmic electrons, were however shown not to fit the data. At least two power laws were needed with a break at ~20 keV.

Cowsik and Kobetich (1972) suggested that the data could be fit with an exponential spectrum as would be expected from thermal bremmstrahlung emission from an optically thin gas at a temperature of several hundred million degrees. In an elegant subsequent paper, Field and Perrenod (1977) gave the most complete and critical treatment of this model. They pointed out several difficulties connected with the existence of the postulated high temperature gas. First the gas energy content would have to be higher than any other form of energy in the universe except in the 3°K or in the rest mass. Second there exists a problem on how the gas could be reheated after the original cooling to the microwave background temperature. Third, they pointed out that, depending on the clumping factor, the required density in baryonic matter could be quite large, of order of closure density and possibly in conflict with the upper limits derived from the Deuterium abundance. It is also interesting to point out that apart from these theoretical difficulties, the observed spectrum (Figure 9) never really fit an exponential shape over the entire range of measured energies. This point was clearly made by Schwartz (1978) prior to the Einstein launch.

It is difficult to explain a posteriori why the enthusiasm for a diffuse XRB origin continued. In part, perhaps because thermal bremmstrahlung models were easy to compute. In part, because an XRB model based on the superposition of individual sources appeared to have its own difficulties. As pointed out by Schwartz (1978), none of the known extragalactic sources could provide the required number count without evolution. The point that, of course, even with very mild evolution one could provide an overabundance of sources was apparently neglected. Furthermore, there appeared to be difficulties in reproducing the observed XRB spectral shape by synthesis of the then known extragalactic source spectra. This argument turns out in retrospect also to have been rather simplistic. It was based on the apparent incongruity between early measurements of AGN spectra which appeared to be consistent with power laws with a small range of exponents (Mushotsky 1984) and the XRB. It is now well understood theoretically that more complex AGN spectra could well reproduce the XRB spectrum (Schwartz and Tucker 1988) and such spectra have in fact been observed by the Ginga experiment (Pounds et al. 1990).

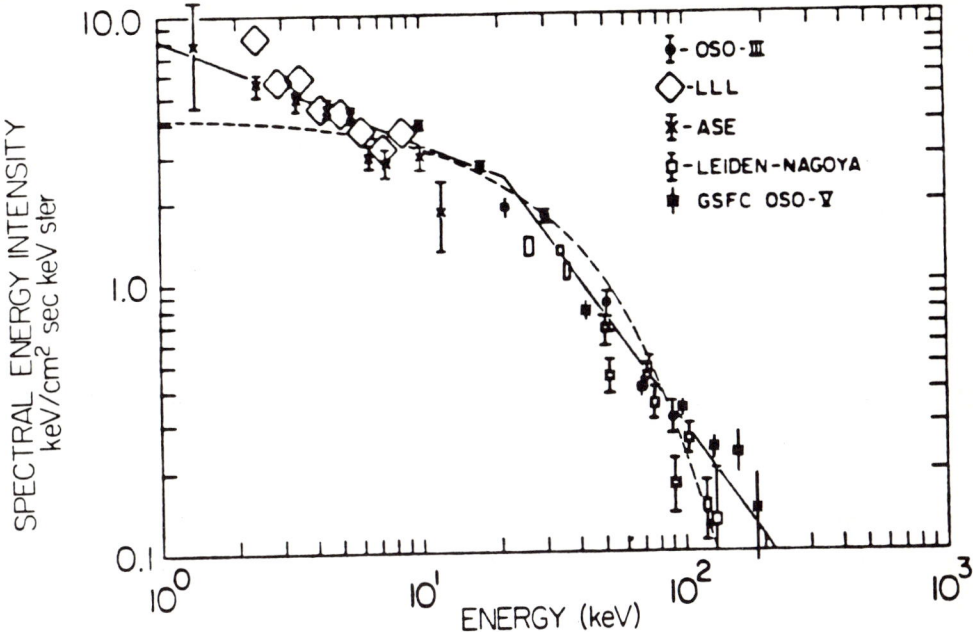

Figure 9: Spectrum of XRB. (Schwartz, 1978)

It is difficult today to recollect the impression that was generated by the finding of HEAO A-2 in 1980 by Marshall et al., Figure 10. It appeared obvious to a number of scientists that such an apparently good fit to a single temperature bremmstrahlung model had to be significant. What had been neglected in the enthusiasm of the moment was that the fit was good only over a decade of energies. Furthermore, given that a substantial fraction of the XRB had to be due to sources (Giacconi et al. 1979) then the very perfect agreement with a thermal bremmstrahlung model spectrum meant that either the individual sources had themselves the required spectrum and therefore assuming an additional diffused component was unnecessary, or that the residual XRB spectrum (after subtraction of the sources) had to be different from thermal bremmstrahlung. Figure 11 follows the calculation by Giacconi and Zamorani of the "residual" spectrum after subtraction of the discrete source component, under a number of assumptions. The strength of the argument was not appreciated except by a few scientists (Cowie 1989).

It required the stringent upper limit set by the COBE on the microwave background fluctuations (Mather et al. 1990) to convince most astronomers that the large amounts of hot gas required to produce the observed XRB by thermal bremmstrahlung could not exist. We must conclude that the diffuse hypothesis was interesting as a theoretical construct which was never supported by observational evidence.

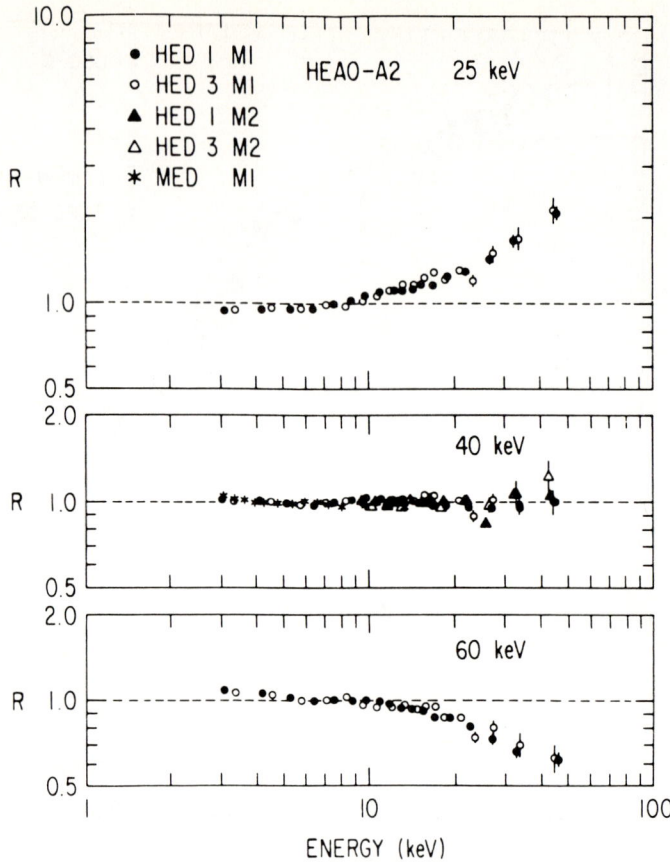

Figure 10: Fit to thermal bremmstrahlung model. (Marshall et al. 1980)

4. SOURCES

Given the rich phenomenology that has been revealed to date, it seems clear that we shall continue to study the properties of individual sources which contribute to the XRB at fainter and fainter fluxes first with ROSAT and then with ASTRO-D, Spectrum X, AXAF, XMM and WFXT. The relative contribution of the different classes of objects will have to be defined.

AGN

We know the quasar counts flatten out too fast to make up the bulk of the XRB (Koo and Kron 1988). On the other hand, a number of authors (see for example Setti and Woltjer 1990) have pointed out that Seyferts can make up the background with moderate density evolution of the type which in fact is being observed (Zitelli et al. 1992). Should the XRB turn out to be largely due to low level AGN this might explain the absence of a Gunn-Peterson effect and contribute to the understanding of

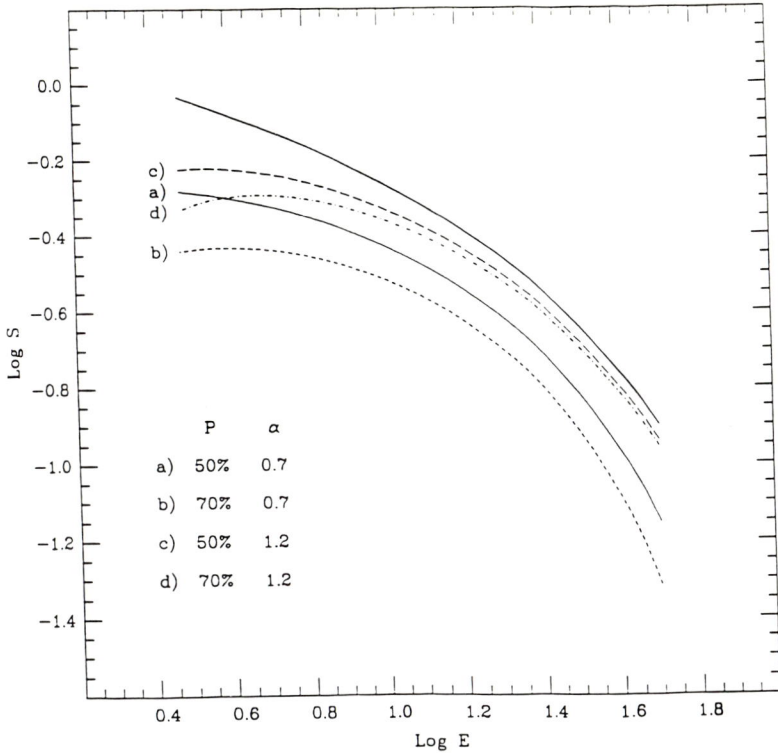

Figure 11: Residual spectrum of XRB after subtracting point sources (Giacconi and Zamorani, 1987)

the missing mass problem in present day Seyferts (Padovani et al. 1990, Collin-Souffrin 1991). As to the shape of the AGN spectrum at high energy, the Ginga spectra can be explained by the existence of large amounts of cold matter in the proximity of the nucleus, which would modify the original spectrum through reflection effects (Terasawa 1990, Rogers and Field 1991, Lightman and White 1988, and Fabian et al. 1991).

Clusters

A very interesting and as yet unresolved issue has to do with the contribution of clusters of galaxies to the XRB. In the most straightforward assumption of hierarchical clustering, Kaiser (1986) has derived scaling laws for density and luminosity evolution of clusters. In this view, cluster density would increase as $(1 + z)^3$ providing perhaps a significant contribution at low temperatures. A very interesting recent hydrodynamical model of Cen et al. (1991) implies late cluster formation whose X-ray emission could also contribute significantly to the low energy XRB.

Starburst and Normal Galaxies

Several authors (Weedman 1986, Griffiths and Padovani 1990) have suggested that normal or starburst galaxies with mild evolutionary properties could make up a substantial fraction (>15%) of the XRB.

While current ROSAT data are too preliminary to permit us to decide on the question of the contribution by individual classes of objects, additional ROSAT surveys and new missions will in the next decade provide observations of even increasing angular and spectral resolution and sensitivity to study the XRB. Table 1 summarizes some of the relevant information about future missions.

TABLE 1. CURRENT AND FUTURE MISSIONS

MODERATE SPATIAL RESOLUTION
- ROSAT all sky survey
    - flux limit of ~2 10^{-13} erg s^{-1} cm^{-2} with arc-minute resolution
    - at polar caps (36 sq. deg.) as sensitive as the Einstein Deep Surveys (~2 10^{-14} erg s^{-1} cm^{-2})
    - especially powerful for studying clusters and large scale structures
- Astro-D, JET-X (early 1990's)
    - large areas to 10 keV
    - direct measure of 2-10 keV background to roughly Einstein limits
    - fluctuations
- XMM
    - largest throughput - 3 times AXAF
    - detailed spectral information of point sources at moderate fluxes

HIGH RESOLUTION
- ROSAT (1990)
    - .2 - 2.0 keV
    - 5" spatial resolution with HRI; 20" with PSPC
    - 10 times more sensitive than the Einstein Deep Surveys
- AXAF (1998)
    - .1 - 10 keV
    - .5" spatial resolution
    - potentially 50-100 times more sensitive than the Einstein Deep Surveys
    - energy resolution of at least 20 at Fe emission line
- WFXT
    - .3 - 3.0 keV
    - 5" over 1° field
    - high sensitivity (5 • 10^{-15} ergs s^{-1} cm^{-2}) surveys
    - 100's of sq. deg. of contiguous survey AREA
    - clusters, AGN, large scale structures
- Large surface density of sources -- high resolution a must

---

5. CONCLUSION

The attempt to resolve the sources of the XRB has been the "Holy Grail" of X-ray astronomy. It has provided an extremely useful goal for the development of X-ray instrumentation. If uniform on the arc minutes scale, the XRB would provide a surface brightness of 10^{-15} erg cm^{-2} sec^{-1} arcmin square^{-1}, close to the design sensitivity of AXAF. In the search for this distant objective, X-ray survey's sensitivity has improved by more then 20 magnitudes in the last 30 years. It seems clear that the study of the XRB will continue with great interest because it provides a powerful tool to investigate the formation and evolution of discrete objects and structures at cosmological distances.

REFERENCES

Barcons, X. and Fabian, A. C. 1989, *M.N.R.A.S.*, **237**, 119.

Cen, R. Y., Jameson, A., Liu, F. and Ostriker, J. P. 1990, *Ap.J.*, **362**, L41.

Collin-Souffrin, S., Invited Talk given at the Moriond Conference on Diffuse Backgrounds, March 1991, eds. DeHarveng, J. M. and Rocca-Volmerange, B. , Editions Frontieres.

Cowsik, R., and Kobetich, E., 1972, *Ap. J.*, **211**, 135.

Fabian, A. C., George, I. M., Miyashi, S. and Rees, M.J. 1991, *M.N.R.A.S.*, in press.

Felten, J. E., and Morrison, P., 1966, *Ap. J.*, **146**, 686.

Field, G., and Perrenod, S.C., 1977, *Ap. J.*, **215**, 717.

Giacconi, R., Gursky, H., Paolini, F. R. and Rossi, B. B. 1962, *Phys.Rev.Letters*, **9**, 439.

Giacconi, R. et al. 1979, *Ap. J.*, **234**, L1.

Giacconi, R., Gursky, H., Paolini, R., and Rossi, B., 1962, *Phys. Rev. Letters,* **9**, 439.

Giacconi, R. and Gursky, H. (eds.) 1974, X-ray Astronomy, D. Reidel Publishing Co. Dordrecht, Holland.

Giacconi, R., Bechtold, J., Branduardi, G., Forman, W., Henry, J. P., Jones, C., Kellogg, E., van der Laan, H., Liller W., Marshall, H., Murray, S. S., Pye, J., Schreier, E., Sargent, W. L. W., Seward, F., Tananbaum, H., and Wright, A., 1979, *Astrophs. J.*, **234**, L1.

Griffiths, R. E. and Padovani, P. 1990, *Ap.J.*, **360**, 483.

Griffiths R. E. et al. 1983, *Ap. J.*, **269**, 375.

Hamilton, T. T. and Helfand, D.J. 1987, *Ap.J.*, **318**, 93.

Hoyle, F., 1963, *Ap. J.*, **137**, 993.

Kaiser, N. 1986, *M.N.R.A.S.*, 204, 33.

Koo D. C.and Kron R. G. 1988, *Ap. J.*, **325**, 92.

Lightman, A. P. and White, T. R. 1988, *Ap. J.*, **335**, 57.

Lockman, F.J., Jahoda, K. and McCammon, D. 1986, *Ap. J.*, **302**, 432.

Maccacaro, T., Feigelson, E. D., Fener, M., Giacconi, R., Gioia, I. M., Griffiths, R., Murray, S. S., Zamorani, G., Stocke, J., and Liebert, J., 1982, *Ap. J.*, **253**, 504.

Marshall, F. et al. 1980, *Ap. J.*, **235**, 4.

Mather, J., et al. 1990, *Ap. J.*, **354**, L37.

Matilsky, T., Gursky, H., Kellogg, E., Taananbaum, H., and Giacconi, R. 1973, *Ap.J.*, **181**, 753.

Mushotsky, R. F. 1984, *Advances in Space Research*, **3**, no. 10-13, 157.

Padovani, P., Burg, R., and Edelson, R. 1990, *Ap.J.*, **353**,438.

Pounds, K.A., et al. 1990, *Nature*, **344**, L32.

Primini, F. A. et al. 1990, *Ap. J.*, **374**, 440.

Rogers, R. D. and Field, G. B. 1991, *Ap. J.*, **378**, L17.

Schwartz, D. A. and Tucker, W. H. 1988, *Ap. J.*, **332**, 157.

Schwartz, D. A., and Gursky, H., 1974, The Cosmic X-Ray Background, X-Ray Astronomy, Giacconi and Gursky, eds., D. Reidel Publishing Co.

Schwartz, D. A. 1978, Proceedings of the IAU/COSPAR Symposium on X-ray Astronomy, Innsbruck, Austria.

Setti, G., and Woltjer, L., 1970, *Astrophy. Space Sci.* **9**, 185.

Setti, G. and Woltjer, L. 1989, *Astron. & Astrophysics*, **224**, L21.

Terasawa, N. 1991, Ap. J., 378, L11.

Weedman, D. W. 1986, in Star Formation in Galaxies, NASA CP-2466, ed. C. J. Lonsdale, 351.

Zitelli, V., Mignoli, M., Zamorani, G., Marano, B and Boyle, B. J. 1992, *M.N.R.A.S.*, in press.

Chapter II

The spectrum

The Soft X-ray Diffuse Background: How Much of it is Extragalactic?

DAN McCAMMON

University of Wisconsin - Madison

1 INTRODUCTION

For the purposes of this paper we will take the "Soft X-ray" Background to include photon energies from 0.1 to 10 keV. It is true that more of the energy of the diffuse background lies in the 10 to 100 keV range, but there are two reasons it is worthwhile to understand what is going on at the lower energies. First, of course, a measurement or upper limit on the cosmological background intensity at any wavelength is interesting in its own right, since it can place useful restrictions on models of the structure and past history of the universe. A second, probably temporary, reason for special interest in the lower energies is that the spectra of the extragalactic objects that seem to be the source of most if not all of the cosmological background have been well determined only below 2-4 keV. As discussed below, it is just at these energies that the extragalactic background flux is most uncertain. Direct calculation of the contribution from even the observed point sources therefore requires a highly uncertain extrapolation of one spectrum or the other.

A recent review by McCammon and Sanders (1990; hereafter M&S) presents the observational data in some detail and includes a discussion of limits on the extragalactic intensity at different energies that will only be summarized here. The observed diffuse background shows such markedly different characteristics at different energies that it seems unlikely to be due to a common source over the entire range. The energy range can conveniently be divided into three regions that show more uniform characteristics. These extend from 0.1 to 0.3 keV, from 0.5 to 1.0 keV, and upwards from 3 keV. The interval from 0.3 to 0.5 keV is effectively unobserved for instrumental reasons, providing a distinct break between the behaviors observed at lower and higher energies. The transition between the 0.5 - 1.0 keV behavior and the high energy limit takes place somewhere over the 1 - 3 keV range, but there is some uncertainty even about the observed intensity in this interval (see M&S), and the nature of this transition will probably not be clear until the origins of the 0.5 - 1 keV flux are better understood.

None of these regions is understood, but the high end is easiest to explain, so I will start there. Above 3 keV, the entire Galaxy is transparent, and the observed isotropy of the X-ray background at these energies provides strong evidence that its origin is primarily extragalactic. The spectrum from 3 to 40 keV can be fit quite precisely by optically thin bremsstrahlung with kT = 40 keV (Marshall et al. 1980). It now appears that this fit is entirely accidental (the first of the misleading coincidences in this area), but the spectrum can be represented adequately in the 3 - 10 keV range by an $8 E^{-0.4}$ keV cm^{-2} s^{-1} keV^{-1} sr^{-1} power law. There is a galactic component that contributes about 1% of the total background up to intermediate galactic latitudes (Iwan et al. 1982, Koyama 1989). It is too faint to add significantly to the spectral uncertainty of the extragalactic component, but its origin and spatial distribution are poorly understood, and it is a major limitation for studies of large-scale fluctuations in the extragalactic background.

The 0.5 - 1.0 keV energy range shows several bright regions that are readily identified with known galactic features such as Loop I and the Eridanus superbubble. Aside from these, and possibly the galactic plane near the galactic center, the X-ray intensity appears quite isotropic. In particular, it shows essentially no variation with galactic latitude. This is quite surprising, since the average interstellar mean free path in this energy range is about 900 pc in the galactic plane. An extragalactic source should therefore drop almost to zero in the plane, while a galactic source should be brighter near the plane than it is at high latitudes. A very local source could be inherently isotropic, but no suitable models exist. As described below, there is now good evidence that a significant fraction of the observed intensity at 1 keV is extragalactic. If this is true, then the observed isotropy must be coincidental, since a galactic source of some sort will be required to fill in the expected absorption near the plane. One candidate for at least part of this is X-ray emission from dM stars which, with large uncertainties, may approach the required intensity (Schmitt and Snowden 1990, Kashyap et al. 1992). This is the energy region where the origin of the bulk of the observed intensity is most uncertain.

In the 0.1 - 0.3 keV range, the interstellar mean free path varies from 6 to 100 pc at an average density of 1 atom cm^{-3}. The intensity varies by a factor of five in different directions, showing little correlation with other identifiable features. There is a strong anticorrelation with total neutral hydrogen column density -- the expected signature of an absorbed extragalactic flux. However, the quantitative relationship is not right for absorption, and the expected strong energy dependence is entirely absent. The existence of a substantial flux in the galactic plane shows that much of the observed radiation must originate within a few hundred parsecs of the Sun. An acceptable but probably incomplete model is that most of the HI-deficient volume known to extend 100 pc or more in most directions from the Sun is filled with low-density gas at a temperature near 10^6 K, and that thermal emission from this gas provides the bulk of the observed X-ray flux (see M&S).

2 EXTRAGALACTIC LIMITS

Figure 1, largely taken from M&S, summarizes the available spectral information on the diffuse background. The data points show the all-sky average intensity at each energy, with the upper "error bar" indicating the average intensity at high latitudes and the lower "error bar" the average intensity near the plane. The heavy bars show the best current upper limits to an extragalactic or uniform halo emission at each energy. Between 0.5 and

2 keV, these were derived simply by taking the lowest intensity observed at moderately high latitude, and correcting this for interstellar absorption, assuming that it originates beyond the entire galactic hydrogen distribution. The limits below 0.3 keV were obtained from shadowing experiments that showed that at most only a small fraction of the observed X-ray flux could be extragalactic, but at these low energies the absorption corrections are substantial, and the resulting limits on the extragalactic intensity are comparable to or exceed the observed intensities.

The long-dash line on Figure 1 shows the model of Schwartz et al. (1989) that was adjusted to provide all of the 3 - 40 keV diffuse background through a superposition of discrete sources. The various versions of the Compton backscatter model that do such a good job of providing the right spectral form in this energy range (e.g. Field, this volume) make similar predictions at low energies. It is apparent that the limit at 1/4 keV ("C band") currently comes closest to constraining these models, and it would be quite interesting if it could be reduced, even by a small factor.

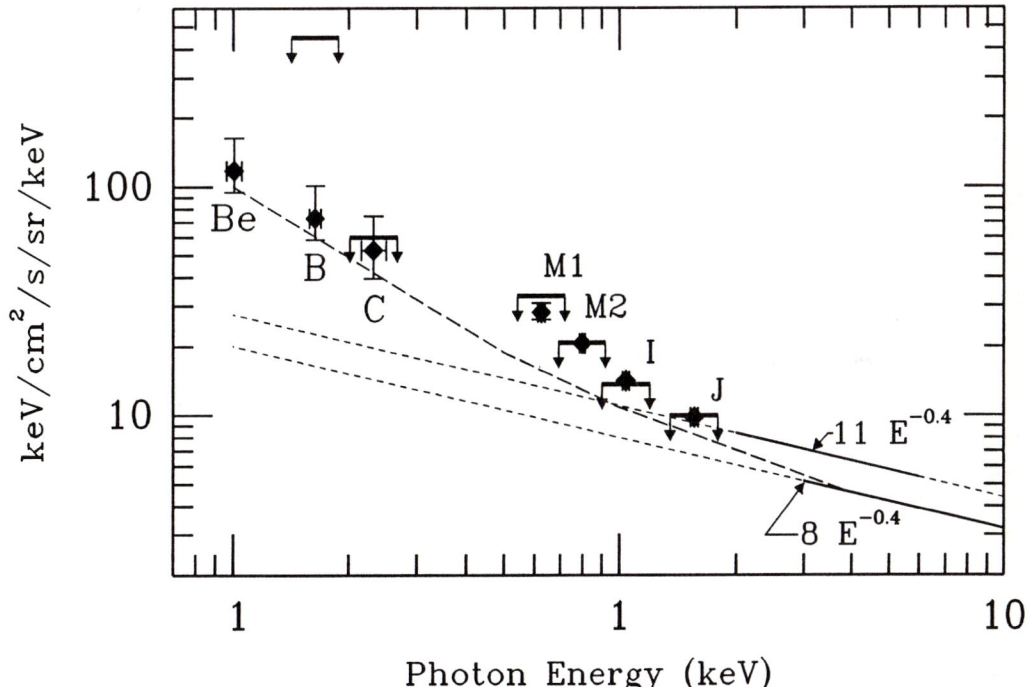

Figure 1. -- Average observed diffuse background intensity vs. energy. The plotted points are the all-sky averages. The upper "error bar" is the average for |b| > 60°, while the lower "error bar" is for |b| < 20°. Be-band points assume that the unobserved part of the sky also tracks the B band. The heavy bars represent the best upper limits that can be placed at each energy for the flux incident on the Galaxy from an extragalactic or uniform halo source. The long-dash line is the extragalactic source model of Schwartz et al. (1989).

3 RECENT DEVELOPMENTS

One of the most exciting early results from ROSAT is the demonstration that deep exposures at high galactic latitudes show as much as 50% of the diffuse background at 1 keV resolved into presumably extragalactic sources (Hasinger, this volume). This already requires a substantial galactic source at the same energy with the right latitude dependence to rather accurately fill in the expected absorption of this extragalactic component. A very preliminary analysis indicates that the "spectral paradox" still exists: the composite spectrum of the resolved sources seems to be considerably steeper than that of the diffuse background in the 0.6 - 2 keV range.

ROSAT also offers the opportunity to do much more sensitive shadowing experiments, potentially making significant reductions in both the 1/4 keV and 0.5 - 1 keV extragalactic limits. However, the very first shadowing results have been positive (Snowden et al. 1991, Burrows and Mendenhall 1991). These results not only don't reduce the existing limits, they show such intense 1/4 keV emission beyond the absorbing target (the Draco molecular cloud) that it is inconsistent with older limits in other directions, and therefore must represent neither a true extragalactic flux nor an isotropic halo.

4 FUTURE PROSPECTS

Shadowing experiments with ROSAT could potentially reduce the current 1/4 keV upper limit on the extragalactic flux by a factor of seven, which would obviously have major repercussions on some existing models. However, the new evidence for a patchy galactic halo makes it less likely that this kind of shadowing experiment will be able to significantly lower the existing limits, unless we are lucky and find a target in a direction where the halo emission happens to be very small.

The emission seen behind the Draco cloud has a temperature near 10^6 K and should not produce significant intensity above 0.5 keV. We can therefore still hope to improve extragalactic limits in the 0.5 - 1 keV range. However, shadowing targets for these energies require column densities of at least 10^{21} atoms cm^{-2} with reasonably sharp edges. Such targets are rare at high latitudes, and since clouds with such high column densities usually contain a substantial molecular component, they are also difficult to characterize.

The use of galactic absorbers for shadowing experiments has the fundamental drawback that if there is a positive result, we cannot distinguish between a halo source and a true extragalactic flux. Until very recently, of course, the complete lack of any detection of a halo flux allowed us to hope that the results might remain negative, at least to the point where very useful limits could be placed on the extragalactic intensity. Use of extragalactic shadowing targets offers the distinct advantage that it is possible to make a direct measurement of just the extragalactic intensity. The major difficulties here are the shortage of suitable, well-characterized targets, and the probability that the targets are themselves diffuse X-ray emitters. This problem is unlikely to be solved except through an extensive program of observing many targets and trying to understand them thoroughly.

The direct resolution of diffuse flux into extragalactic point sources in ROSAT deep survey exposures will of course continue to provide lower limits on the extragalactic intensity. Unless *all* of the observed intensity is resolved, however, this will always beg the important question of how much extragalactic flux is left to be accounted for by fainter sources or some other mechanism. We can potentially close in on this number by determining lower limits to the galactic contribution. One possibility is that study of the very high quality maps being produced from the ROSAT all sky survey will help us understand the galactic diffuse background well enough to predict its high latitude intensity. Another, very direct route is to use high *spectral* resolution observations: all plausible galactic emission mechanisms are thermal, and thermal emission at temperatures below several million degrees is almost entirely in lines. A true extragalactic flux, on the other hand, must be red-shifted into at least a quasi-continuum. Figures 2 and 3 show models prepared by Jiahong Zhang, showing what the diffuse background integrated over a steradian field of view centered in Hercules should look like if observed with a spectral resolution of 5 eV FWHM. The assumption in Figure 2 was that an extragalactic power law of 11 $E^{-0.4}$ keV cm^{-2} s^{-1} keV^{-1} sr^{-1} could be extrapolated down through the 0.1 - 1 keV range. The total intensity in different parts of this range is known from broad-band measurements, and the

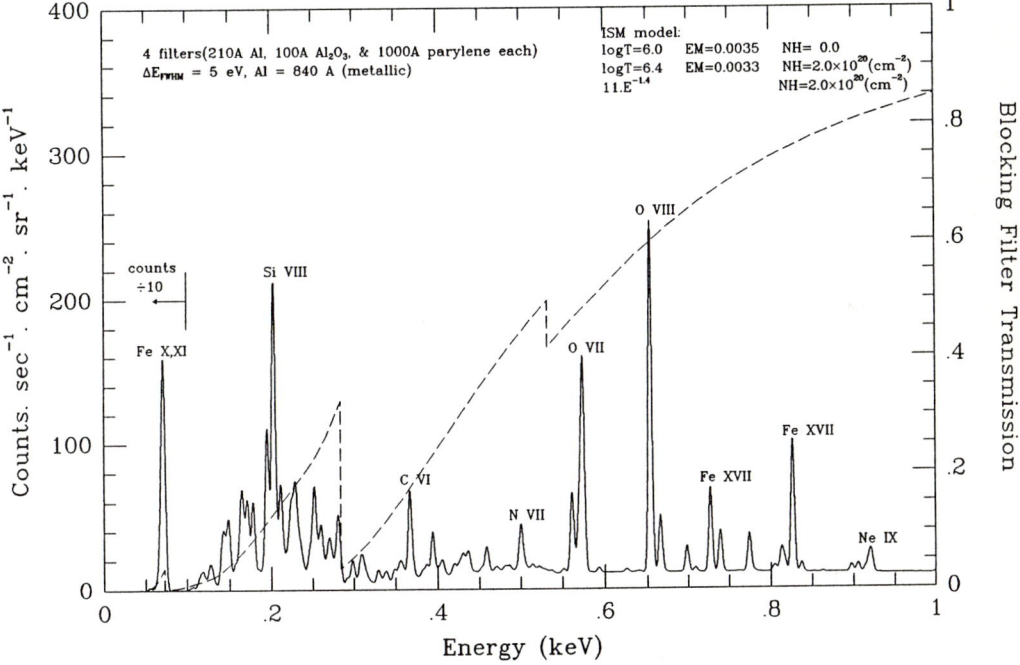

Figure 2. -- Instrument response for a diffuse spectrometer with 5 eV FWHM resolution to a model in which the observed diffuse background background intensity in broad bands has been fit by an extragalactic 11 $E^{-0.4}$ keV cm^{-2} s^{-1} keV^{-1} sr^{-1} power law plus two galactic thermal components. One of the thermal components has T = 1 x 10^6 K and provides most of the observed flux below 0.3 keV. The other has T = 2.5 x 10^6 K and provides the additional intensity required in the 0.5 - 1 keV range.

remainder of this total is assumed to be provided by two thermal components, one at 1 x 10^6 K, which appears primarily in the 0.1 - 0.3 keV range, and one at 3 x 10^6 K, which provides the rest of the flux in the 0.5 - 1 keV interval. In Figure 3, it was assumed that there is no extragalactic contribution below 1 keV, and the emission measure of the higher temperature thermal component was increased to make up the necessary total intensity. This we now know is probably incorrect, but one could go the other direction and assume that the extragalactic power law gets steeper, as both current models and the early ROSAT results imply, and the thermal emission must therefore be considerably reduced.

This spectral resolution has recently been obtained with cryogenic microcalorimeters (McCammon et al. 1991), and we, in conjunction with the X-ray group at the Goddard Space Flight Center, are currently constructing a sounding rocket payload to do just this experiment sometime next year. Figures 4 and 5 show Monte Carlo simulations of results for a sounding rocket experiment that correspond to the models of Figures 2 and 3, respectively. It is clear that the statistics should be adequate to determine the relative line and continuum contributions between 0.7 and 1.0 keV. Although there are potential experimental uncertainties about the continuum level, the intensity resolved into lines should provide a firm lower limit to the galactic contribution. The CUBIC experiment from Penn State (Burrows, this volume) will have poorer energy resolution, but its far better statistical precision may enable a similar analysis to be done. At the least, it should provide a definitive answer to the question of the absolute spectrum of the total diffuse

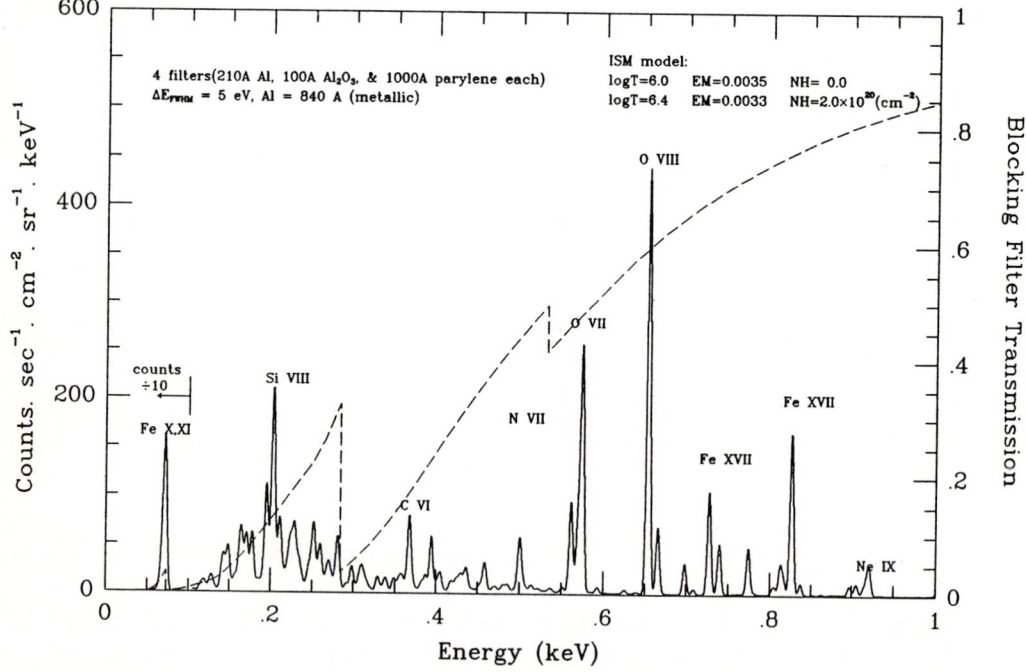

Figure 3. -- As in Figure 2, but the extragalactic component is assumed to be negligible below 1 keV, and the emission measure of the high-temperature galactic component has been increased to make up the total observed flux.

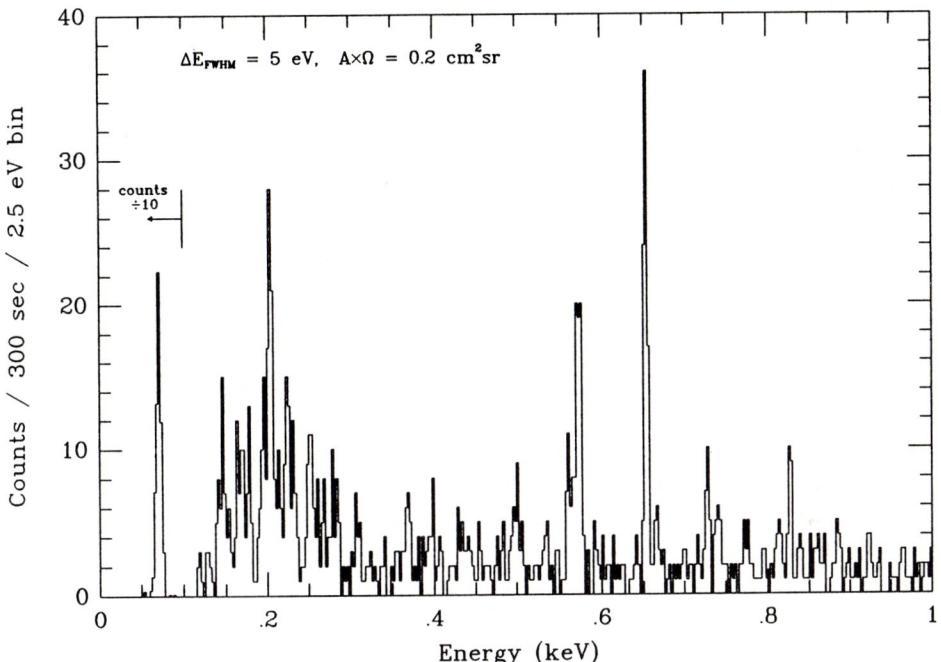

Figure 4. -- Monte Carlo simulation of the spectrum in Figure 2 for an observation that could be made during a sounding rocket flight with a 1 sr field of view.

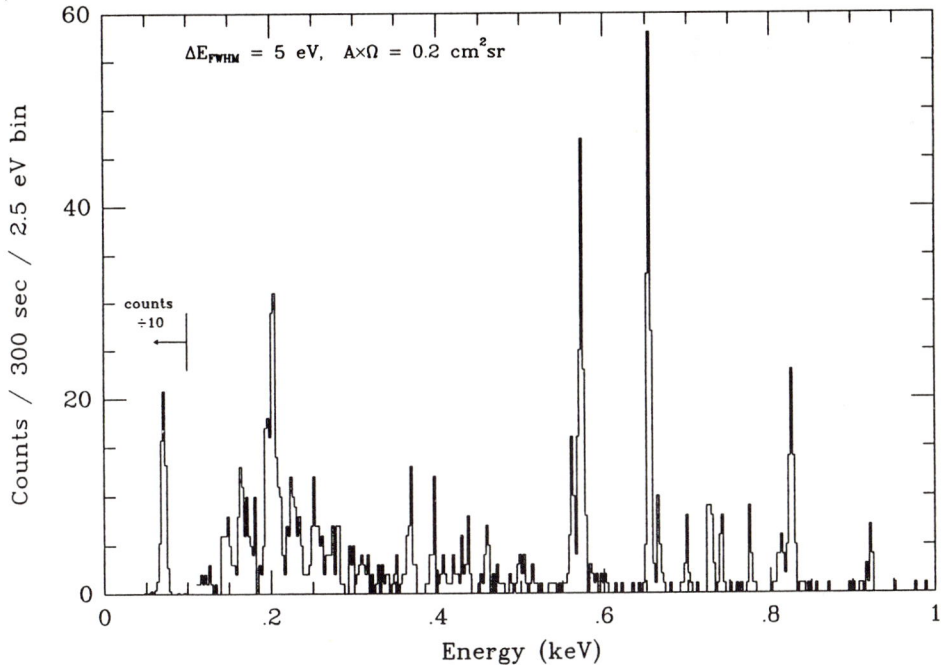

Figure 5. -- As in Figure 4, but for the assumed spectrum of Figure 3.

background in the 1 - 3 keV range, making it possible to tie together the low energy portion of the spectrum with the well-determined 3 - 40 keV flux.

5 CONCLUSION

We do not yet understand very well either the galactic or the extragalactic components of the diffuse soft X-ray background at any energy. ROSAT is almost ideally suited for low energy diffuse observations, and it seems our knowledge should soon be improving rapidly. High spectral resolution observations offer another route to determining the galactic portion of the diffuse intensity between 0.7 and 1 keV. ASTRO-D will have the ability to make accurate measurements of the spectra of reasonably distant extragalactic sources at energies up to 10 keV, hopefully solving or at least shedding light on the "spectral paradox."

This work was supported in part by the National Aeronautics and Space Administration under grant NAG 5-629.

REFERENCES

Burrows, D. N., and Mendenhall, J. A. 1991, *Nature*, 351, 629

Iwan, D., Marshall, F. E., Boldt, E. A., Mushotzky, R. F., Shafer, R. A., and Stottlemyer, A. 1982, *Ap. J.*, 260, 111

Kashyap, V., Rosner, R., Micela, G., Sciortino, S., Vaiana, G. S., and Harnden, F. R., Jr. 1992, *Ap. J.*, 389, in press

Koyama, K. 1989, *Publ. Astron. Soc. Jpn.*, 41, 665

Marshall, F. E., Boldt, E. A., Holt, S. S., Miller, R. B., Mushotzky, R. F., et al. 1980, *Ap. J.*, 235, 4

McCammon, D., Cui, W., Juda, M., Plucinski, P., Zhang, J., Kelley, R. L., Holt, S. S., Madejski, G. M., Moseley, S. H., and Szymkowiak, A. E. 1991, in *PANIC XII: Particles and Nuclei*, J. L. Matthews, T. W. Donnelly, E. H. Fahr, L. S. Osborne, eds., (North Holland), 821c

McCammon, D., and Sanders, W. T. 1990, *Annu. Rev. Astron. Astrophys.*, 28, 657

Schmitt, J. H. M. M., and Snowden, S. L. 1990, *Ap. J.*, 361, 207

Schwartz, D. A., Qian, Y., and Tucker, W. H. 1990, in *23rd ESLAB Conference on X-Ray Astronomy* (ESA SP-296), N. White ed., 1043

Snowden, S. L., Mebold, U., Hirth, W., Herbstmeier, U. and Schmitt, J. H. M. M. 1991, *Science*, 252, 1529

The Spectrum of the Cosmic X-Ray Background

STEPHEN S. HOLT

NASA/Goddard Space Flight Center

1 INTRODUCTION

From its initial discovery in the same rocket-borne investigation in which the first discrete extra-solar source of cosmic X-rays was found (Giacconi, *et al.* 1962), the origin of the cosmic X-ray background (hereafter, CXB) has been recognized to be both immensely important and astonishingly enigmatic. The identification of 3C273 as an X-ray source from another early rocket-borne investigation "explained" the CXB quite elegantly (*e. g.* Setti and Woltjer, 1979): the CXB will be overproduced by more than an order of magnitude if all known quasars have X-ray luminosities similar to that of 3C273, so that quasars must clearly be responsible for the CXB and the only problem is how to make most of them X-ray underachievers relative to 3C273! Subsequent investigations have found apparent consistencies and inconsistencies with the quasar hypothesis for the origin of the CXB, but a single serious inconsistency is sufficient to nullify all the consistencies.

The electromagnetic radiation that permeates the local universe is dominated by the cosmic microwave background, with an enormous cosmic radiation density of approximately 1/4 eV/cm^3 that is widely interpreted as the 3-degree remnant radiation of the hot big bang. We cannot observe the background radiation in the ultraviolet shortward of the Lyman limit directly, because galactic neutral hydrogen presents an impenetrable barrier. When the galactic gas starts becoming optically thin at energies >1 keV, the CXB is rising with energy and peaks at about 30 keV, with a radiation energy density some two orders of magnitude less than that of the microwave background. If it were related to the cooling of a uniform hot intergalactic gas, the particle kinetic energy density would be enormous since its cooling via bremsstrahlung is so inefficient ($t_{cool} >> t_{Hubble}$), and its mass would exceed all that observed in condensed matter.

The evidence that can be brought to bear to explain the origin of the CXB includes direct CXB measurements (*e. g.* X-ray spectrum, small-scale fluctuations, large-scale isotropy), as well as indirect measurements that can provide constraints on potential scenarios (*e. g.* discrete source candidates and the microwave background). This review will concentrate on information that might be relevant to the explanation of the spectrum of the CXB.

2 THE MEASUREMENT OF THE SPECTRUM OF THE CXB

Figure 1 is a characterization of the best available data on the spectrum of seemingly diffuse electromagnetic radiation >1 keV, displayed as *EF(E)*, where *F(E)* is the differential energy spectrum (see equation 1 below) measured in the units keV/(cm^2-s-sr-keV). This form of presentation demonstrates the dominance of the 30 keV CXB peak, and also indicates that there is a lesser "MeV bump" that will not be discussed here in any detail (but see the review by Gruber in this volume).

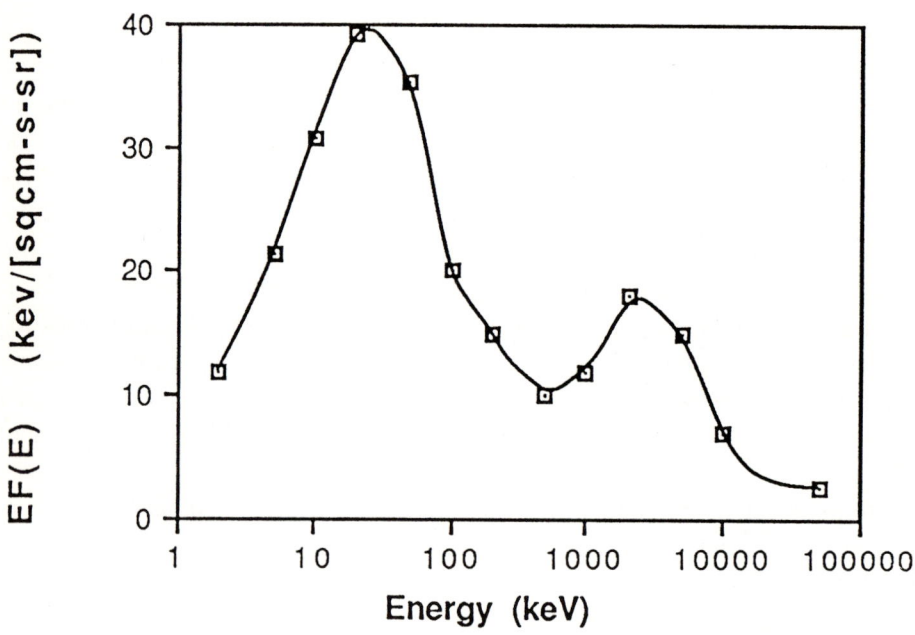

Figure 1 The CXB >1 keV, where the data points approximate the raw data displayed in Rothschild *et al.* (1983), plotted as *EF(E)* versus log*E*.

The 3-50 keV data that currently exist will not be supplanted by better data for a long time. This is not because they have better spectral or spatial resolution than can be obtained from other instruments (indeed, they have neither). Instead, it is because they were obtained with an instrument which had the ability to extract a systematics-free signal proportional to solid angle. The low-internal-background proportional counters of *HEAO-1 A2* had the unique feature of having different portions of exactly the same detecting volume with differing solid angles (Rothschild *et al.* 1979). Even if the internal back-ground varied, the signal that was proportional to the solid angle could be extracted unambiguously.

These data, from "source-free" regions at high galactic latitude, were carefully fit by Marshall *et al.* (1980), as demonstrated in Figures 2 and 3. Figure 2 attempts to fit the data with simple power laws. The indicated power-law index Γ refers to that for the differential photon spectrum , *i. e.* $(dN/dE) \propto E^{-\Gamma}$ $(cm^2\text{-s-sr-keV})^{-1}$ where Γ is related to the index α of the differential energy spectrum $F(E)$ via $\alpha = \Gamma\text{-}1$). In the upper trace of Figure 2, it is clear that $\Gamma=1.4$ ($\alpha=0.4$) fits the <10 keV data reasonably well, but utterly fails to match the data at higher energies. $\alpha=0.7$ is a better fit to all the data in an average sense, but it actually fails to match the data anywhere (the error bars on the data points are typically smaller than the symbols). No single power-law can match the data over the range 3-50 keV.

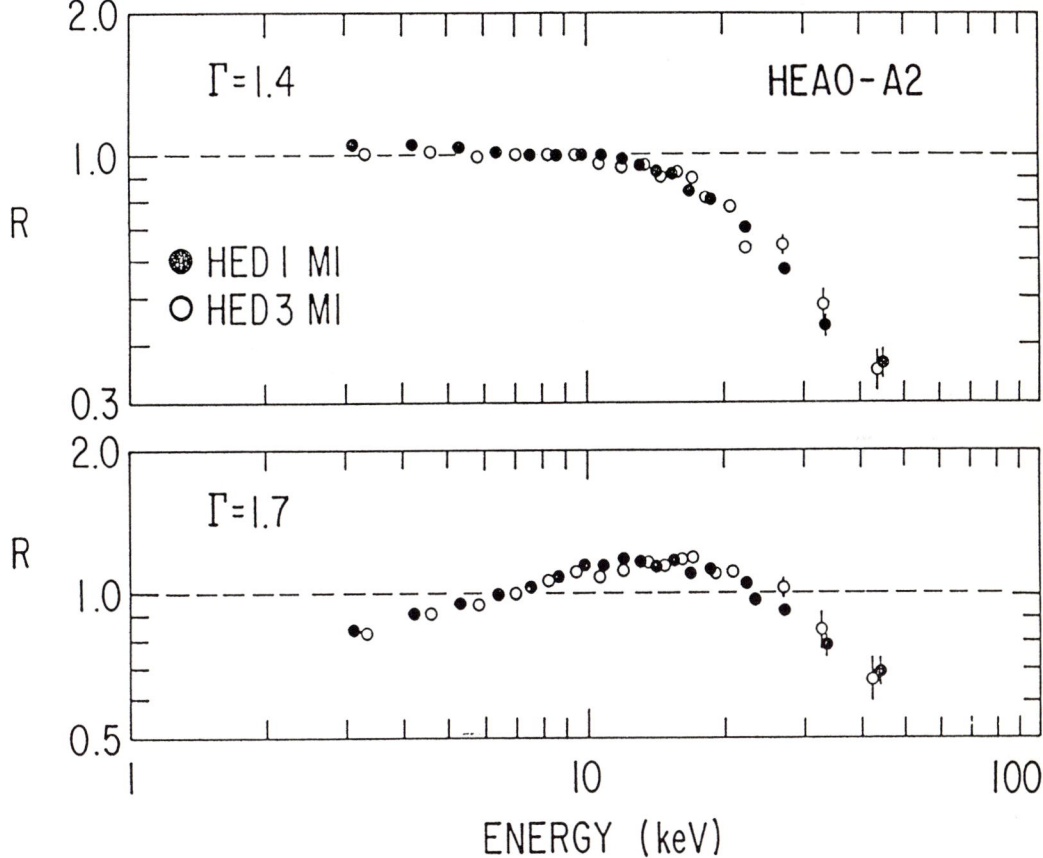

Figure 2 The CXB data from two of the several independent counter volumes (open and closed circles) of the *HEAO-1 A2* instrument, plotted as the ratio *R* of the observed counts from the CXB to that predicted for power-law incident spectra . Statistical errors are shown when larger than the sizes of the symbols. (*from Marshall et al. 1980*)

The same data are fit with coronal (optically thin bremsstrahlung) photon spectra, close to the form $(dN/dE) \propto E^{-.3} \exp(-E/kT)$, in Figure 3. Paraphrasing a discussion in the children's story *Goldilocks and the Three Bears*: "60 keV is too hot, and 25 keV is too cold, but 40 keV is just right." In fact, the fit seems to be so astonishingly good that departures from it have probably been taken much too seriously.

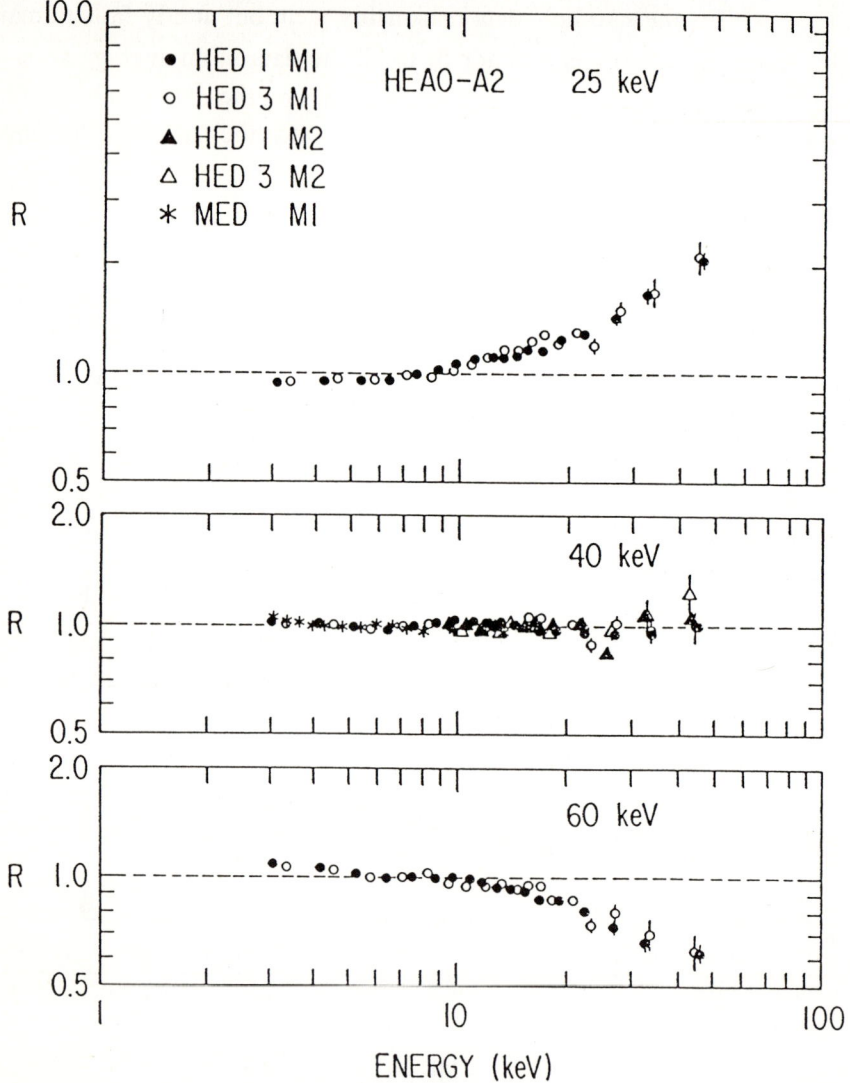

Figure 3 The CXB data from several independent counter volumes of the *HEAO-1 A2* instrument, plotted as the ratio R of the observed counts from the CXB to that predicted for thermal bremsstrahlung incident spectra, as indicated. Statistical errors are shown when larger than the sizes of the symbols. (*from Marshall et al. 1980*)

There are new data <2 keV available from *Rosat* that largely supplant the older *Einstein* data (which had considerably more systematics contamination). At energies <1/2 keV, the x-ray sky is dominated by galactic absorption and emission. At 1 keV, however, where the radiation can penetrate the galactic hydrogen at high galactic latitude, the majority of the incident X-rays arise from resolved quasars consistent with power law spectra of index $\alpha \approx 1.2$. Further, the remaining unresolved emission at 1 keV has a similar steep spectrum (Hasinger, this volume). It would appear, therefore, that such quasars can explain the great majority of the CXB at 1 keV. It would also appear, however, that such steep-spectrum sources may be irrelevant in explaining the 30 keV CXB.

3 CONTINUUM SPECTRA FROM DISCRETE SOURCES

Even though the *Rosat* and *Einstein* Observatories are not capable of measuring X-rays ≥ 3 keV, they are responsible for the preponderance of evidence that has been brought to bear in support of AGN (active galactic nuclei, *i. e.* quasars and Seyfert I galaxies) as the origin for the CXB. With $\alpha \approx 1.2$, the new *Rosat* spectra of AGN are somewhat steeper than those reported from the *Einstein Observatory*, which have been most recently characterized as $\alpha \approx 1$ (*e. g.* Canizares and White 1989; Maccacaro *et al.* 1991).

At energies that are more directly relevant to the CXB, data from *HEAO-1* indicate that the spectra of Seyfert I galaxies seem to be well-fit by $\alpha \approx 0.7$ power laws over 2-40 keV (Mushotzky 1984), and perhaps out to at least 100 keV (Rothschild *et al.* 1983). *Exosat* 2-20 keV data confirm this "canonical" spectral form (Turner and Pounds 1989), and additionally demonstrate that AGN sometimes exhibit much steeper spectra <2 keV. Interestingly, an extension of the $\alpha = 0.7$ spectrum to higher energies can reconcile the "MeV bump" if approximately 20% of Seyferts maintain this form beyond 1 Mev before they steepen (equivalently, that the log$N$-log$S$ distribution of local AGN reported by Piccinotti *et al.* (1982) flattens for $L < 10^{43}$ erg/s).

Clusters of galaxies, having coronal spectra with typical $kT = 6$ keV, cannot make a meaningful contribution to the 30 keV CXB. Nor can BL Lacs or galaxies with X-ray emission dominated by normal stars, both of which usually exhibit very steep spectra (and, like clusters of galaxies, do not exhibit the evolutionary characteristics of quasars which would make them much more numerous in the past). The X-ray binaries in starburst galaxies (Bookbinder *et al.* 1980; Griffiths and Padovani 1990) have spectra that are generally similar in character to the 30 keV bump, but not similar enough: few X-ray binaries have cutoffs >20 keV (White, Swank and Holt, 1983), and the combination of such binaries and supernova remnants cannot replicate the sharpness of the fall of the CXB spectrum >30 keV.

Boldt (1987) has characterized arbitrary spectral forms with the parametric form:

$$F(E) = E \, (dN/dE) = C \, E^{-\alpha} \exp(-E/B) \tag{1}$$

If we can identify the brightest discrete contributors to the CXB, *i. e.* if we know about the sources that are just below the threshold of the *HEAO-1 A2* experiment, we can subtract out their contribution to the Marshall *et al.* (1980) CXB spectrum. A prescription for this spectral analysis is elegantly treated in Figure 4, from Boldt (1987). Here the approximate residual spectrum required to produce the CXB, in the form of equation (1), is exhibited for an arbitrary contribution to the "foreground" from sources with spectra like those of Seyfert galaxies. For a foreground >30% (at 3 keV), $\alpha<0.2$ and $B<30$ keV. Since α approximates the Gaunt factor for thermal spectra (which decreases as a function of temperature), this would imply candidate thermal sources for the residual CXB with $kT>200$ keV in their proper frame at $z>6$.

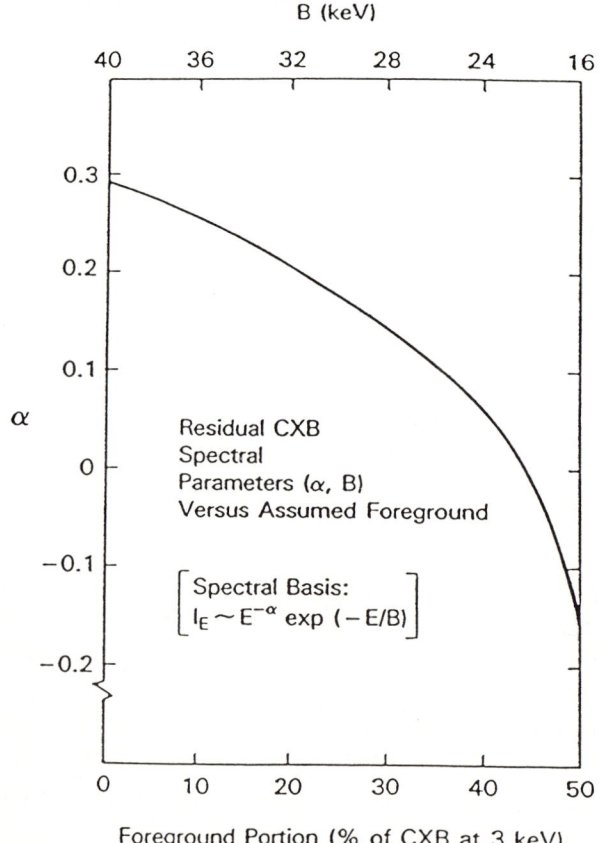

Figure 4 Spectral parameters (α and B, from Equation 1) for the residual CXB as a function of the assumed percentage of AGN foreground at 3 keV.

Table 1 Approximate Spectral Forms

Type	α	B
Seyfert I Galaxies (as per Turner and Pounds 1989)	0.7	>100 keV
Quasars (as per Hasinger, this volume)	1.2	>20 keV
Clusters of Galaxies (as per Smith *et al.* 1979)	0.4	6 keV
Pulsar-driven Supernova Remnants (as per the Crab nebula)	1.1	>100 keV
Neutron Star Binaries (as per White, Swank and Holt 1983)	~ 1	~ 20 keV
CXB (as per Boldt 1987)	0.29	40 keV

In fact, there are no discrete sources whose spectra can match the CXB. Further, virtually all the discrete sources that might substantially contribute to the CXB exhibit spectra that are steeper than the CXB, so that there is no simple way to imagine how some combination of such sources can work, either. On the other hand, the excellent fit of the CXB spectrum to that of a hot thin gas deserves a careful examination.

4   TRULY DIFFUSE X-RAYS AND THE SPECTRAL COINCIDENCE

The suggestion that the CXB might arise in an intergalactic hot gas actually pre-dates the detailed spectral measurement of Marshall *et al.* (1980) (*e. g.* Cowsik and Kobetich 1972). Additional impetus to take the spectral coincidence seriously is afforded by the fluctuation data. Hamilton and Helfand (1987) and Barcons and Fabian (1990), from analyses of the "source-free" portions of images in the *Einstein Observatory* data base, determined that the amplitudes of small-scale fluctuations were much smaller than expected on the basis of the CXB arising from discrete objects like quasars. These data demand at least thousands of sources per square degree, which is not compatible with quasar counts (*e. g.* Schmidt and Green 1986) but could be trivially satisfied by a truly diffuse origin.

Field and Perrenod (1977) carefully considered the implications of such a hot intergalactic medium (IGM), and there have been many subsequent scenarios that invoke a thin, hot gas to match the 3-50 keV spectrum. A perfectly smooth distribution of hot gas in the IGM can explain both the spectral and fluctuation data and, as a bonus, will provide fully 1/3 the closure density of the universe (considerably more mass than has been observed directly by any other means). This required mass can be reduced by clumping (since the bremsstrahlung emissivity is proportional to the square of the particle density), but the fluctuation data severely limit random clumping, so that additional *ad hoc* assumptions are required. For example, Daly (1987) can match both the spectral and fluctuation data with a scenario that features a uniformly spaced grid of diffuse clumps of gas condensed around decayed seeds of unspecified dark matter.

The most serious difficulty that a diffuse origin for the CXB must face is the recent *COBE* (*Cosmic Background Explorer*) measurement of the spectrum of the cosmic microwave background. The lack of sub-mm deviations from a perfect blackbody (Mather *et al.* 1990) requires that the Comptonization parameter $y=(kT_e/m_ec^2)\tau_e <10^{-3}$ (where τ_e is the optical thickness to Thomson scattering in the IGM), in contrast to the $y\approx10^{-2}$ that is expected for a uniform IGM. This totally precludes a hot smooth IGM which produces more than a few percent of the CXB, since $y\propto n_e$ via τ_e and the X-ray emissivity $\propto n_e^2$. In principle, there can still be a substantial contribution to the CXB from clumped diffuse matter arranged to satisfy the X-ray data and the *COBE* y-parameter limits.

Surprisingly, the limits imposed by X-ray measurements of discrete sources are not very restrictive, in spite of reports that large percentages (~25-50%) of the CXB are identified as arising from point sources (*e. g.* Giacconi *et al.* 1979; Maccacaro *et al.* 1991; Burg, this volume). It would appear that such reports, even though they are based upon <3 keV data, would devastate the possibility of substantial CXB arising from diffuse emission, since the subtraction of the contribution of such sources would seem to destroy the spectral goodness-of-fit at low energies, as absorption by cold gas local to the source cannot be invoked for diffuse emission. This argument seems intuitively obvious, but it actually does not substantially reduce the fraction of the CXB that can arise from hot gas, as demonstrated in Figure 5 (which reprises a discussion from Holt, 1980).

The "fraction of the CXB" at energies <3 keV has usually referred to the low-energy extension of the Marshall *et al.* (1980) spectrum (using the form of Boldt (1987), or a power-law approximation $8E^{-0.4}$ that is actually 7% in excess of that extension at 1 keV). We know from *Einstein Observatory* data that there are important contributions to the 1 keV CXB from quasars, Seyfert galaxies and clusters of galaxies. Figure 5 assumes that each of these three classes contribute exactly 10% of the extension of the Marshall *et al.* (1980) spectrum at 1 keV. The "standard" spectra of Table 1 have been used for quasars ($\alpha=1.2$), Seyfert galaxies ($\alpha=0.7$) and clusters of galaxies ($kT=6$ keV), and the remainder of the CXB is synthesized from a 40 keV hot gas (*i. e.* no attempt is made to find a better temperature for the residual hot gas component).

As can be seen from the figure, the apparent 40 keV coronal spectrum of the CXB between 3-50 keV can be reasonably well-matched by the sum of the three 10%-at-1 keV discrete components and almost 90% of exactly the same 40 keV bremsstrahlung spectrum. The synthesized spectrum is therefore 20% in excess of the extension of the CXB spectrum at 1 keV, although it never differs by more than about 5% from the measured CXB spectrum within the 3-50 keV band. It also exceeds the extension of the 40 keV spectrum at high energies, where local Seyferts dominate.

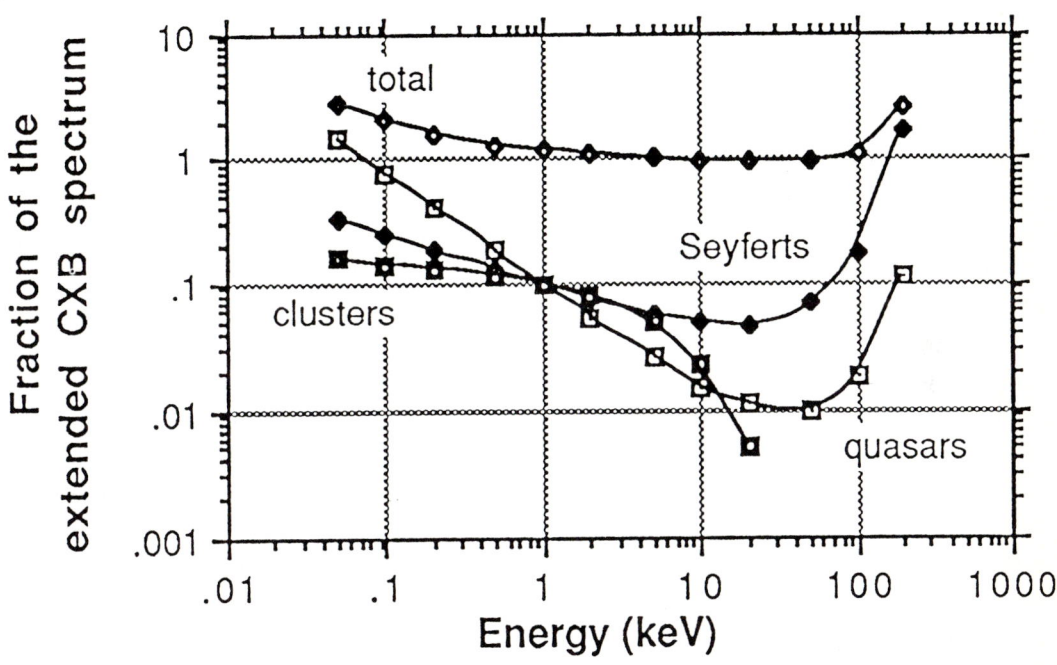

Figure 5 A synthesis of the CXB from clusters of galaxies, quasars, Seyfert I galaxies (each of which contributes exactly 10% of the extension of 3-50 CXB spectrum at 1 keV), and the CXB spectral form itself making up the rest (approximately 89%). The data are expressed as fractions of the extension of the CXB spectral form of Boldt (1987). The trace labelled "total", the sum of the four contributions, is similarly expressed as a ratio relative to the Boldt spectrum.

It is interesting to note that the addition of a fourth 10%-at-1-keV component of "steep spectrum quasars" with $\alpha=2$ will not substantially change the amount of CXB-like component allowed such that the synthesis will always be within 5% of the 3-50 keV CXB. Here the excess relative to the extension of the measured 3-50 keV spectrum would be almost 30% at 1 keV, >60% at 0.5 keV, and >200% at 0.2 keV. In terms of the standard jargon that has been used in the literature, the previous sentence might be restated: "discrete sources make up 30% of the CXB at 1 keV, >60% of the CXB at 0.5 keV, and overproduce the CXB at 0.2 keV by a factor of two!").

This exercise is not meant to demonstrate the necessity of a large fraction of the CXB arising from diffuse hot gas. Instead, the importance of the preceding argument is twofold. First, we cannot rule out a significant contribution to the CXB from hot gas (clumped, to be sure, in order to satisfy the *COBE* limits on the y-parameter), but even <10% would be "significant." Second, and even more important, we should never allow conventional wisdom to influence the manner in which data are utilized to exclude unpopular interpretation. Although there are lots of good arguments against it, the fact remains that clumped thin hot gas can still be a major contributor to the CXB on the basis of the available hard evidence.

5   AGN SPECTRA REVISITED

The deep exposures of *Ginga* have yielded some interesting results on the spectra of AGN. There had been previous indications of partial (~10%) covering of AGN by optically thick clouds (Holt *et al.* 1980), and Morisawa *et al.* (1990) found instances in the *Ginga* data of Seyfert galaxies that are slightly more covered. These authors concluded that they could, with the addition of some *ad hoc* assumptions (including an even higher covered fraction), synthesize a fair approximation to the CXB spectrum, since partial covering can produce an apparent hardening of the source spectrum. Their synthesis involves a population of sources with moderately strong exponential luminosity evolution that are uniformly distributed out to $z_{max} \approx 4$, ~60%-covered with log $N_{Hcov} > 24$, and each with a sharp cutoff at 110 keV in its rest frame. The manner in which partial covering can simulate spectral hardening is shown in Figure 6.

Another way to produce such hard spectra is via Compton scattering in compact sources. There have been many such suggestions in the literature. One scenario for the production of gray-body spectra of arbitrary "grayness" that has some semi-quantitative connection to a physical model arises naturally in the suggestion of Boldt and Leiter (1987) that AGN can be compact enough to be optically thick to Compton scattering in their "precursor" stages at $z>5$. Somewhat more arbitrarily, Morisawa and Takahara (1989) postulated that AGN might have flatter spectra with increasing z (spectral evolution), while Schwartz and Tucker (1988) suggested that if AGN have spectra that flatten with increasing energy, they would appear to have flatter spectra with increasing z without any actual spectral evolution. Note that these ideas all require an unexplained high energy cutoff >100 keV in order to match the CXB deficiency >40 keV.

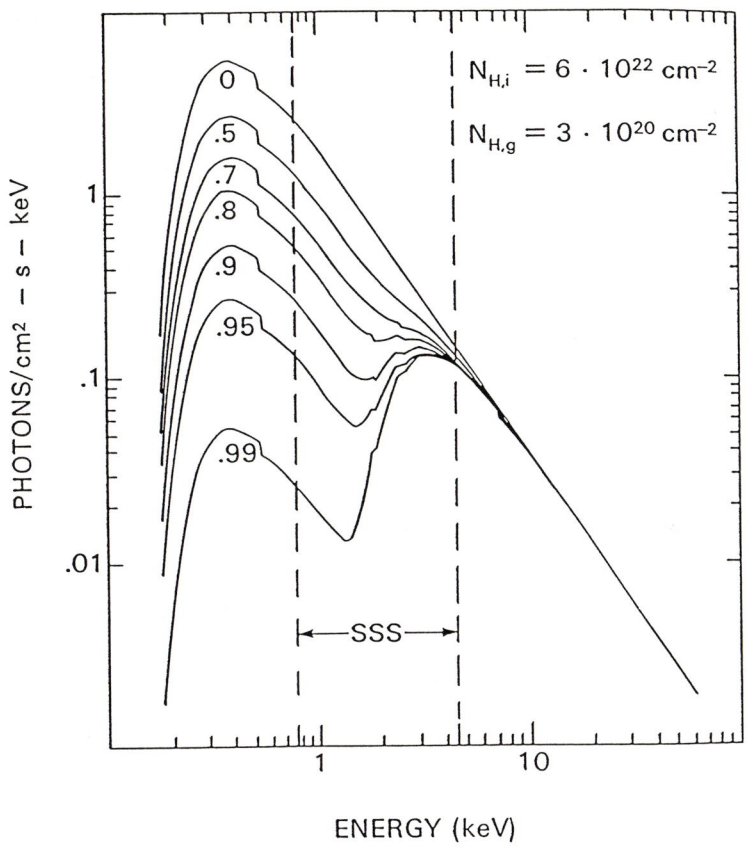

Figure 6 The spectral hardening effect of partial covering of an X-ray source in the energy range of the *SSS (Solid State Spectrometer)* onboard the *Einstein Observatory*. $N_{H,g}$ is the local neutral gas column density in our galaxy in the direction of the source, and $N_{H,i}$ is the intrinsic column density in the source, illustrated for values of covering fraction from 0 to .95 *(from Reichert et al. 1985)*.

Further analysis of *Ginga* spectra of AGN (Pounds *et al.* 1990) has elicited some interesting evidence for Compton scattering, with implications for the origin of the CXB. If we assume that the X-rays that reach us are the combination of those emitted primarily from the source plus those that are scattered or fluoresced into the line of sight as the products of the interaction of other primary X-rays with circumsource material, we can produce spectral hardening that is (almost) indistinguishable from that which can be produced via partial covering. Low energy X-radiation that is incident on cold (10^5 K) material (an accretion disc, for example) will be photo-absorbed, but higher energy X-rays will scatter and can re-emerge from the scattering medium, as illustrated in Figure 7. In principle, the relative roles of absorption, partial covering and reflection may be distinguishable from the details of the Fe emission lines and absorption edges that will be present in the emergent spectrum. It is interesting to note that such an analysis by Pounds *et al.* (1990) suggests that the $\alpha\approx0.7$ that are observed for most AGN >3 keV can arise from the combination of reflected radiation with somewhat steeper primary spectra ($\alpha\approx1$), which are similar to those that have been observed from AGN <3 keV.

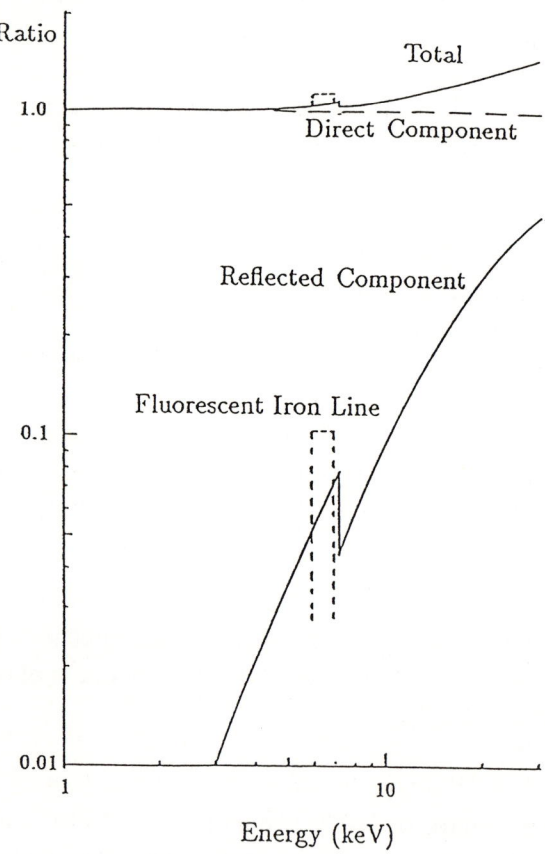

Figure 7 The ratio of total emergent X-rays from AGN with a 2π cold reflector, to the primary radiation that would be observed with no such reflector. The details of the Fe line emission are a function of the reflector parameters, and are included here as a 1 keV-wide band (*from Inoue 1989*).

6 SUMMARY

In spite of the remarkable fit to a 40 keV bremsstrahlung spectrum over a dynamic range of almost 20, the extent to which diffuse hot gas can be a major contributor to the CXB is severely limited by the *COBE* limit on the *y*-parameter. Even with the clumping that is then required, however, any substantial (>10%) contribution can have major implications on the total baryon mass in the universe, so that even minor contributions to the CXB from hot diffuse gas can be of fundamental importance, and cannot be discounted.

The measured spectra of discrete sources are not consistent with that of the CXB, and cannot be simply reconciled with a model that will superpose them to synthesize the CXB spectrum. A prime difficulty with such models is that they require an *ad hoc* high energy cutoff in each source to reproduce the relatively sharp cutoff of the CXB above 40 keV. The problem is not that each source must have a high energy cutoff, but that they all conspire to have cutoffs such that their superposition displays a cutoff that is, itself, relatively sharp. It would seem that this necessary coincidence must have an origin which is fundamental rather than accidental.

A logical candidate for this "fundamental" cutoff energy is the annihilation of electron-positron pairs at photon energies $E=m_ec^2$ (Fabian, Done and Ghisellini 1988). The difficulty with this suggestion is that the radiation must originate at a mean redshift $z\approx20$, *i. e.* such that Compton scattering will smooth out precisely the the sharpness that we seek. There is no obvious natural energy $E<200$ keV that can alleviate this difficulty. Or is there?

Compton reflection in AGN spectra supplies such a natural, if not immediately obvious, intermediate energy that depends solely upon physical constants. The Klein-Nishina cross section for the Compton scattering of photons as they near energies at which electrons become relativistic peaks strongly in the forward direction, so that the <u>backscattered</u> photons that constitute the reflected component described in Figure 7 exhibit a natural peaking at ~120 keV (Lightman and White 1988). Fabian *et al.* (1990) have suggested that AGN, with the scattered component an order-of-magnitude in excess of the direct emergent $\alpha=0.7$ component, and with volume emissivity varying as $(1+z)^4$ out to $z_{max}=5$ ($z_{mean}\approx1.5$), can come very close to matching the CXB spectrum. Rogers and Field (1991) have embellished the details to accommodate an $\alpha\approx1$ primary spectrum that can make the spectral fit even better at low energies, but this may not be necessary in view of the fact that there are other sources that can make up the difference. These authors also provide a possible explanation for the otherwise bizarre geometry that might be required to produce an order of magnitude more scattered component than direct component: they suggest that the primary X-ray source arises from magnetically confined electrons that would necessarily produce radiation which could be directed at a thick cold disc instead of being emitted isotropically (Rogers and Field, this volume). As attractive as this model for the X-ray spectrum is, however, it is a bit disconcerting that not a single such 90%-obscured source has been observed directly.

For the past thirty years, all reviews of the CXB spectrum have looked to future space-borne instruments for the unambiguous resolution of the mystery of its origin. This review concludes with a respectful reprise of that tradition.

7 REFERENCES

Barcons, X. & Fabian, A. C. 1990, *MNRAS* **243**, 366

Boldt, E. 1987, *Phys. Rept.* **146**, No. 4, 215

Boldt, E. & Leiter, D. 1987, *Ap. J. (Letters)* **322**, L1

Bookbinder, J., *et al.* 1980, *Ap. J.* **237**, 647

Burg, R. (this volume)

Canizares, C. R. & White, J. L. 1989, *Ap. J.* **339**, 27

Cowsik, R. & Kobetich, E. J. 1972, *Ap. J.* **177**, 585

Daly, R. 1987, *Ap. J.* **322**, 20

Fabian, A. C., George, I. M., Miyoshi, S. & M. J. Rees 1990, *MNRAS* **242**, 14p

Fabian, A. C., Done, C. & Ghisellini, G. 1988, MNRAS **238**, 729

Field, G. B. & Perrenod, S. C. 1977, *Ap. J.* **215**, 717

Giacconi, R., Gursky, H., Paolini, F. & B. Rossi 1962, *Phys. Rev. Lett.* **9**, 439

Giacconi, R., *et al.* 1979, *Ap J. (Letters)* **234**, L1

Giacconi, R. & Zamorani, G. 1987, *Ap. J.* **313**, 20

Griffiths, R. E. & Padovani, P. 1990, *Ap. J.* **360**, 483

Gruber, D. (this volume)

Hamilton, T. T. & Helfand, D. J. 1987, *Ap. J.* **318**, 93

Hasinger, G. (this volume)

Holt, S. S. 1980, in *X-Ray Astronomy* (R. Giacconi & G. Setti, eds), Reidel, p. 327

Holt S. S., *et al.* 1980, *Ap. J. (Letters)* **241**, L13

Lightman, A. P. & White, T. R. 1988, *Ap. J.* **335**, 57

Maccacaro, T., *et al.* 1991, *Ap. J.* (in press)

Marshall, F. E. *et al.* 1980, *Ap. J.* **235**, 4

Mather, J. C. *et al.* 1990, *Ap. J. (Letters)* **354**, L37

Morisawa, K., Matsuoka, M., Takahara, F. & Piro,L. 1990, *Astron Ap.* **236**, 299

Morisawa, K. & Takahara, F. 1989, *P. A. S. J.* **41**, 873

Mushotzky, R. F. 1984, *Adv. Space Res.* **3**, 157

Piccinotti, G.*et al.* 1982, *Ap. J.* **253**, 485

Pounds, K. A. *et al.* 1990, *Nature* **344**, 132

Reichert, G. A., Mushotzky, R. F., Petre, R. & Holt , S. S.1985, *Ap. J.* **296**, 69

Rogers, R. D. & Field, G. B. 1991, *Ap J (Letters)* **370**, L57

Rogers, R. D. & Field, G. B. (this volume)

Rothschild, R. *et al.* 1979, *Space Sci. Instr.* **4**, 269

Rothschild, R. E. *et al.* 1983, *Ap. J.* **269**, 423

Schmidt, M. & Green, R. F. 1986, *Ap. J.* **305**, 68

Schwartz, D. A. & Tucker, W. H. 1988, *Ap. J.* **332**, 157

Setti, G. & Woltjer, L. 1979, *Astron. Ap.* **76**, L1

Smith, B. W., Mushotzky, R. F., & Serlemitsos, P. J. 1979, *Ap. J.* **227**, 37

Turner, T. J. & Pounds, K. A. 1989, *MNRAS* **240**, 833

White, N. E., Swank, J. H. & Holt , S. S. 1983, *Ap. J.* **270**, 711

Wilkes, B. J. & Elvis, M. 1987, *Ap. J.* **323**, 243

The Hard X-Ray Background

D. E. GRUBER

CASS, University of California, San Diego

1 ABSTRACT

The observations over the last 20 years of the x-ray background in the "hard" x-ray range above 10 keV are reviewed, with special emphasis on those from the HEAO-1 mission. Near 10 keV the spectrum is very well determined, near 100 keV fairly well, and above one MeV barely determined. Dipole anisotropy of the background, marginally measured at 90-165 keV, is at least consistent with the 2-10 keV dipole and also the microwave dipole. Fluctuations, and hence implied source density, up to 160 keV are roughly consistent with 2-10 keV results, and may indicate a spectral index smaller than 0.7.

2 INTRODUCTION

The question of the origin of the diffuse X-ray background continues to challenge us. Counts of high-latitude sources near one keV, most recently with ROSAT (Hasinger 1992), indicate that active galaxies (AGN) can account for a large fraction of the background at this energy. Moreover, measured graininess of the background is strongly suggestive of a point source origin for at least most of the background. All the spectral information so far obtained, however, indicates that the prime candidates, Seyfert galaxies and QSO's, or at least those close to the present epoch, have spectra which are markedly different from the background. Recently the possible importance of Compton reflection in AGN x-ray spectra has been recognized (Lightman and White 1988), and the corresponding effect on the background spectrum (Fabian et al. 1990). Supporting observational evidence for reflection features in a composite AGN spectrum has been obtained (Pounds et al. 1990) with Ginga data. The reflection features in the Ginga data are somewhat faint, and if AGN's are to make up most of the background, then a larger ratio of reflected to unreflected flux, or "covering factor", is required in these sources (Rogers and Fields 1991). Two possibilities suggest themselves: either we preferentially see present-epoch sources with low covering factor, or earlier epoch sources had on average higher covering factor. Since definitive observations may not be quick in coming, it is worthwhile at the present to make a best assessment of the available data, particularly that in the "hard" x-ray range at energies above 10 keV or so. Hard X-ray and soft gamma-ray observations of the diffuse background are thus reviewed here.

While measurements of the diffuse background properties can be made with reasonable precision and accuracy near the lower threshold of this energy range, with increasing energy photon fluxes and instrument sensitivities decline rapidly, until in the energy range of 1-10 MeV it is a bit of a triumph merely to make a significant observation. This being the case, one might ask whether the weaker observations, obtained by different techniques, are in such agreement that the collective result may be used with confidence. For the spectral results reviewed here, such confidence appears warranted. Results on spatial structure are so far significant up to 160 keV.

Emphasis is placed here on results from the HEAO-1 mission. Both the Cosmic X-Ray Experiment (A2) and the Hard X-Ray/Low-Energy Gamma Ray Experiment (A4) had as primary goals the measurement of the diffuse background. They have indeed produced the most solid observations to date. The A2 experiment (Rothschild et al. 1979) was made up of large-area proportional counters in three energy ranges: Low energy detectors (LED) for 0.2-2 keV, Medium Energy Detectors (MED) for 2-10 keV, and High Energy Detectors (HED) sensitive from 3-60 keV. Apertures were specially designed to map the spectrum and structure of the background. Spectral results were obtained from 3 to 60 keV, structure (dipole term and fluctuations) results pertain mostly to the 2-10 keV band. The A4 experiment (Matteson 1978) consisted of scintillators also optimized for three energy bands. Using the same nomenclature, these has ranges 13-160 keV for the two LED's, 80-2000 keV for the four MED's, and 120 keV to 10 MeV for the HED. Rather large apertures of 114 sq. deg. (LED), 855 sq. deg. (MED), and 4170 sq. deg., reflected a tradeoff between maximum signal-to-noise and source confusion limitations.

3 THE SPECTRUM

Measurements of the cosmic background spectrum in the hard x-ray range from 10 keV up to 10 Mev have been obtained with a variety of techniques: proportional counters at the lower end, scintillators in the middle range, and Compton telescopes at the high end. Observations have been conducted from balloons and space platforms. Balloon observations are limited both in observing time and coverage of the sky. Experiments on space platforms generally suffer from a variable ambient radiation environment and consequent variable internal background, which complicates analysis (although Mazets et al. (1975) made this the key to their analysis). Results reported in the last 20 years are listed in Table 1. Only the most recent report from each group is listed.

Observing techniques are varied. The A2 experiment (Rothschild et al. 1979, Marshall et al. 1980) used apertures of differing solid angles and measured the difference in counting rates. Daniel et al. (1972) and Mandrou et al. (1979) extrapolated to zero atmospheric contribution through a curve of growth analysis. The two Compton telescope teams also made use of an observed curve of growth to determine atmospheric contribution. Trombka et al. (1977) applied large calculated corrections for induced internal radioactivity in the detector. Mazets et al. (1975) used observed dependences with geomagnetic latitude to correct for counting rates from atmospheric fluxes and internal background induced by cosmic rays. Fukuda et al. (1975) and the A4 team (Gruber et al. 1985) relied primarily on a shutter, although the A4 result also employed a modest degree of correction for variable internal background (Gruber et al. 1989, Jung 1988, Briggs 1991). For the most part the results are remarkably coherent. Only the Kosmos result is clearly discordant; its very high reading at energies below 200 keV is probably due to the influence of the brighter x-ray sources in this omnidirectional detector. The A4 data have undergone some refinement since 1985, and the results from the LED and MED detectors in the region of overlap at 80-160 keV are no longer in conflict. At the lowest end of the range, where results are precise to a few percent, the close agreement between A2 and the A4 LEDs is especially remarkable (and comforting) in view of the differing observational techniques.

Table 1

Observations of the Diffuse Background

Energy Band	Detector	Platform	Reference
.1-8.5 MeV	scintillator	balloon	Daniel et al. (1972)
28-4100 keV	scintillator	Kosmos	Mazets et al. (1975)
.1-4 MeV	scintillator	balloon	Fukuda et al. (1975)
2-10 MeV	Compton tel.	balloon	White et al. (1977)
.3-10 MeV	scintillator	Apollo	Trombka et al. (1977)
0.3-6 MeV	scintillator	balloon	Mandrou et al. (1979)
3-60 keV	prop. counter	HEAO-1	Marshall et al. (1980)
1.1-10 MeV	Compton tel.	balloon	Schönfelder et al. (1980)
13-4000 keV	scintillator	HEAO-1	Gruber et al. (1985)

In Figure 1 are displayed a selection of the observed spectra. The results with smallest reported errors were chosen for display, with attention given to representing results obtained by different techniques. Empirical functional fits to the data in Figure 1 have been determined, and the most successful is shown below and also in Figure 1. The energy flux in units of keV/(cm²-s-keV-sr) is represented in piecewise fashion:

$$7.877E^{-.29} * exp(-E/41.13 keV) , \quad 3keV < E < 60keV$$

$$1652 * E^{-2.00} + 1.754 * E^{-.70} \quad , \quad 60keV < E < 6000keV.$$

Below 60 keV the bremsstrahlung form which fits the A2 data so well is employed. The Gaunt factor $E^{-.29}$ (Boldt, private communication) was used unaltered, and it is seen that an e-folding energy slightly larger than 40 keV is required by the combination of data. Above 60 keV the sum of two power laws is needed: one essentially continues as an extension of the bremsstrahlung form, and the other has index 0.7, which characterizes the spectra of nearby AGN's. If the index of the second power law is allowed to vary, its best-fit value is 0.56 ± 0.20, which does not differ significantly from 0.7. An interesting rejected model consists of the sum of the bremsstrahlung form and one power law; this falls considerably below the measurements in the region 100-300 keV. Although the fit above is the best in a least-squares sense, it is not formally acceptable, having reduced chi-square around two. The scatter of the data from this function increases by about a factor of two each half decade, from around 2% at 3-120 keV to 60% RMS 1-3 MeV. It is a curious fact that the variance on each interval is roughly double that from stated errors. The data from the different experiments are somewhat more consistent with each other than they are with the function. Even so, the observational picture is clear: from an initial index of 0.4 at the lowest energies, the spectrum follows a bremsstrahlung form until by 60 keV it has an index of about 1.6, then between 200 keV and 1 MeV undergoes a second transition to an index near 0.7. Above a few MeV the spectrum almost certainly steepens in order to connect smoothly with spark chamber fluxes at tens of MeV, but the 1-10 MeV data by themselves cannot indicate such a break.

4 SPATIAL STRUCTURE

In the hard x-ray range systematic studies of spatial structure of the extragalactic sky were performed first with an experiment on the OSO-3 satellite, although only upper limits were obtained (Schwartz 1970). Schwartz examined his data for field-to-field fluctuations, presumably arising from fluctuations in the number of sources, and also for various forms of global structure: galactic emission; emission associated with the supercluster; 12 and 24 hour components arising for various cosmological reasons, but most notably the 24-hour ($\cos\theta$) component resulting from our peculiar velocity with respect to this radiation field.

More recent studies have been performed with the HEAO-1 satellite. With the A2 experiment comprehensive studies have been performed of the galactic emission (Iwan et al. 1982) and of the extragalactic sky (Shafer 1983). Shafer limited his data set to the most stable HED, which was nominally sensitive to 60 keV, although it should be noted that only about 10% of its diffuse background counting rate came from photons with energy above 10 keV, and 5% above 14 keV. His results are thus more characteristic of the traditional 2-10 keV band than of the hard band. Field of view was roughly pyramidal, 3 by 3 degrees FWHM for half the cells of the detector and 3 by 6 degrees FWHM for the others. The sensitivity of the HEAO-1 A4 experiment begins at 13 keV and extends nominally to several MeV, but useful sensitivity for spatial structure ended at 165 keV. The Low Energy Detectors, sensitive from 13 to 160 keV, had a fan beam aperture, 1.43 by 20 degrees FWHM, while the Medium Energy Detectors, whose threshold was about 90 keV, has a circular field of view with FWHM 16.5 degrees. The A4 results have not yet been published, although papers are now in preparation.

4.1 The Dipole

Phenomenologically, the first order terms in a multipole expansion of a radiation field on the sphere are simply terms obtained as dot products with the three unit vectors of a cartesian coordinate system. Equivalently, the first order deviation from spherical uniformity can be expressed as an amplitude, I_1, times $cos\theta$, where θ is the angle to an apex direction. This *dipole* variation is of special interest, because it arises naturally from peculiar motion of the observer with respect to an isotropic radiation field. Thus a measured dipole amplitude, $\delta I/I_0$, is normally related to a peculiar velocity v with

$$\delta I/I_0 = (2 + \Gamma)(v/c)cos\theta,$$

where c is the speed of light, Γ is the power law spectral index of the radiation field, and $(2 + \Gamma)$ is the well-known Compton-Getting factor. It is useful to keep in mind that a measured dipole anisotropy is not necessarily kinematic in origin, but may measure (in part) a local enhancement of x-ray intensity.

Shafer (1983) measured a dipole component at 95% confidence with amplitude 0.5% and apex at $(l = 282°, b = 30°)$. With kinematic interpretation a value of 475 ± 165 km/sec is obtained. The 90% confidence region for the Shafer result is rather large ($\pm75°$ for b, $\pm60°$ for l), nevertheless the result is in remarkable agreement with the microwave peculiar velocity and apex (Smoot et al. 1991).

With the A4 experiment a dipole term has been measured on the band 95-165 keV with the MED detectors. A 5-month section of the mission was chosen during which the shutter was alternated between two detectors. The counting rate while shuttered thus provided a control experiment during which no dipole term should appear. Without the control any result from the open detectors could be compromised by fluctuations of the detector internal background, which varied on all time scales. Indeed, no dipole term was seen in the control experiment, while the two detectors while open detected a dipole anisotropy $\delta I/I_0$ of 2.0%. By referencing to the control experiment, a net amplitude of $(2.18 \pm .66)\%$ is obtained, with apex $(l = 304°, b = 26°)$. This direction is consistent, within errors, with both the Shafer apex and the microwave apex. However, the inferred velocity, 1450 ± 440 km/sec, is very much larger. The one-sigma error bars quoted here are deceptive. Using the technique of Lampton, Margon, and Bowyer (1976) for three interesting parameters, the null hypothesis (no dipole component) is excluded at only 90% confidence. The 95-165 keV dipole is thus not inconsistent with the A2 result, but more importantly, it cannot yet be said to be securely measured. If this dipole anisotropy can be confirmed at its measured amplitude, a kinematic explanation would seem less likely than a concentration of hard x-ray emitting sources in the general vicinity of the apex.

4.2 Fluctuations

Probably the most solidly determined footing for our understanding of the graininess of the x-ray sky comes from the joint work with A2 by Shafer (1983) on the fluctuations and by Piccinotti et al. (1982) on the log N – log S relation and luminosity functions for active galaxies. Shafer shows that the observed A2 fluctuations are closely consistent with a Euclidean extension of the Piccinotti sample downward in observed flux by more than two orders of magnitude. This is to say, it very much appears that the Piccinotti population is responsible for the observed fluctuations. Ginga observations of fluctuations (Hayashida 1990, Warwick and Stewart 1989) and of faint sources (Kondo 1991) are both very consistent with the A2 results. The question here is whether hard x-ray results are also consistent with A2.

Hard X-ray Fluctuations have been studied with the LED and MED detectors of A4. The sky outside a $\pm 20°$ band on the galactic equator was partitioned into roughly 1000 full-response pixels for the LEDs. After elimination of pixels contaminated by the 12 high-latitude A4 sources (Levine et al. 1984) there remained about 750. Data from the entire mission were accumulated in these pixels and then examined for variance. For the MED a similar procedure was followed, although this MED set consisted of only 99 approximately square pixels of area 225 square degrees. In addition to sky fluctuations, variance in these pixels contained contributions from counting statistics and from fluctuations of the detector internal background. The latter term was largely suppressed through averaging of enough data, but a residual correction, typically of the order of 0.3% of the total counting rate and 0.6% of the sky flux, was accomplished with a time dependent model of the detector internal background (Gruber et. al 1989). The LED data were divided into four factor-of-two energy bands and the MED data into a single band from 95 keV to 165 keV. On these bands the variance for the pixels was about twice that from counting statistics. Variances were then subtracted in quadrature, corrected for pixel smoothing, and referenced to the mean cosmic background intensity in the band. A final scaling by the square root of solid angle was made to correct to the RMS fluctuation levels in Table 2.

Boldt (private communication) has normalized these fluctuation statistics to the Piccinotti sample, and the result is seen in Figure 2. In it, the dashed line at R=1 indicates the expected fluctuation level if the source population extends to the hard x-ray band with power law index α for energy flux. The upper panel shows the result for the customary index; the lower panel shows rather closer agreement with a harder index of 0.4. Although perhaps not conclusive in itself, this data set indicates the sort of harder spectrum for the source population required by the numerous compton reflection models and by the more phenomenological model of Schwartz and Tucker (1988).

Table 2

A4 Sky Fluctuations

Energy Band (keV)	Obs. RMS %/ster
15-25	0.29 ± 0.05
25-40	0.34 ± 0.04
40-80	0.39 ± 0.02
80-160	0.97 ± 0.03
95-165	0.73 ± 0.11

5 SUMMARY

Reported observations of the background spectrum over the last 20 years are generally consistent, given a factor 1.4 adjustment to the stated errors, but this is a very strong statement at a few tens of keV, where agreement is a few percent, and a weak statement above one MeV, where many of the measurements represent marginal detections. Nevertheless, the well-known bremsstrahlung shape is closely verified from its 0.4 index regime below 10 keV to its transition to an asymptotic power law with index 1.6 above 60 keV. Also the spectral shelf around one MeV, having been measured many times by several techniques, appears real. An empirical function of fairly simple form has been generated from the data.

The only hard x-ray dipole measurement, although marginally significant, is consistent in magnitude and direction with the 2-10 keV dipole, and also with the corresponding microwave values. It may indicate, however, a much higher cosine anisotropy, which may point to a concentration of hard x-ray sources.

Fluctuations have been measured to 165 keV, their level is consistent with that in the 2-10 keV band, and they may indicate a mean spectrum harder than $\alpha = 0.7$ for the source population, presumably AGN's.

6 REFERENCES

Briggs, M. S., 1991, unpublished dissertation Univ. Calif. San Diego.

Daniel, R. R., Joseph, G., Lavakare, O., 1972, *Astrophys. Space Sci.*, **18**, 462.

Fukuda, Y., Hayakawa, S., Kasahara, I., Makino, F., Tanaka, Y., 1975, *Nature*, **254**, 398.

Fabian, A. C., George, I. M., Miyoshi, S., & Rees, M.J. 1990, *M.N.R.A.S.*, **242**, 14p.

Gruber, D. E., Matteson, J. L., Jung, G. V., 1985, *Proc. 19th Intl. Cosmic Ray Conf.*, ed. F. C. Jones, **OG-1**, 349.

Gruber, D. E., Jung, G. V., Matteson, J. L., 1989, in *Aip Conf. Proc nr. 106, High Energy Radiation Background in Space*, eds. A. C. Rester and J. I. Trombka (AIP:New York), 232.

Hasinger, G., 1992, these proceedings.

Hayashida, K., 1990, unpublished dissertation, Univ. Tokyo

Iwan, D., Marshall, F. E., Boldt, E. A., Mushotzky, R. F., Shafer, R. A., and Stottlemyer, A., 1982, *Ap. J.*, **260**, 111.

Jung, G. V., 1988, unpublished dissertation, Univ. Calif., San Diego.

Kondo, H., 1991, unpublished dissertation, Univ. Tokyo.

Lampton, M., Margon, B., and Bowyer, S., 1976, *Ap. J.*, **208**, 177.

Levine, A. M., *et al.*, 1984, *Ap. J. Suppl.*, **54**, 581.

Lightman, A. P. and White, T. R., 1988, *Ap. J.*, **335**, 57.

Mandrou, P., Vedrenne, G., Niel, M., 1979, *Ap. J.*, **230**, 97.

Marshall, E. F., Boldt, E., Holt, S. S., Miller, R., Mushotzky, R. F., Rose, R. E., Rothschild, R. E., & Serlemitsos, P. 1980, *Ap.J.*, **235**, 4.

Matteson, J. L. 1978, *AIAA Conference* paper 78-35.

Mazets, E. P., Golenetskii, S. V., Il'inskii, V. N.,Gur'yan, Y. A., Kharitonova, T. V., 1975, *Astrophys. Space Sci.*, **33**, 347.

Piccinotti, G., Mushotzky, R. F., Boldt, E. A., Holt, S. S., Marshall, F. E., Serlemitsos, P. J., Shafer, R. A., 1982, *Ap. J.*, **253**, 485.

Pounds, K. A., Nandra, K., Stewart, G. C., George, I. M., and Fabian, A. C. 1990, *Nature*, **344**, 132.

Rogers, R. D. and Field, G. B. 1991, *Ap. J. (Letters)*, **370**, L57.

R. E. *et al.* 1979, *Space Sci. Instr.*, **4**, 269.

Schönfelder, V., Graml, F., Penningsfeld, F.-P., 1980, *Ap. J.*, **240**, 350.

Schwartz, D. A., 1970, *Ap. J.*, **162**, 439.

Schwartz, D. A., & Tucker, W. H., 1988, *Ap.J.*, **322**, 157.

Shafer, R. A., 1983, unpublished dissertation Univ. Maryland.

Smoot, G. F., *et al.*, 1991, *Ap. J. (Letters)*, **371**, L1.

Trombka, J. I., Dyer, C. S., Evans, L. G., Bielefeld, M. J., Seltzer, S. M., Metzger, A. M., 1977, *Ap J.*, **212**, 925.

Warwick, R. S., and Stewart., G. C., 1989, in *Proceedings 23rd ESLAB Symp. on two Topics in X-Ray Astronomy*, (ESA:Noorwijk), 727.

White, R. S., Dayton, B., Moon, S. H., Ryan, J. M., Wilson, R. B., Zych, A. D., 1977, *Ap. J.*, **218**, 920.

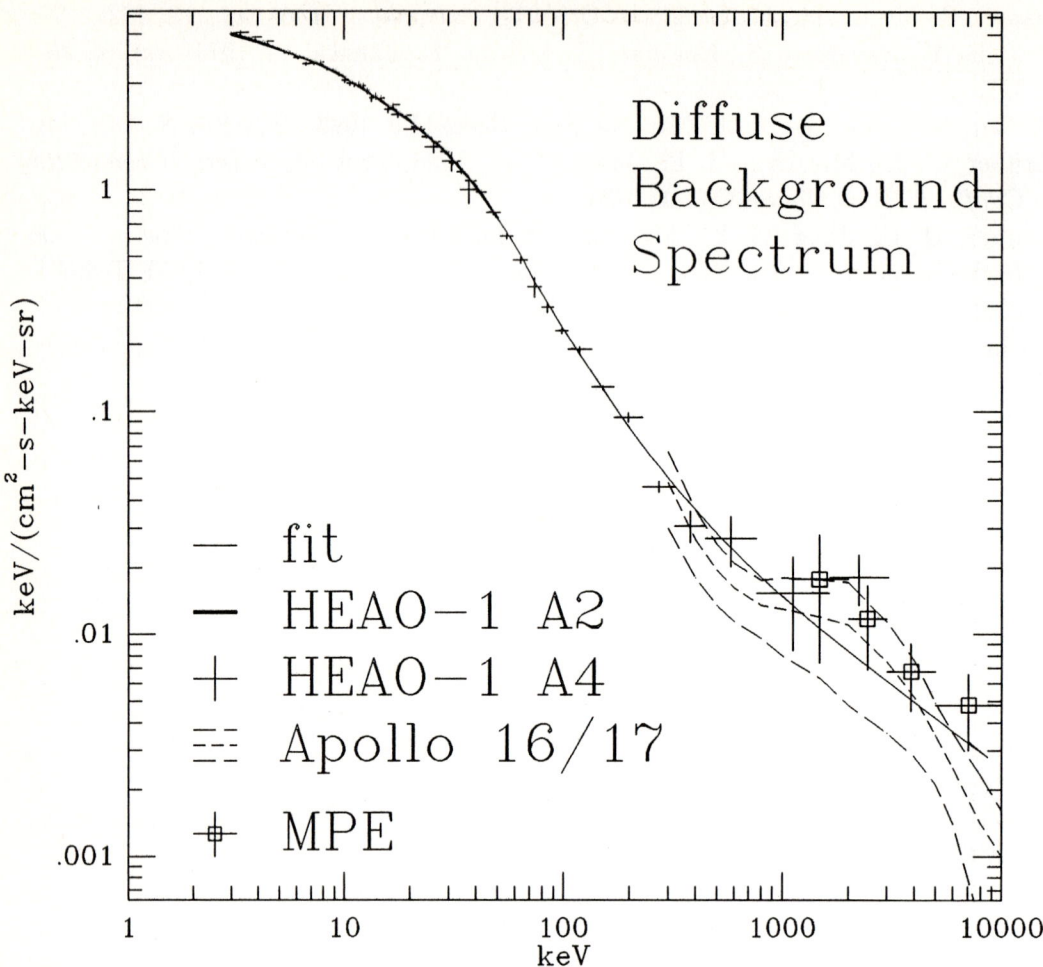

Figure 1. Energy spectrum of the diffuse background. Data from experiments with smallest quoted errors have been selected for display. The thin line is an empirical fit, given in the text. The spectrum makes two smooth transitions between three approximately power-law segments with spectral indices of 0.4, 1.6, and 0.7, in increasing order of energy. The index at highest energy is poorly determined.

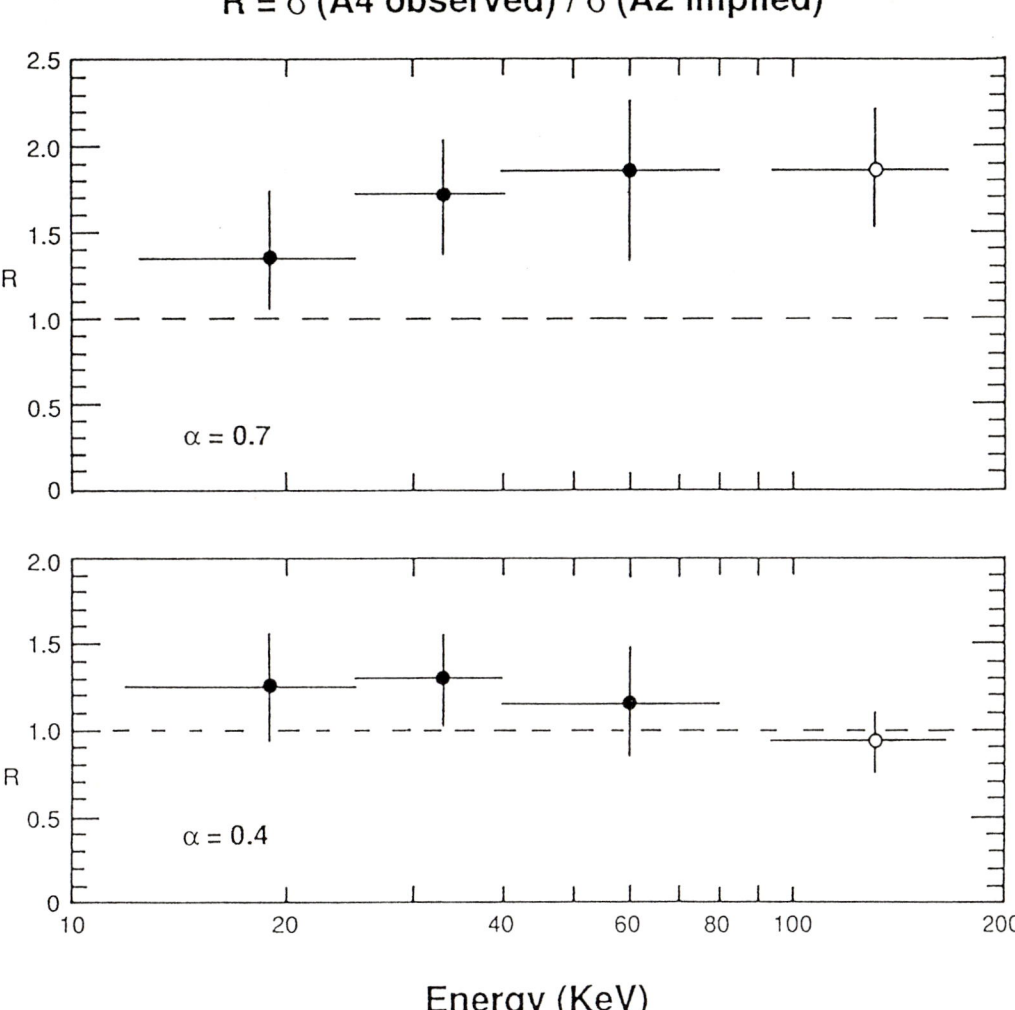

Figure 2. Observed sky fluctuation levels at higher energies compared to levels expected by extrapolation of the Piccinotti et al. (1982) sample. The upper panel shows a general disagreement with the expected R=1 for an assumed spectral index for the earlier-epoch sources which make up the fluctuations. The lower panel shows agreement for an assumed index of 0.4. Thus more distant Seyferts, too faint to observe individually, may have spectral index harder than 0.7, thus easing the problem of explaining the "spectral paradox".

Shadows in the Soft X-ray Diffuse Background

S. L. SNOWDEN

University of Wisconsin – Madison

Max Planck Institute for Extraterrestrial Physics

ABSTRACT

The study of shadows cast by discrete interstellar objects against the soft X-ray diffuse background (0.1−2.0 keV) will provide a sensitive tool for the separation of emission components responsible for the observed flux. This will have extensive implications for our understanding of subjects ranging from local galactic structure to cosmology. This is particularly true in the 0.5−1.0 keV energy range where some models attribute nearly all of the observed diffuse background to galactic sources while other models attribute nearly all to extragalactic sources.

1 INTRODUCTION

1.1 The Soft X-ray Diffuse Background

The spatial structure of the soft X-ray diffuse background (SXRB) in the 0.1−2.0 keV energy range varies very strongly as a function of energy. At 1/4 keV (0.07−0.284 keV), the SXRB exhibits a strong negative correlation with the column density of galactic H I. However, there remains a significant flux in the galactic plane requiring the existence of galactic emission. It is generally accepted that the bulk of the observed flux is galactic in origin though there is still much discussion about the exact distribution of that emission (cf., Jakobsen and Kahn 1986; Snowden et al. 1990; Hirth et al. 1991). In the 0.5−1.0 keV band, the SXRB exhibits greater isotropy suggestive of an extragalactic origin. However, the sky is still dominated by discrete galactic features of varying size up to ∼ 90° (Loop I). An interesting aspect of this band is the general lack of absorption in the galactic plane. This implies the existence of diffuse galactic emission which "fills in" for the absorption of the extragalactic flux. Presumably there is an extension of the emission required in the plane to higher galactic latitudes such as in the case of the collective emission from late-type stars (cf., Schmitt and Snowden 1990). The local cavity, which is apparently filled with hot gas that emits at 1/4 keV, is also a possible source for X-rays observed in the 0.5−1.0 keV band (Arnaud and Rothenflug 1986). In the 1.0−2.0 keV band, galactic features have faded though there is still relatively strong emission towards the

North Polar Spur, the galactic center, and supernova remnants such as Vela/Puppis, the Cygnus Loop, and the Cygnus Superbubble. The general character, however, is isotropic and undoubtedly extragalactic in origin. In the higher energy range, there is obviously some doubt about the appropriateness of the word "diffuse" to describe the X-ray background. I use it in the context of "unresolved."

1.2 Observations with the *ROSAT* XRT/PSPC

With the launch and operation of the *Röntgen Satellite* (*ROSAT*, Trümper 1983), we have by far the most sensitive instrument ever flown to observe the SXRB in the 0.1−2.0 keV energy range. The X-ray Telescope (XRT, Aschenbach 1988), coupled with the Position Sensitive Proportional Counter (PSPC, Pfeffermann *et al.* 1987), provides a large effective area, fast optics, a large field of view ($\sim 2°$), and exceptionally low instrumental background (Snowden *et al.* 1991b).

The quality and utility of the XRT/PSPC for the study of shadows was manifestly demonstrated by the detection of a serendipitous shadow during the survey-verification phase of *ROSAT* operation (Snowden *et al.* 1991a) and the Guest Observer (pointed observation) detection of a shadow by a high-latitude molecular cloud during the pointed-observation-verification phase (Burrows and Mendenhall 1991). Both shadows were cast by clouds associated with the Draco Nebula. The results of the observations were consistent with each other strengthening the confidence in the scientific analysis of data from this new telescope.

ROSAT also carries the Wide Field Camera (WFC, Sims *et al.* 1990), a telescope which covers the energy range 0.070−0.188 keV. The data provided by the WFC are complimentary to the XRT/PSPC and will provide sensitivity to interstellar clouds of lower column density (Harris, Sumner, and Walker 1989). The WFC has higher instrumental background levels than the XRT/PSPC making the analysis of shadow targets considerably more difficult and there is much work in progress.

2 THE SEARCH FOR SHADOWS

2.1 Observation Modes

As demonstrated by the Snowden *et al.* (1991a) and Burrows and Mendenhall (1991) papers, *ROSAT* XRT/PSPC data from both survey and pointed observations can be used in the search for shadows. However, the applications and expected results for the two cases are markedly different due to the significant difference in exposure times. Typical survey exposures are on the order of 0.5 ks while typical pointed observations of shadow targets are over an order of magnitude longer. For a shadow target of degree size, the survey exposure is sufficient only to study shadows in the 1/4 keV energy range and then only at the 5−10% statistical level. The longer exposure of a pointed observation not only improves the statistics of a 1/4 keV observation but allows the search for shadows in the 0.5−2.0 keV energy range (providing the shadow target is sufficiently optically thick, i.e., greater than roughly 2×10^{21} H cm^{-2}). On the other hand, the number of pointed observations of shadow targets is obviously going to be limited.

Data collected during the survey phase of *ROSAT* operations will allow the search for shadows over nearly the entire sky. While, in general, the search will be limited to the 1/4 keV band, optically thick targets of large angular scale can also be investigated in the 0.5−2.0 keV band. The Ophiuchus molecular clouds provide an excellent example of a survey shadow in the higher energy band. Also, with the survey data, analysis is not limited to targets of small angular extent. There are at least two discrete H I clouds which show shadows in the 1/4 keV band and have diameters of $\sim 10°$ (Snowden 1991).

2.2 Pointed Observations

By 1991 October, roughly 11 shadowing targets had been observed by *ROSAT*. The bulk were high latitude molecular clouds identified by Magnani, Blitz, and Mundy (1985). Roughly another 15 targets were scheduled for observation during the remainder of the AO-2 observation period (through 1992 April 15). A number of these targets are particularly interesting since there exist reliable distances to the clouds (cf., Blitz 1991). The prototypical example is MBM 12 with a distance of 60 to 70 pc (Hobbs, Blitz, and Magnani 1986) and a maximum column density of roughly 3×10^{21} H cm^{-2}.

2.3 Results to Date

Unfortunately, at the time of this writing, only the two papers on the shadows in the Draco Nebula have been published, although much work is in progress. It is also unfortunate that the Draco Nebula results will probably not prove to be typical of the sky at high galactic latitudes. The shadows were observed to be at the 50% level (in the 1/4 keV band) which, when coupled with the distance to the clouds (> 210 pc, Blitz 1991; most likely beyond the low velocity H I, Lilienthal *et al.* 1991), implies very strong galactic halo or extragalactic emission. However, preliminary results from the "Lockman's Hole" region (the direction of minimum galactic H I column density; $\sim 40°$ separation on the sky from Draco) are considerably at odds with the Draco results. If the background emission in the direction of the Draco Nebula was also found behind Lockman's Hole, the maximum X-ray intensity would be at least a factor of 3.3 higher than observed (Snowden *et al.* 1991c). Taking the Lockman's Hole result for an estimate of the nominal flux of distant origin, at most 25% of the observed 1/4 keV flux originates beyond the low and intermediate velocity galactic H I (for a typical high latitude column density of 1.5×10^{20} HI cm^{-2}). Even this residual 25% can not be unconditionally attributed to an extragalactic origin since X-ray emission in the halo of our galaxy cannot be ruled out.

While both the Draco Nebula and the Lockman's Hole region show detailed shadows in the 1/4 keV band, this is not a general feature of molecular cloud observations. Preliminary analysis of the MBM 12 data show no shadow at all at 1/4 keV though a probable shadow at 0.5−1.0 keV is observed. The same is true for MBM 7 (Harris 1991). A final example is provided by the Ophiuchus molecular clouds where a very strong shadow is seen in the 0.5−2.0 keV band with no shadow in the 1/4 keV band. The obvious conclusion is that *all* of the observed 1/4 keV flux originates on the near side of these objects.

3 THE FUTURE
Even with the limited results thus far, a complicated picture of the diffuse X-ray emission in our galaxy is beginning to arise. In some directions (e.g., MBM 12, $\sim$ 65 pc; Ophiuchus, < 150 pc, de Geus, Bronfman, and Thaddeus 1990), all of the observed 1/4 keV flux originates locally while in other directions (e.g., Draco), half of the observed flux originates at distances greater than a few hundred parsecs. Given sufficient shadowing targets with a variety of distances and directions, the galactic distribution of X-ray emission can be mapped out. This is also true, but to a lesser extent, in the 0.5−2.0 keV band.

A byproduct of the understanding of galactic emission is, of course, the specification of the extragalactic flux. This knowledge will constrain extragalactic emission models in an energy range where previously there has been significant controversy.

This work was supported in part by the National Aeronautics and Space Administration under grant NAG 5-1438 and the Max Planck Institute for Extraterrestrial Physics.

REFERENCES
Arnaud, M., and Rothenflug, R. 1986, *Adv. Sp. Res.*, **6**, No. 2, 119.
Aschenbach, B. 1988, *Appl. Opt.*, **27**, No. 8, 1404.
Blitz, L. 1991, in *Proceedings of IAU Symposium No. 144, The Interstellar Disk-Halo connection in Galaxies*, ed. J. G. B. M. Bloemen, in press.
Burrows, D. N. and Mendenhall, J. A. 1991, *Nature*, **351**, 629.
de Geus, E. J., Bronfman, L., and Thaddeus, P. 1990, *Astr. Ap.*, **231**, 137.
Harris, A. W. 1991, private communication.

Harris, A. W., Sumner, T. J., and Walker, H. J. 1989, in *Proceedings of IAU Symposium No. 139, Galactic and Extragalactic Background Radiation*, in press.

Hirth, W., Mebold, U., Dahlem, M., and Müller, P. 1991, *Astr. Sp. Sci.*, in press.

Hobbs, L. M., Blitz, L., and Magnani, L. 1986, *Ap. J. (Letters)*, **306**, L109.

Jakobsen, P., and Kahn, S. M. 1986, *Ap. J.*, **309**, 682.

Lilienthal, D., Wennmacher, A., Herbstmeier, U., and Mebold, U. 1991, *Astr. Ap.*, in press.

Magnani, L., Blitz, L., and Mundy, L. 1985, *Ap. J.*, **295**, 402.

Pfeffermann, E., Briel, U. G., Hippmann, H., Kettenring, G., Metzner, G., Predehl, P., Reger, G., Stephan, K.-H., Zombeck, M. V., Chappell, J., and Murray, S. S. 1987, *Proc. SPIE Int. Soc. Opt. Eng.*, **733**, 519.

Schmitt, J. H. M. M. and Snowden, S. L. 1990, *Ap. J.*, **361**, 207.

Sims, M. R. *et al.* 1990, *Opt. Eng.*, **29**, No. 6, 649.

Snowden, S. L. 1991, in *Proceedings of the RAL Workshop on Astronomy and Astrophysics, Hot Gas in the Galaxy*, in press.

Snowden, S. L., Cox, D. P., McCammon, D., and Sanders, W. T. 1990, *Ap. J.*, **354**, 211.

Snowden, S. L., Mebold, U., Hirth, W., Herbstmeier, U., and Schmitt, J. H. M. M. 1991a, *Science*, **252**, 1529.

Snowden, S. L., Plucinsky, P. P., Briel, U. G., Hasinger, G. R., and Pfeffermann, E. 1991b, *Ap. J.*, submitted.

Snowden, S. L., Sanders, W. T., Lockman, F. J., McCammon, D., *et al.* 1991c, in preparation

Trümper, J. 1983, *Adv. Space Res.*, **2**, No. 4, 241.

Chapter III

The isotropy

Small scale anisotropies of the XRB

L. DANESE[1], G. DE ZOTTI[2] & P. ANDREANI[1]

[1] Universita' di Padova, Padova, Italy

[2] Osservatorio Astronomico di Padova, Padova, Italy

1 INTRODUCTION

Studies of fluctuations, small scale anisotropies and AutoCorrelations (ACs) of the X-ray Background (XRB) are useful in constraining counts, emissivity and clustering properties of X-ray extragalactic sources; therefore they furnish relevant clues to the solution of the XRB problem.

In particular source counts can be extrapolated well below the detection limits by studying the small scale fluctuations, with a technique known as P(D) analysis (Scheuer 1957, 1974).

The emissivity of a population can be investigated even in the absence of a significant detection rate, comparing the distribution of deflections of the counts (negative signals allowed) from positions coincident with the sources under study with that found for control fields (Worrall, Marshall and Boldt 1979; Worrall and Marshall 1984; Persic et al 1989).

Depending on the source detection limits and the angular scales used in the analyses, the spatial distribution of the extragalatic XRB is a useful tool to explore the large scale structure of the Universe in nearby portions of the Universe (Miyaji and Boldt 1990; Jahoda and Mushotzky 1989) as well as the clustering of the sources at epochs when galaxy formation took place.

In this paper we review the main topics concerning the P(D) and the ACs analysis. Particular attention will be paid to the ACs analysis, showing that the data already obtained with the Einstein Observatory and ROSAT and even more those obtainable exploiting the full capabilities of ROSAT, give unique insights into the problem of the development of the clustering in the high redshift Universe.

In section 2 we will briefly present the observations so far done on fluctuations and on AutoCorrelations. The main theoretical relations on the P(D) and AutoCorrelation Functions (ACFs) will be illustrated in section 3 and 4 respectively. In section 5 the main results of the P(D) analysis will be presented. Section 6 is devoted to confrontation of the ACF data with the predictions. In section 7 the conclusions are summarized.

A value $H_0 = 50$ will be used in the following ($h_{50} = H_0/50$).

2 THE OBSERVATIONS

Small scale fluctuations of the XRB have been studied by many authors, exploiting the data bases of the X-ray satellites. In particular the probability density distribution of the deflection from the mean [P(D), where D=I-$\langle I \rangle$] is used to infer the counts beyond the detection limits. The intensity I and its spatial average $\langle I \rangle$ are referred to the residual background, once the detected sources have been subtracted. Sometimes the spectral characteristics of the *positive* fluctuations are derived for sake of comparison with the spectra of detected sources and XRB.

The angular distribution of the XRB, usually investigated through the autocorrelation function

$$W(\theta) = \frac{\langle I(\mathbf{n})I(\mathbf{n}') \rangle}{\langle I \rangle^2},$$

(1)

where θ is the angle between the directions $\mathbf{n}$ and $\mathbf{n}'$, is used to constrain the clustering properties of the sources contributing to the XRB.

2.1 P(D) distributions

Hard X-ray bands

The P(D) analysis is extremely important in hard X-ray bands, because the direct source counts are still confined to quite bright fluxes.

The analysis of the small scale fluctuations of the UHURU survey has been performed by Fabian (1975) and by Schwartz et al (1976). The energy band was between 2 and 7 keV, the limiting flux was about 4.8×10^{-11} $ergcm^{-2}s^{-1}$, and the FWHM $\simeq 5°$.

The fluctuations in the Ariel V survey have been studied by Pye and Warwick (1979) and by Warwick, Pye and Fabian (1980), in the energy band 2-18 keV with limiting flux 3×10^{-11} $ergcm^{-2}s^{-1}$, and field of view $0.75° \times 10.6°$.

The HEAO1-A2 all sky survey has been analyzed by Shafer (1983) and Shafer and Fabian (1983) in the energy band 2-10 keV with limiting flux 3×10^{-11} $ergcm^{-2}s^{-1}$. A futher analysis of the HEAO1-A2 data have been done recently by Mushotzky and collaborators (see Mushotzky these proceedings).

Recently measurements of small scale fluctuations of the XRB have been done with the large area counter (LAC) on GINGA, using dedicated background observations (Warwick and Stewart, 1989); the energy band is 4-11 keV and field of view is $1° \times 2°$. Futher analyis of GINGA data on XRB fluctuations have been done by Stewart (these proceedings) and Hayashida (these proceedings).

Soft X-ray bands

As it is well known, in the soft bands the results obtained with Einstein Observatory have boosted the X-ray astronomy in the last decade. Quite important results have

been obtained with the Extended Medium Sensitivity Survey (EMSS) (Gioia et al 1990), the Deep Survey (DS) and the Extended Deep Survey (EDS) (Giacconi et al. 1979, Griffiths et al. 1983, Primini et al. 1991).
Small scale fluctuations have been investigated using the IPC exposures of the EDS by Hamilton and Helfand (1987), by Barcons and Fabian (1990) and by Soltan(1991). The energy band is 1-3 keV for former authors, while Soltan (1991) used the 0.8-3.5 keV energy band. The analysis have been done in the central part (0.25 sq. deg.) of the IPC exposures and with pixel sizes ranging from 1' to 4'. The source detection limit changes from exposure to exposure and ranges between 3 to 5 $\times$ 10^{-14} $erg cm^{-2} s^{-1}$.
Shanks et al (1991) have analyzed the P(D) distribution for a deep ROSAT exposure with a source detection flux of 1×10^{-14} $erg cm^{-2} s^{-1}$ in the 0.5-2 keV energy band.

2.2 AutoCorrelation Functions

The ACs have been studied both in hard X-ray and in soft bands. The data base used are almost the same as for the P(D) analysis. The instruments in the hard and soft X-ray bands have quite different spatial resolutions and sensitivities. Therefore we expect that the ACs of hard XRB (HXRB) and soft XRB (SXRB) probe the clustering of sources of different luminosities in rather different regions of the Universe.

Hard X-ray bands

The HEAO1-A2 all sky survey data have been exploited by Persic et al. (1989), and Martin-Mirones et al (1991) to determine the ACs on angular scales from 0° to 27°. Mushoztky and collaborators (see Mushoztky these proceedings) using the same data base in a more extensive way have found a signal in the autocorrelation of HXRB $W \simeq 3 \times 10^{-5}$ on scales ranging from 10° to 18°.
A total sample of 132 exposures with the LAC of the GINGA satellite allowed Carrera et al (1990) to investigate the ACs from 2° to 23° angular scales.

Soft X-ray bands

In softer bands the ACs have been studied by Barcons and Fabian (1989), on the angular scales in the range 1' to 15', using EDS exposures taken with the Einstein Observatory IPC. In particular, they found $W(5') \leq 0.088$ at the 95 per cent confidence level.
Analyzing 6 EDS fields, Soltan (1991) has found very stringent upper limits $W(2' - 5') \leq 3 \times 10^{-3}$.
The angular distribution of the background measured by the ROSAT Medium Sensitivity Survey (RMSS) down to a limiting flux $S \simeq 8 \times 10^{-15}$ $erg cm^{-2} s^{-1}$ (0.5-1.6

keV) is compatible with $W(2' - 5') \leq 2 \times 10^{-3}$ (Hasinger, these proceedings).

3 THE P(D) DISTRIBUTION OF THE FLUCTUATIONS, SOURCE COUNTS AND CLUSTERING

The contribution of undetected sorces to the small scale fluctuations of background radiations has been extensively discussed in the literature. The statistical analysis of the "graininess" has been developed in particular by radioastronomers (Scheuer 1957, 1974; Condon, 1974).

If $f(\vartheta, \varphi)$ is the response of the detector to a source of flux S located at angular coordinates (ϑ, φ) in the beam and N(S) are the differential counts per steradian, then the mean number responses of intensity x in the beam is

$$R(x) = \int_{\Omega_b} N\left(\frac{x}{f(\vartheta, \varphi)}\right) \frac{d\omega}{f(\vartheta, \varphi)}, \qquad (2)$$

where Ω_b is the solid angle over which the response function $f(\vartheta, \varphi)$ is appreciably different from zero; then the effective solid angle is given by $\omega_e = \int_{\Omega_b} f \, d\omega$.

In the case of a Poissonian distribution of sources in the sky, there is a direct relationship between the probability distribution P(D) of the total deflection D contributed by all responses x in the beam area Ω_b and the $R(x)$ distribution. Scheuer (1957) has shown that the Fourier transform of P(D) is a simple function of $R(x)$:

$$p(\omega) = \int_{-\infty}^{\infty} P(D) \, exp(2\pi i \omega D) \, dD = exp[r(\omega) - r(0)], \qquad (3)$$

where

$$r(\omega) = \int_{-\infty}^{\infty} R(x) \, exp(2\pi i \omega x) \, dx.$$

The above relationships hold for a pencil beam telescope, but have been generalized to interferometers.

For a general N(S) distribution, the expected P(D) can be computed as the real part of the Fourier antitransform of equation (3):

$$P_{theor}(D) = \int_0^{\infty} exp\left[R(x)cos(2\pi\omega x)dx - \int_0^{\infty} R(x)dx\right] \times$$
$$\times cos\left[\int_0^{\infty} R(x)sin(2\pi\omega x)dx - \omega D\right] d\omega. \qquad (4)$$

Under the hypothesis of power-law differential conts $[N(S) = \kappa S^{-\gamma}]$ Scheuer (1957, 1974) and Condon (1974) have derived simple formulae relating P(D) to the parameters κ and γ.

To compare the predicted distribution to the observed ones, the $P_{theor}(D)$ must be combined with the noise distribution and with other possible contributions to the deflections, such as those deriving from the clustering of the sources contributing to the background.

The differential counts below the detection threshold are inferred by minimizing the differences between the observed and predicted fluctuation distributions.

A useful estimate of the fluctuation level generated by randomly distributed sources is provided by the second moment σ of the $R(x)$ distribution (Condon, 1974):

$$\sigma^2 = \int_0^{D_c} x^2 R(x)dx, \tag{5}$$

where D_c is a cutoff, which is usually assumed to be the flux S_{lim} of the faintest detectable source.

Again in the case of power-law counts, we can integrate and obtain

$$\sigma = \left(\frac{\kappa \omega_e}{3 - \gamma}\right)^2 D_c^{3-\gamma}, \quad 2 < \gamma < 3, \tag{6}$$

where ω_e is the effective solid angle. To get significat probability of detecting real sources, threshold fluxe $S_{lim} = 4 - 5\sigma$ are usually adopted.

In presence of power-law counts and defining $S_{lim} = q\sigma$, it is interesting to note that the number of sources/beam is $N_b(> S_{lim}) = \frac{3-\gamma}{\gamma-1}q^{-2}$, while $N_b(> \sigma) = \frac{3-\gamma}{\gamma-1}q^{3-\gamma}$. In the case $q \simeq 4$ and $\gamma = 2.5$, the threshold flux corresponds to one source every 50 beams, while at $S = \sigma$ there is about 1 source/beam. Because the P(D) distribution is dominated by 1σ deflections, the fluctuations give poor information on sources much below the level of 1 source per beam.

So far we have discussed the case of a Poissonian distribution of sources, but clustering is expected to increase the fluctuations, modifying the P(D) distribution.

Barcons and Fabian (1988) suggested that the effect of the clustering can be taken into account, at least at first order, by assuming that it generates a gaussian noise. This noise must be combined with the instrumental noise and the P(D) expected from a poissonian distribution, and then compared with the observed P(D). With the assumption of gaussian clustering Barcons (1991) and collaborators (see Toffolatti and Barcons, these proceedings) have shown that source clustering significantly affects the shape of the P(D) distribution.

As we shall see later (section 5.2), clustering of X-ray sources could significantly increase the fluctuations in the case of the HEAO1-A2 experiment.

4 THE AUTOCORRELATION FUNCTION. BASIC RELATIONS

Cell-to-cell fluctuations of randomly distributed unresolved sources as well as source clustering produce intensity fluctuations δI of the observed background. The angular correlation of the intensity fluctuations can be computed under general hypotheses (see for details Martin-Mirones et al 1991).

The general equation of the ACF is cumbersome, but it remarkably simplifies under the assumption that the maximum scale r_{max} of the clustering is much smaller than Hubble radius, c/H_o, and that the maximum value of angular separation of lines of sight in the two beams is much less than one radiant. In this case the proper separation r between points on two lines of sight separated by an angle θ can be approximated by:

$$r = [(c\delta t)^2 + (d_A \theta)^2]^{1/2}, \tag{7}$$

where $d_A = d_L(1 + z)^{-2}$ is the angular diameter distance, and $\delta t = (dt/dz)\delta z = -H_o^{-1}(1 + z)^{-2}(1 + \Omega z)^{-1/2}\delta z$. The luminosity distance is given by $d_L = \frac{2c}{\Omega^2 H_o}\{\Omega z + (\Omega - 2)[-1 + (\Omega z + 1)^{1/2}]\}$.

With the above simplification for r and the additional hypothesis that the beams [response function $f(\vartheta, \varphi)$] do no overlap, Martin-Mirones et al (1991) have shown that the ACF as function of the angular scale $\theta_\star$ is:

$$W(\theta_\star)_c = \left(\frac{c}{4\pi H_o \langle I \rangle}\right)^2 \int d\omega f(\vartheta, \varphi) \int d\overline{\omega} f(\overline{\vartheta}, \overline{\varphi}) \cdot$$
$$\cdot \int_{z_m(L_{min}, S_l)}^{z_{max}} dz \frac{j_{eff}^2(z)}{(1 + z)^4(1 + \Omega z)} \int_{max[z_m - z, -\Delta(r_{max})]}^{min[z_{max} - z, \Delta(r_{max})]} d(\delta z)\xi(r, z), \tag{8}$$

where $\Delta(r_{max})$ is the value of δz corresponding to the maximum scale of clustering, and

$$j_{eff}(z) = \int_{L_{min}}^{min[L_{max}, L(S_l, z)]} d\log L \, L \, n_c(L, z)K(L, z) \tag{9}$$

is the effective volume emissivity; $L(S_l, z)$ is the luminosity of a source at the redshift z that yields a flux equal to the detection limit S_l, whereas $z_m(L_{min}, S_l)$ is the redshift at which a source of minimum luminosity L_{min} has the limiting flux S_l. The average $\langle I \rangle$ is performed over the blank sky after subracting sources brighter than S_l. An upper limit z_{max} to the redshifts of sources is fixed.

The epoch-dependent luminosity function of the clustered sources is $n_c(L, z)dlogL$, $\xi(r, z)$ is the epoch-dependent spatial correlation function and the K-correction factor is given by

$$K(L, z) = \frac{\int_{E_1}^{E_2} L[E(1 + z), z] \, dE}{\int_{E_1}^{E_2} L(E, 0) \, dE}, \tag{10}$$

E_1 and E_2 bounding the relevant energy band.

Available data suggest that epoch-dependent spatial correlation function can be represented as (see Peebles 1980; Bahcall and Soneira 1983; Sebok 1986)

$$\xi(r, z) = D^2(z)\xi(r) \tag{11}$$

$$\xi_o = (r_o/r)^\gamma \tag{12}$$

with $\gamma \simeq 1.8$ for a number of populations of extragalactic sources.

Actually in numerical calculations more refined models are used, that make the correlation function flat at small radii and explicitely incorporate a maximum radius r_{max}, such as $\xi = 0$ if $r > r_{max}$.

The factor $D^2(z)$ allows for clustering evolution. A largely used parametrization is

$$D^2(z) = (1+z)^{-(3+\epsilon)} \tag{13}$$

The case $\epsilon = -3$ corresponds to r_0 constant in physical coordinates, while $\epsilon = \gamma - 3$ implies r_0 constant in comoving coordinates. A self-similar evolution of the correlation function (requiring $\Omega = 1$ and a power-law spectrum of initial density perturbations) yields $\epsilon = 0$. In the latter case, on scales where $\xi \gg 1$ the number of object in a physical volume is constant (statistically stable clustering).

On the other hand, some scenarios for galaxy formation would rather predict correlations at early epochs comparable to, or even larger than, those observed today (i.e. $\epsilon \leq -3$). This is the case for CDM models in which high redshift and luminous QSOs are associated with rare high density peaks and are thus expected to have enhanced clustering (Kaiser 1986; Efstathiou and Rees 1988).

It is worth noticing that the ACF depends on the square of the contribution to the background and on the clustering scale of the population under scrutiny as well as on the sensitivity and *geometry* of the experiment. For instance, hard X-ray observations so far done were limited to a minimum available beam separation of at least few degrees that yet at modest redshifts may encompass many clustering scales. Under these circumstances the contribution to ACs drastically decreases with increasing redshift.

The hard X-ray surveys so far available have bright limiting fluxes (typically few times $10^{-11} erg\ cm^{-2} s^{-1}$) and large fields of view (from few to several degrees) so that they allow to better investigate the *local* clustering of X-ray sources. Contrarywise the soft X-ray surveys are more sensitive to clustering of high redshift sources even on relatively small scales, because of their fainter detection limits ($S_l \sim few\ 10^{-14} erg\ cm^{-2} s^{-1}$) and higher resolution that permit to get good measurements of the ACs even on small angular scales ($\theta_* \sim$ few arcminutes).

A relevant simplification of equation (8) is possible in the case that the angular scales we are dealing with (θ_* and θ_{beam})are much smaller than the minimum angle $\theta(r_{circ}, z)$ subtended by the physical clustering scale. Then the innermost integration in equation (8) can be extended from $-\infty$ to ∞ and we get (De Zotti et al 1990):

$$W(\theta_*)_c = H_\gamma \left(\frac{c}{4\pi H_o \langle I \rangle}\right)^2 \left(\frac{H_o}{cr_o}\right) \theta_*^{1-\gamma}.$$

$$\cdot \int_{z_m(L_{min}, S_l)}^{z_{max}} dz\ j_{eff}^2(z) \frac{D^2(z)\left[(H_o/c)\,d_A(z)\right]^{1-\gamma}}{(1+z)^2 (1+\Omega z)^{1/2}}, \tag{14}$$

with

$$H_\gamma = \Gamma\left(\frac{1}{2}\right)\Gamma\left(\frac{\gamma-1}{2}\right)/\Gamma\left(\frac{\gamma}{2}\right). \tag{15}$$

($H_\gamma \simeq 3.7$, if $\gamma = 1.8$).

If the ACF is mainly contributed by small redshifts ($z \ll 1$), then, for the case of a class of non evolving sources with statistical stable clustering ($\epsilon = 0$) and $\gamma = 1.8$, the ACF is well reproduced by (Martin-Mirones et al 1991)

$$W(\theta_\star) \simeq 1.2 \times 10^{-4} \left(\theta_\star^\circ\right)^{-0.8} \left(\frac{r_0}{50\,\mathrm{Mpc}}\right)^{1.8} \left(\frac{(j_{sources}/j_{XRB})_{z=0}}{0.1}\right)^2. \tag{16}$$

At the same approximation, the contribution of the clustering to small scale fluctuations ($\theta_\star \to 0$) is then given by (Martin-Mirones et al 1991)

$$W(0) \simeq 2 \times 10^{-4} \left(\frac{\omega_{eff}}{\mathrm{sq.\ deg}}\right)^{-0.4} \left(\frac{r_0}{50\,\mathrm{Mpc}}\right)^{1.8} \left(\frac{(j_{sources}/j_{XRB})_{z=0}}{0.1}\right)^2, \tag{17}$$

where ω_{eff} is the effective solid angle of the detector.

Equation (17) provides reasonable estimates also in the case of evolving sources (errors smaller than 30 % with widely accepted evolution models), provided that the ratio of local volume emissivities $(j_{sources}/j_{XRB})_{z=0}$ is replaced by the global contribution of the given class of sources to the XRB.

The the second moment of the P(D) distribution expected from randomly distributed sources, that is well approximated by equation (5), may be compared to

$$\sigma_c^2 = W(0) \cdot \langle I \rangle^2, \tag{18}$$

to understand the relevance of the clustering in respect of small scale fluctuations.

5 RESULTS OF P(D) ANALYSES OF FLUCTUATIONS

5.1 The results in the soft bands

The integral counts of the EMSS (Gioia et al 1990) and of the DS (Giacconi et al 1979) are well described by a euclidean power law $N(> S) \propto S^{-1.5}$. The main point addressed by the authors who have used the Einstein Observatory IPC data (Hamilton and Helfand 1987; Barcons and Fabian 1990; Soltan 1991), has been the possible flattening of the counts below the DS limiting flux and the contribution to the background from the inferred counts. Actually such a discussion is partly overridden by the results of the ROSAT MSS (Hasinger et al 1991), which show that the observed counts are flatter than euclidean counts at fluxes a few times fainter than the Einstein Observatory DS limit. Nonetheless it is worth giving a quick review of the arguments and results, because of the uncertainties still present in the counts at faintest fluxes.

Hamilton and Helfand (1987) derived their P(D) distribution using pixel sizes of 1'. However in their discussion they make use of the second moment rather than of the global distribution.

Using 1'x1' pixel they have 1 source about every 160 pixels, and, extrapolating the counts with -1.5 slope, there is 1 source per beam at $S \simeq few \ 10^{-15} \ ergcm^{-2}s^{-1}$. At that level the photon noise would be larger than the fluctuations.

However their main conclusion is probably not affected by this problem. They found that the observed "granularity" of the radiation requires that the counts become flatter than euclidean in the range $0.5 - 1 \times 10^{-14} \ ergcm^{-2}s^{-1}$.

They have also shown that the integral counts extrapolated below the point of inflection with a slope of 1.2 down to very faint limits ($\sim 2 \times 10^{-15} \ ergcm^{-2}s^{-1}$) are consistent with the observed "graininess" and would account for 100% of the XBR in the band 1-3 keV.

Barcons and Fabian (1990) have analyzed 5 deep IPC fields. They choose to use a pixel size of 4 arcmin, because it is large enough to avoid that the fluxes in adjacent pixels are not independent. They found that a pixel-size of 4 arcmin is a better choice in analysing the data. With this pixel size they have about 1 source every 14 pixels and they expect one-source-per-pixel at $S \sim S_{DS}/4$ (with euclidean counts).

They computed the P(D) distributions for five IPC fields and then compared the results to the expected distributions. The comparison has been done in a somewhat intriguing way, in the sense that first they fixed the normalization of the counts assuming a minimum cutoff and an euclidean slope, then, with this normalization, they investigated the limit beyond which the counts must flatten, in order to reproduce the P(D) distributions.

Their conclusion is that the counts have to bend down below $S \sim S_{DS}/(2-4)$ and that probably less than 40% of the background can be produced by sources brighter than this flux.

The analysis of Soltan (1991) has led to a rather different conclusion. He has analyzed 5 IPC fields, carefully modelling the instrumental effects and using different pixel sizes.

In the fluctuations analysis he used as a reliable evaluation of the variance of the P(D) distribution, the second moment of the distribution of the responses (see equation 5) with pixel size of about 1 arcmin. He claimed that the flux of inflection of the counts may range from 0.3 to $1 \times 10^{-14} \ ergcm^{-2}s^{-1}$. As a consequence the contribution of all the discrete sources may reach 70% of the SXRB, as computed extrapolating the Bremsstrahlung representation of the HXRB.

Although the conclusions of the three papers look different, however they are not incompatible, when allowance is made for the uncertainties in the counts above the detection limit and in the extrapolation to lower fluxes using the P(D) or its second moments.

Soltan results also allow to investigate the influence of the clustering on P(D) distributions. The variances of the P(D) distributions of five IPC fields computed by Soltan (1991) range from 1 to 3×10^{-26} (cgs). and we can compare these values to σ_c^2 as given by equation (17). Specializing to the case under scrutiny, we get ($\gamma = 1.8$)

$$\sigma_c^2 \simeq 2 \times 10^{-26} \left(\frac{\omega_{eff}}{0.25 \ sq \ deg} \right)^{-0.4} \left(\frac{r_0}{50 \ Mpc} \right)^{1.8} \left(\frac{f}{0.5} \right)^2. \qquad (19)$$

Therefore the clustering has negligible effect on the extrapolation of the counts in the case of IPC fields, unless unresolved sources give a relevant contribution to the XRB and cluster on large scales. We shall see later on that this circumstance is excluded by the limits on the ACs of SXRB.

Shanks et al (1991) have analyzed the P(D) distribution of a long exposure done with the PSPC of ROSAT. The predicted source counts below $S \simeq 10^{-14} \ ergcm^{-2}s^{-1}$ are above any reasonable extrapolation of AGN counts. Because AGNs and QSOs are dominating the counts above $S \simeq 10^{-14} \ ergcm^{-2}s^{-1}$, then the same authors have envisaged the existence of a new population of sources, although they also warn that the observed P(D) could be affected by instrumental effects.

5.2 The results in the hard bands

P(D) analyses are even more interesting in the hard bands, where the counts are limited to bright fluxes. A good, concise summary of the results has been presented by Warwick and Stewart (1989). In particular from their figure 2 it is apparent that the counts as inferred by their analysis of GINGA background observations are in good agreement with those derived by Shafer (1983) from HEAO1-A2 data and are compatible with an euclidean extrapolation of the Piccinotti et al (1982) counts. Pye and Warwick's (1979) analysis of the Ariel V survey gave similar results. These studies significantly constrain the counts down to fluxes $S \sim 1 \times 10^{-12} \ ergcm^{-2}s^{-1}$ (2-10 keV).

At corresponding fluxes in the 0.3-3.5 keV band of EMSS the counts are rather well established. To compare the counts in the two bands, we can define an average spectral conversion factor $F = S(2 - 10 \ keV)/S(0.3 - 3.5 \ keV)$. Assuming a "canonical" AGN spectrum, $\alpha_x \simeq 0.7$, and no absorption, we expect $F \simeq 1$. With this flux conversion the EMSS counts fall a factor of about 3 below the P(D).

A possible explanation has been presented by Warwick and Stewart (1989), who pointed out that a substantial photoelectric absorption ($N_H \sim 10^{22} \ cm^{-2}$) would make low luminosity AGN's to be under-represented in the EMSS. However analyses of AGN spectra in the soft bands have shown that steep spectra $\langle \alpha_x \rangle \simeq 1$ and soft excesses are common features, in contrast with Warwick and Stewart (1989) suggestion.

On the other hand, because X-ray spectra of AGN's are rather complex and diverse, it is expected that surveys done in different bands select objects with different spectral properties.

This is confirmed by the fact that the F ratios of 18 AGNs in the Piccinotti sample show an average value $\langle F \rangle \simeq 1.8$, while for a sample of bright AGNs in the MSS the average is $\langle F \rangle \simeq 0.6$, and the QSOs of the PG sample and the objects of a sample of optically selected Seyferts have $\langle F \rangle \simeq 0.75$ and $\langle F \rangle \simeq 1$ respectively (Martin-Mirones et al 1991). These results may help in reconciling the soft and hard X-ray counts.

Martin-Mirones et al (1991) analysing HEAO 1 A-2 data base have found evidence for fluctuations in excess of those expected from a Poissonian distribution of extragalactic sources, in the case that their counts follow a euclidean power law at fluxes fainter than the limit of the Piccinotti et al (1982) survey.

A possible explanation is that the integral counts have a slope $\simeq 1.65 \pm 0.06$, steeper than euclidean, still compatible with the limits on the counts inferred from the P(D) analysis.

On the other hand the excess in the fluctuations can also be explained in term of clustering of sources.

For instance clustering of AGNs can do the job. Using luminosity evolution models (Danese et al 1986) which fit the counts down to the DS limit, values of clustering radii r_0 ranging from 20 to 30 Mpc are required in the case of pure luminosity evolution and with stable clustering ($\epsilon = 0$); in the case of luminosity evolution confined at high luminosity ($L_x \geq 3 \times 10^{43} \ ergs^{-1}$) then r_0 ranges from 30 to 45 Mpc. Slightly smaller value of r_0 are required in the case of a correlation function constant in physical coordinates (see Table 1).

Table 1. Constraints on the clustering radius r_0 (Mpc) of AGN's ($H_0 = 50$).

	$\epsilon = 0$		$\epsilon = -3$	
	r_{max}/r_0		r_{max}/r_0	
	2	6	2	6
PLE	20–30	15–20	10–20	10–15
log $L_s \geq 43.5$	30–45	30–35	20–40	15–25

On the other hand upper limits on ACs of SXRB practically exclude such a level of clustering for AGNs (see section 6.2). Thus slopes steeper than euclidean in the counts are the most plausible explanation of the fluctuation excess (if real).

6 RESULTS FROM ACS

Significant information on clustering of X-ray sources can be obtained from the Auto-Correlations of the XRB. Due to the difference in limiting fluxes, angular resolutions and beam separations of hard and soft X-ray surveys, ACs of the SXRB and HXRB are sensitive to clustering of sources of different luminosities and redshifts.

6.1 The results in the hard bands

As for the data, there is good agreement between the constraints on ACFs obtaineded by Martin-Mirones et al (1991) and by Carrera et al (1991) on scales from few to several degrees, once upper limits are scaled to the same significance level. The comparison is meaningful because the data used by the two groups are from experiments with similar energy bands and limiting fluxes.

At larger scales a positive autocorrelation $W(10° - 18°) \simeq 3 \times 10^{-5}$ has been found by Mushotzky and collaborators (see Mushotzky these proceedings).

AGNs and clusters of galaxies are expected to give relevant contributions to ACFs, because of their X-ray volume emissivity.

As for rich clusters, the optical observations show that they are strongly clustered, although the amplitude of their correlation function is still matter of debate (Bahcall and Soneira 1983; Sutherland 1988; Bahcall et al 1988; Batuski et al 1989; Olivier et al 1990).

Martin-Mirones at al (1991) have computed the expected contribution of galaxy clusters to ACs at various scales. They used the 2-10 keV luminosity function derived by Piccinotti et al (1982) (excluding Virgo) and the temperature-luminosity dependence (Mushotzky 1988)

$$T \simeq 2.7 \left(\frac{L}{L_0} \right)^{0.421} keV \tag{20}$$

where $L_0 = 10^{44} \ ergs^{-1}$.

In the framework of hierarchical theories with approximately scale-free primordial density fluctuations (CDM or "cosmic string" scenarios) within a flat universe ($\Omega = 1$ and $\Lambda = 0$), Kaiser (1986) has shown that luminosity and temperature evolve as

$$L \propto (1 + z)^{\frac{(7n+5)}{(2n+6)}} \tag{21}$$

$$T \propto (1 + z)^{\frac{(n-1)}{(n+3)}}, \tag{22}$$

where n is the post-recombination spectral index of the perturbation field.

The maximum contribution to the ACs from clusters is obtained in the case n=0 and $\epsilon = -3$ with a cut-off of the clustering at $r \geq 300$ Mpc and is well below the observational limits and detections on scales below 5°. On the other hand the same clustering produces an autocorrelation $W(10°) \simeq 1.5 \times 10^{-5}$ lower than but close to the observed value.

The only way to make the clusters important contributors to the ACs on scales of few degrees would be to increase the clustering scale well beyond the optical limits, with the consequence of producing autocorrelations on the scale of $10°$ larger than observed.

On the other hand the contribution of the AGNs to the HXRB is large and therefore we can expect that their contribution to the ACs is substantial, even in the case of relatively small clustering

To model the AGN contribution to counts and HXRB, Martin-Mirones et al (1991) have used luminosity evolution models parameterized as follows (Danese et al 1986; Maccacaro et al 1991):

$$L(t) = \begin{cases} L_S + (L_0 - L_S)\, exp(\kappa\tau) & \text{if } L_0 > L_S \\ L_0 & \text{if } L_0 \leq L_S, \end{cases} \tag{23}$$

where $L_0 \equiv L(t_0)$ is the present luminosity and τ is the look-back time. For $L_S = 0$ we have Pure Luminosity Evolution (PLE). Models consistent with the available X-ray surveys range from PLE to $L_S \leq 10^{43.5}\ ergs^{-1}$. The spectrum of the sources has been assumed to be a power-law with spectral index $\alpha_x = 0.7$. As it can be seen in equation (8) the ACF depends on the product of the *geometrical* factor by the square of the contribution to the background of the sources fainter than the limiting flux.

Martin-Mirones et al (1991) have explored various possible values of the clustering radius r_0, the clustering evolution parameter ϵ and ratio r_{max}/r_0, under the conditions that they produce ACs compatible with the observations on angular scales smaller than $3°$. The values of r_0 reported in Table 1 are upper limits to the clustering scale of AGNs contributing about 45 % of the HXRB in the case of PLE and only 18 % in the case with $log\ L_s \simeq 43.5$. With the same parameters it turns out that $W(10°) \leq 2 \times 10^{-5}$. On the other hand, if the clustering is still significant at $r \geq 180$ Mpc, the predicted $W(10°)$ exceeds the value found by Mushotzky and collaborators (Mushotzky these proceedimgs).

Optical evidence for substantial clustering of QSOs with an average redshift $z \simeq 1$ has been found by Shanks et al (1988), Shaver (1988), Kruszewski (1988), Shaver, Iovino and Pierre (1989). More recently Boyle (1991) using a sample of 400 objects has found that QSOs cluster with a typical scale $r_0 \simeq 12h_{50}^{-1}$Mpc constant in comoving coordinates. On the other hand Iovino, Shaver and Cristiani (1991) have studied the clustering using 5 homogeneous samples, totalling about 1000 QSOs. They confirm the reality of the clustering of the QSOs on a proper scale $r_0 \simeq 10 - 15h_{50}^{-1}$Mpc, at least for objects at substantial redshift ($z \geq 1$). A further analysis (Andreani et al 1991) of a larger sample suggests that the clustering scale keeps almost constant in comoving coordinates (i.e. $\epsilon = -1.2$) at least down to $z \simeq 0.5$, with $r_0 \simeq 20h_{50}^{-1}$Mpc. It is worth noticing that these studies are limited to bright objects (QSOs), because the optical selection pratically excludes low luminosity AGNs.

How does the clustering observed for optically selected QSOs compare to the limits imposed by the smoothness of the HXRB? It can be seen that a large contribution to $W(3°)$ comes from relatively low luminosity AGNs. Thus we can conclude that optical studies and hard X-ray ACs are in some sense complementary. However we can make the hypothesis that the optical clustering applies to AGNs in general. From Table 1 it is apparent that in the case of PLE, i.e. substantial contribution to HXRB, the clustering revealed in the optical bands is just consistent with X-ray limits.

6.2 The results in the soft bands

In the soft X-ray band the AC based on the MSS of ROSAT has a value of $W(2'-5') \leq 2 \times 10^{-3}$ in the energy band from 0.5 to 1.6 keV (Hasinger, 1991 these proceedings). The telescope resolution is about 25", much less than the AC scales, and the limiting flux is $S_{lim} \simeq 1 \times 10^{-14} \ ergcm^{-2}s^{-1}$.

The AC derived by Soltan has an upper limit of $W(2'-5') \leq 3 \times 10^{-3}$ in the 0.8-3.5 keV band. The observed fields have different flux limits and $S_{lim} \simeq 4 \times 10^{-14} \ ergcm^{-2}s^{-1}$ is the average value quoted by Primini et al (1991).

Note that also objects at substantial redshifts contribute to the ACs at small angular separations. In particular when computing the ACFs on arcmin scales, we can use equation (14), because we can assume that the angular separation is much smaller than the projected clustering scale at any redshift.

Small autocorrelation scales and faint limiting fluxes make the soft ACF sensitive to the clustering of high redshift and high luminosity objects. High redshift and high luminosity ($L_x \geq 10^{44} \ erg \ s^{-1}$) sources give large contributions to the ACs.

The typical X-ray luminosity of optically selected QSOs detected by Shanks et al (1991) in their ROSAT survey is $L_x \simeq 3 \times 10^{44} \ erg \ s^{-1}$ in 0.1-2.4 keV band, that translates to $L_x \simeq 10^{44} \ erg \ s^{-1}$ in the bands used by Hasinger and by Soltan in their analyses.

Therefore the limits on clustering derived from ACs of the SXRB are directly comparable with QSO clustering properties derived from the optical observations and complementary to those from the hard X-ray observations.

Again to predict the autocorrelations of the SXRB we have used the models described in section 6.1, constrained by the condition of reproducing the available Einstein Observatory and ROSAT extragalactic source counts and related stastistics.

Although it is also possible to study the autocorrelations assuming the contribution to the background from unresolved sources as a free parameter (see e.g. Barcons and Fabian 1989), the use of models fitting the available counts can clarify many aspects of the problem. One of the results of the models is the prediction of the background produced by AGNs at various limiting fluxes.

Using 0.5-1.6 keV as a reference band and following Hasinger et al (1991) and Shanks et al (1991), the background produced by extragalactic sources brighter than 8×10^{-15} $ergcm^{-2}sec^{-1}$ is around 35% of the total, that in turn is a factor of about 1.7 larger than the background extrapolated from higher energies. The models predict that the total contribution of AGNs to SXRB is not larger than 40-45 %, then allowing unresolved AGNs to give only about 10%. It is worth noticing that part of the SXRB could be originated in, or just around our Galaxy (McCammon and Sanders 1990; Wu et al 1991). When computing the ACFs we have taken into account both that the total background is higher than the extrapolation of the HXRB bremsstrahlung approximation.

If we allow all the unresolved sources to cluster as optical QSOs do, the predicted autocorrelations $W(5')$ exceed the values found both by Hasinger (1991) as well as by Soltan (1991).

For instance, if QSOs have a present day clustering scale of about 50 Mpc with an evolution parameter $\epsilon = 0$, as suggested by Iovino et al (1991), then the ACF computed matching the conditions of ROSAT Deep Survey, is $W(5') \simeq 1 \times 10^{-2}$, five times larger than the observational limit. On the other hand, Bahcall and Chokshi (1991) suggested that QSO clustering is intermediate between that of galaxies and rich clusters, with $r_0 \simeq 24$ Mpc and $\epsilon = -1.8$ as possible clustering parameters. However this clustering also produces a value of $W(5')$ that is at least 3 times larger than the observed upper limits. In the case of clustering with $r_0 \simeq 20$ Mpc and $\epsilon = -1.2$ (Andreani et al 1991), then $W(5') \simeq 2.5 \times 10^{-3}$.

Limiting the clustering only to sources of high luminosity $L_x \geq 10^{44}$ $ergs^{-1}$ does not change significantly the predicted autocorrelations ($W(5') \simeq 2 \times 10^{-3}$ in the latter case). This result stems from the fact that the main contribution to ACs comes from high redshift and high luminosity AGNs (QSOs), because both of the small angular scale and of the faint limiting flux.

The upper limit derived by Soltan (1991) is even more effective in constraining the clustering, because it has almost the same value as that found by Hasinger (1991) but at a brighter limiting flux. Thus it allows for a larger fractional contribution by unresolved sources, that implies higher values of autocorrelations, keeping the same clustering.

For instance, in the case of QSO clustering such as found by Andreani et al (1991) the AC computed matching the conditions of IPC fields is $W(5') \geq 4 \times 10^{-3}$.

Only the case of QSO clustering similar to that of bright galaxies, as suggested by Boyle (1991), is compatible with the smoothness of the SXRB. To keep a clustering significantly stronger than that of bright galaxies, as found in most of the optical analyses, we have to decrease the fractional contribution to the SXRB from unresolved QSOs. Moreover in the models we have used the contribution is already as small as 3%. It is hard to imagine a smaller contribution, if we take into account that QSOs with fluxes $S(0.5 - 2keV) \geq 1 - 2 \times 10^{-14}$ $ergcm^{-2}s^{-1}$ give at least 20% (and possibly

30%) of the total background (Shanks et al 1991).

Therefore it is possible that QSOs are more clustered than galaxies, but *i)* the cluster-ing scale should not exceed $r_o \simeq 20$ Mpc, *ii)* clustering evolution should be significant (i.e. $\epsilon \geq -1.2$), and *iii)* the contribution to the SXRB of QSOs not detected in the RMSS should be only a few percent of the total.

Clustering at level of galaxy clusters is still acceptable only for very special classes of QSOs (for example brighter than $L_x(0.5 - 1.6 keV) \geq 6 \times 10^{44}$ $erg cm^{-2} s^{-1}$), if they give a negligible contribution to the SXRB.

Faint ($S(0.5 - 1.6 keV) \leq 8 \times 10^{-15}$ $erg cm^{-2} s^{-1}$), low luminosity AGNs at high redshifts (say $z \geq 0.8$) and clustered like galaxies and with substantial clustering evolution ($\epsilon = 0$) can add only about 20% to the SXRB. Therefore the smoothness of the background seems to require an upper limit of about 60% to the contribution to the SXRB by extragalactic sources clustered like galaxies. Interestingly enough this would correspond to the whole background extrapolated from higher energies on the basis of the bremsstrahlung representation of the spectrum. Actually the relationship between HXRB and SXRB is quite complicated, and depends on the detailed spectral properties of X-ray sources (see Setti 1991).

Analysing the P(D) distribution of their long exposure with PSPC of ROSAT, Shanks et al (1991) have found that the fluctuations are in excess respect to those predicted on the basis of reasonable extrapolations of the AGN counts (see above sect 5.1). They have noticed that the excess of fluctuations could be attributed to instrumental effects. Indeed models fitting EMSS (Gioia et al 1990) and RMSS (Hasinger et al 1991) counts, show a clear convergence at a flux level of about 1×10^{-15} $erg cm^{-2} s^{-1}$ and predict source counts at least a factor 2 below those inferred from fluctuations. The convergence in the counts are particularly constrained by Hasinger et al (1991) results.

Sources brighter than $S \simeq 1 \times 10^{-15}$ $erg cm^{-2} s^{-1}$ following the counts predicted by the above mentioned P(D) analysis would produce $80 - 90\%$ of the SXRB, with about 50% produced by sources fainter than the RMSS limit. The smoothness of the background implies that either the typical correlation length of these sources is as small as $r_o \leq 5$ Mpc, half of that of bright galaxies, or that their clustering evolution is much more rapid ($\epsilon \geq 1$) than predicted by gravitational instability. There are indeed extragalactic sources that show weak clustering. For instance HI dwarf galaxies are likely to be less clustered that the bright ones (Weinberg et al 1991). Low luminosity AGNs could also be weakly clustered objects.

It is quite interesting that Efstathiou et al (1991) have found that the faint blue galaxies detected in deep CCD images (Tyson 1988) show weak clustering. The same authors also suggest that either these galaxies belong to a population of weakly clustered and intrinsically faint objects or their clustering evolution is strong ($\epsilon \geq 0$).

7 CONCLUSIONS

From the above discussion it is apparent that studies of ACs of SXRB and HXRB are extremely powerful in exploring the clustering of galaxy clusters and AGNs.

The smoothness of the SXRB excludes that QSOs have clustering properties similar to those of galaxy clusters. Unless QSOs fainter than the ROSAT Medium Sensitivity Survey produce less than a few percent of the SXRB, QSOs are bound to be weakly clustered ($r_0 \leq 20\text{Mpc}$).

As is well known, low luminosity AGNs are good candidates to give a substantial fraction of both the HXRB and the SXRB. For a contribution to the HXRB larger than 50%, the smoothness of the background constrains their typical clustering scale to be $r_0 \leq 20$ Mpc.

The smoothness of the SXRB sets an upper limit of about 60% to the contribution to the background by sources clustered like galaxies.

In general very little room is left for contributions to the XRB from sources more clustered than galaxies. If sources contribute up to 80% of the SXRB, as suggested by Shanks et al (1991), then their typical correlation length must be $r_0 \leq 5$ Mpc.

The limits imposed by ROSAT and Einstein Observatory data on autocorrelations are interesting challenges for theories of galaxy and QSO formation. For instance in the CDM scenario the luminous high redshift QSOs and rich clusters are associated with rare high density peaks of the density perturbation field and thus are expected to show enhanced clustering at high redshifts, rather in contrast with the above findings. This again illustrates the relevance of the studies of ACs of the XRB. Particularly the information contained in deep exposures obtained by ROSAT are extremely important to study the development of the clustering at very crucial epochs, where theories of structure formation give rather well defined predictions.

Work supported in part by MURST, GNA/CNR and ASI.

REFERENCES

Andreani, P. et al 1991. In preparation

Bahcall, N.A., & Soneira, R.M. 1983 ApJ 270, 20

Bahcall, N.A., Batuski, D.J., & Olowin, R.P. 1988, ApJ (Letters) 333, L13

Bahcall, N.A., & Chokshi 1991, ApJ (Letters) 380, L9

Batuski, D.J., Bahcall, N.A., Olowin, R.P., Burns, J.O. 1989, ApJ 341, 599

Barcons, X. 1991. In *The Infrared and Sub-mm Sky after COBE*. Eds. Signore, M. & Dupraz, C., in press

Barcons, X., & Fabian, A.C. 1988 MNRAS 230, 189

Barcons, X., & Fabian, A.C. 1989 MNRAS 237, 119

Barcons, X., & Fabian, A.C. 1990 MNRAS 243, 366

Boyle, B.J., 1991. In *Texas-ESO/CERN Conference on Relativistic Astrophysics*, in press

Carrera, F.J., Barcons, X., Butcher, J.A., Fabian, A.C., Stewart, G.C., Warwick, R.S., Hayashida, K. & Kii, T., 1991, MNRAS 249, 698

Condon, J.J. 1974, ApJ 188, 279

Danese, L., De Zotti, G., Fasano, G., & Franceschini, A. 1986. A&A 161, 1

De Zotti, G., Persic, M., Franceschini, A., Danese, L., Palumbo, G.G.C., Boldt, & E.A., Marshall, F.E. 1990, ApJ 351, 22

Efstathiou, G. & Rees, M.J. 1988 MNRAS 230, 5P

Efstathiou, G., Burnstein, G., Katz, N., Tyson, J.A., & Guhathakurta, P. 1991, ApJ (Letters) 380, L47

Fabian, A.C., 1975, MNRAS 172, 149

Giacconi, R. et al ApJ 230, 540

Gioia, I.M., Maccacaro, T., Schild, R.E., & Wolter, A. 1990, ApJS 72, 567

Griffiths, R.E. et al 1983 ApJ 269, 375

Gush, H.P., Halpern, M., & Wishnow, E.H. 1990, Phys.Rev.Lett. 65, 537

Hamilton, T.T., & Helfand, D.J. 1987, ApJ 318, 93

Hasinger, G., Schmidt, M., & Trumper, J. 1991, A&A 246, L2

Iovino, A,n Shaver, P.A., & Cristiani, S. 1991. Preprint

Jahoda, K., & Mushotzky, R.F. 1989, ApJ 346, 638

Kaiser, N. 1986 MNRAS 222, 323

Kruszewski, A. 1989, in *Large Scale Structure and Motions in the Universe*, ed. M. Mezzetti, G. Giuricin, F. Mardirossian, and M. Ramella (Dordrecht: Kluwer), p. 385.

Maccacaro, T., Della Ceca, R., Gioia, I.M., Morris, S.L., Stocke J.T., & Wolter, A. 1991. Preprint

Mather, J.C. et al 1990, ApJ (Letters) 354, LL37

Mc Cammon, D. & Sanders, W. T. 1990, ARA&A 28, 657

Martin-Mirones, J.M., De Zotti, G., Boldt, E.A., Marshall, F.E., Danese, L., Franceschini, A., & Persic, M. 1991 ApJ 379, 507

Miyaji, T., & Boldt, E. 1990, ApJ (Letters) 327, L25

Mushotzky, R.F. 1988, in *Hot Thin plasma in Astrophysics*, ed R. Pallavicini (Dordrecht:Reidel), P. 273

Olivier, S., Blumenthal, G.R., Deel, A., Primack, J.R., & Stanhill, D. 1990, ApJ 356, 1

Peebles, P.J.E. 1980, *The Large-Scale Structure of the Universe* (Princeton: Princeton University Press)

Persic, M., De Zotti, G., Boldt, E.A., Marshall, F.E., Danese, L., Franceschini, A., & Palumbo, G.G.C. 1989 ApJ (Letters) 336, L47

Piccinotti, G., Mushotzky, R.F., Boldt, E.A., Holt, S.S., Marshall, F.E., Serletmitsos, P.J., & Shafer, R.A. 1982, ApJ 253, 485

Primini, F.A., Murray, S.S., Huchra, J., Schild, R., Burg, R., & Giacconi, R. 1991, Center for Astrophysics Preprint Series No. 3170

Pye, J.P.,& Warwick, R.S. 1979, MNRAS 187, 905

Scheuer, P.A.G. 1957, Proc. Cambridge Phil. Soc., 53, 764.

Scheuer, P.A.G. 1974, MNRAS 166, 329

Schwartz, D.A., Murray, S.S.,& Gursky, H. 1976 ApJ 204, 315

Sebok, W.L. 1986, ApJS 62, 301

Setti, G. 1991, Proceedings 28th Yamada Conference, Y. Tanaka and K. Koyama eds.

Shafer, R.A. 1983, Ph. D. thesis, University of Maryland

Shafer, R.A, & Fabian, A.C. 1983, in IAU Symposium 104, Early evolution of the Universe and Its Present Structure, ed. G.O. Abell and G. Chincarini (Dordrecht:Reidel), p. 333

Shanks, T., Boyle, B.J., and Peterson, B.A. 1988, in *Proceedings of a Workshop on Optical Surveys for Quasars*, ed. P.S. Osmer, A.C. Porter, R.F. Green, and C.B. Foltz (San Francisco:Astronomical Society of the Pacific), p. 244.

Shanks, T., Georgantopoulos, I., Stewart, G.C., Pounds, K.A., Boyle, B.J., & Griffiths, R.E. 1991, Nature 353, 315

Shaver, P.A. 1988, in *IAU Symposium 130, Large Scale Structures of the Universe*, ed. J. Audouze, M.-C. Pelletan, and A. Szalay (Dordrecht: Kluwer), p. 359.

Shaver, P.A., Iovino, A, & Pierre, M. 1989, in *Large Scale Structure and Motions in The Universe*, ed M. Mezzetti et al. (Dordrecht:Kluwer), p. 101

Soltan, A.M. 1991, MNRAS 250, 241

Sutherland, W.J., 1988, in *IAU Symposium 130, Large Scale Structures of the Universe*, ed J. Audouze, M.C. Pelletan, and A. Szalay (Dordrect: Kluwer), p. 538

Tyson, J.A. 1988, AJ 96, 1

Worrall, D.M., Marshall, F.E., & Boldt E.A. 1979, Nature 281, 127

Worrall, D.M., Marshall, F.E. 1984 ApJ 276, 434

Wu, X., Hamilton, T., Helfand, D.J. and Wang, Q. 1991, ApJ 379, 564

The Large Scale Surface Brightness Distribution of The X-Ray Background

Richard Mushotzky and Keith Jahoda
Laboratory for High Energy Astrophysics
NASA/GSFC

1 Introduction

The x-ray background and the μ-wave background are the dominant "isotropic" radiation fields available for measurement. While there has been a convergence of opinion on the origin of the μ-wave background and the measurement of its dipole variation and upper limits on higher order multipole terms (Klypin et al. 1987, Strukov et al. 1987, Smoot et al 1990) there has been no such agreement on the x-ray background. However, because of its relative uniformity (Schwartz and Gursky 1974, Schwartz 1979,Turner and Geller 1980) its seems fairly clear that the bulk of the E>2 kev background is also of "cosmological" origin (e.g. due to objects or truly diffuse radiation originating at z>0.5) and as such, is of great interest. At E<1 kev much of the observed flux is likely due to a galactic component. (McCammon this symposium) (For a recent extensive review on the cosmic x-ray background see Boldt 1987 and the many of the papers at this symposium).

There has been extensive work on trying to determine the physical origin of the background. That is, whether it is due to a superposition of numerous faint "well-known" sources such as active galaxies (Giacconi et al. 1979), an early unidentified population of AGN at high redshift (Boldt and Leiter 1987), a new population of objects, or to truly diffuse processes (Guilbert and Fabian 1986) or to a superposition of these. However while of great intrinsic interest these studies have not been aimed at using the XRB to provide the cosmological information that has been gleaned from the μ-wave background. Our group has been attempting an alternate approach, to use the available information on the large, $\theta>5°$, scale distribution of the sky flux to see if the XRB can provide cosmological information .

As opposed to the μ−wave background, much of the x-ray background is presumably due to sources at z<3 (if it is due primarily to active galaxies) or z<10 (as a very general limit) . For the determination of large scale motions the x-ray and μ−wave backgrounds are thus complementary, both defining "distant" reference frames with which one can measure motion with respect to. However, the x-ray background can provide information not contained in the μ-wave background: the distribution of matter perturbations at intermediate (z<1) redshifts; (see the detailed discussion in Rees 1979 of how the optical galaxy counts, μ-wave background measures and x-ray surface brightness measurements complement each other)[1].Warwick, Pye and Fabian 1979, and Rees 1979 have pointed out how measurements of the variation of the x-ray sky surface brightness on moderate angular scales can get the tightest constraints available on density perturbations in the universe on scales from 100-1000 Mpc at redshifts <1.

If the sources of the x-ray background are distributed roughly like matter, variations in the x-ray surface brightness are tracers of the its distribution on large scales. Even if most of the sources responsible for the background are not distributed in such a fashion the best estimates are that ~20-30% of the x-ray background comes from objects such as Seyfert galaxies, normal galaxies and clusters that are distributed like matter.[2] While, potentially, this information resides in the distribution of x-ray sources (in the way that the IRAS counts were used to examine the distribution of matter in the local universe) it is very difficult to obtain all of it from source counts alone because 1/2 of the sky surface brightness comes from sources dimmer than 3×10^{-15} ergs/cm2/sec (Hasinger this symposium) or two orders of magnitude below the Rosat all sky survey limit. Thus very deep surveys are required to "resolve out" the background (see section III). Cataloging and identifying the $\sim 1 \times 10^6$ sources/sr, most of which will be quite faint optically ($M_v > 20$ mag), will be difficult[3]

In addition such surveys have strong selection effects due to source spectra and extent. However, it is clear that the Rosat all sky data can be analyzed in a fashion similar to the IRAS data base to obtain another measure of our local velocity vector.

---

[1]Of course, the x-ray background does not contain the same information as the μ-wave background on the spectrum and amplitude of primoridal perturbations

[2]In particular, Giaconni and Zamorani (1987) have estimated based on the optical-x-ray ratio of low redshift normal galaxies and the deep optical counts of Tyson (1988) that ~13% of the x-ray background at 2 kev is due to normal galaxies, Persic et al 1989b, based on the local luminosity function of Seyfert galaxies, have calculated that they contribute ~ 25% of the background and various authors have calculated that clusters contribute ~5-10 % of the 2 kev x-ray background

[3]The best such use of an all sky catalog to obtain cosmological information has been from the IRAS all sky survey . However there are substantial differences between an x-ray and IR survey. The x-ray sources tend to be much more distant and have fainter optical counterparts. In addition the x-ray source counts at high latitude will be dominated by active galaxies rather than the star forming galaxies which dominate the IRAS counts. Perhaps the most fundamental difference is that the IR background is dominated by local, solar system and galactic effects while the x-ray background is of cosmological interest and its flux is dominated by effects originating at z>0.5 while the fluctuations in it may be dominated by objects at considerably lower redshifts.

The simplest way to see the usefulness of measuring "lumps" in the x-ray surface brightness observation is to assume a Eucledian universe and that the sources of the x-ray background are distributed roughly like matter. Then (Rees 1979), the amplitude variation in the x-ray background $(\Delta I/I)_x$ due to perturbation in matter $(\Delta \rho/\rho)$ on a scale λ compared to the Hubble scale, λ_H is approximately $(\Delta I/I)_x = (\Delta \rho/\rho)(\lambda/\lambda_H)f$; where f is a parameter that describes cosmological effects and x-ray source evolution . Rees (1979) estimates that f~1/2 (see Goicoechea and Martin-Mirones 1990 for a more detailed calculation) . If we assume $(\Delta \rho/\rho)$ ~2, as seems appropriate for the largest scale perturbation claimed to date (the Great Attractor) and similarly (λ/λ_H)~0.02 (~100h$^{-1}_{50}$ Mpc) then the predicted variations in the x-ray surface brightness on angular scales from 1/2 steradian (z~0.01) to 1-2 degrees (z~1) should be of the order of 1-2% (see Shafer 1983 for a detailed discussion)

Similarly interesting results can be obtained by searching for a 24 hour effect in the distribution of the all sky flux (Warwick and Fabian 1979) due to velocities induced by matter perturbations on scales $\lambda \ll \lambda_H$ (the Compton-Getting effect). The expected peak amplitude is $(\Delta I/I)_x \sim (3+\alpha)V_{pec}/c \sim (\Delta \rho/\rho)(\lambda/\lambda_H)\Omega$;or more exactly (Peebles 1980) $(\Delta I/I)_x = (1+\alpha/3)(\Delta \rho/\rho)(\lambda/\lambda_H) \Omega^{0.6}$, where α, the effective energy index in the 2-20 kev x-ray band, is ~0.4 (Boldt 1987). Using the micro-wave dipole velocity of the sun as a scaling parameter we expect $(\Delta I/I)_x{}^{obs}$~0.4% . As opposed to the fluctuation amplitude, which is confined to a small region of the sky, the "dipole-like" term is an all sky effect and thus needs a large solid angle for its measurement. However, the dipole effect may be swamped by smaller scale perturbations if they occupy a large solid angle in toto.

To summarize, the x-ray diffuse background should show a large scale effect (spherical harmonic like) $(3+\alpha)$ times larger in $(\Delta I/I)$ than the μ−wave background and an effect due to the clumping of sources - which does not have a simple dipole like shape, larger yet by a factor $(f/\Omega^{0.6})$. The fact that the amplitude of any 24 hour-like variation in the x-ray background is smaller than 2% (see below; Boldt 1987, Shafer 1983, Warwick, Fabian and Pye 1980) gives a lower bound on the closure ratio and the relative contribution of sources (which are distributed like matter) to the XRB.

2 Recent Results on Structures in the X-ray Background

2.1 HEAO-1 A-2 Capabilities

We have used HEAO-1 A-2 data, which is the only available all sky data base with sufficient sensitivity, to look for effects at <5% level on angular scales $\geq$ 200 square degrees. The limits on this data base are set by Poisson noise, sky fluctuations due to unresolved point sources and residual systematic errors. For the smallest field of view available from this detector (3x1.5 degrees) the sky variance is, using the distribution of the data itself (figure 1) (Persic et al. 1989), ~3.5x10^{-12} ergs/cm^2/sec/5 degrees^2 in the 2-10 kev band for a total variance of ~ 2.6% of the sky flux in the beam. In the absence of variance due to other causes (such as background variations) the absolute value of the noise due to the sky fluctuations scales as solid angle$^{1/2}$ as does the poisson error. Estimates of the variance in the internal background (Shafer 1983) are on the order of 2-3x10^{-12} ergs/cm^2/sec . Thus beam sizes of > 60 square degrees are necessary to examine surface brightness variations of <2% with the HEAO-1 data with greater than 3 σ confidence.

2.2 Galactic Component

Previous studies using this data base (Shafer 1983, Iwan et al. 1983) have shown that the strongest large scale feature in the sky at galactic latitudes greater than 10 degrees is due to the galaxy. This component has an effective scale height of <15 degrees (figure 2 and Iwan et al.). This emission is not symmetric about the galactic plane or the galactic center and has a "softer" spectrum that the "extra-galactic" background . Because it is bright, roughly 7% of the diffuse x-ray background flux at b~10^0 , l ~45^0, it has been mapped by HEAO-1 A-2 with the full resolution of the detector. At lower latitudes this component has been studied in detail by Exosat (Warwick et al. 1985) and Ginga (Koyama 1988) and it is clear from the Tenma spectroscopy that most of the emission is due to thermal processes from plasmas of T<10 kev, but as of yet its physical origin has not been determined (Koyama 1988). It is the presence of this strong spatially complex Galactic component that makes determination of the x-ray surface brightness dipole moment in the direction of the micro-wave dipole difficult .

2.3. The Great Attractor Region

The next strongest feature in the high latitude sky is a "bright" spot at b~20^0, l~330^0 (figure 2) which is located in the direction of the "Great Attractor" (GA) (Lynden-Bell et al. 1988) and the large concentration of clusters (Scaramella et al. 1989) located at v~14,000 km/sec. This region has an enhancement of ~4-5% of the average sky flux (ΔI_{GA}~7.8±1.0x10^{-13} ergs/cm^2/sec/deg^2) (Jahoda and Mushotzky 1989) and subtends a solid angle of ~1000 square degrees for a total flux of ~10^{-9} ergs/cm^2/sec. If it were at the distance of the GA it would have a luminosity of ~3x10^{44} ergs/sec and if were due to the superposition of objects at the distance of the large concentration of clusters L~3x 10^{45} ergs /cm^2/sec. We have visually compared our map of x-ray surface brightness with a smoothed map of the Abell cluster distribution (figure 4) and this region appears to be the only place where an x-ray surface brightness enhancement at this level may be due to "superclusters" at z<0.1. There is another region, a factor of 2 dimmer that maybe associated with supercluster 12A in the list of Bahcall and Soneira (1984) at l=30, b=70 and z~0.07. If this is a real association,

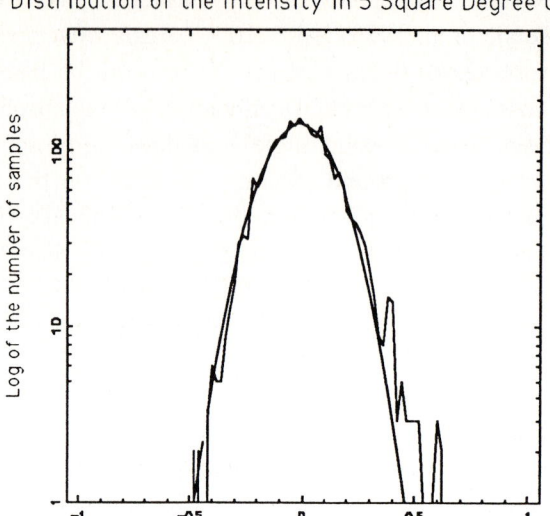

Figure 1: The distribtuion of intensities in 5 square degree cells. The solid line is the fit of a gaussian with FWHM= .35 cts/sec. The mean counting rate ~5.6 cts/sec

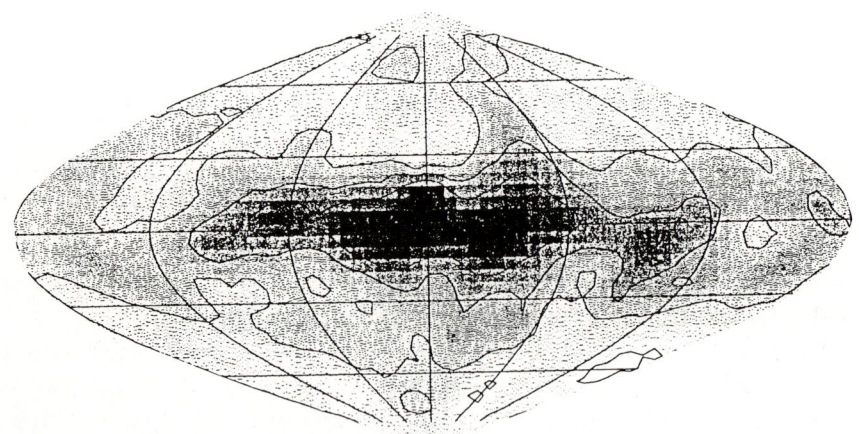

Figure 2a: The 2-20 kev surface brightness of the sky in galactic coordinates boxcar smoothed by 10°. The contour levels represent -2.5%,0,+2.5% and +5% of the mean high galactic latitude surface brightness . Point sources have been removed by clipping all 1/4x1/2 degree cells greater than 5 sigma above the mean.

and the lack of other "detections" places this in doubt , $L_X \sim 10^{44}$, roughly consistent with the Persic et al. upper limits. We may assign a upper limit to $(\Delta\rho/\rho)$ in the GA or similar region from the x-ray data of $(\Delta\rho/\rho) < (\Delta I/I)_x)(\lambda/\lambda_H)^{-1} f^{-1}$: using $(\Delta I/I)_x < 0.05$ and $(\lambda/\lambda_H) \sim 0.02$ we find $(\Delta\rho/\rho) < 2.5 f^{-1}$, comparable to the values obtained from model fitting the optical velocity data for the GA (Lynden-Bell et al. 1988) of $(\Delta\rho/\rho) \sim 0.4$.

Because the GA is located close to the galactic plane it is quite difficult to conduct an optical search for clusters of galaxies. It is thus not clear if, like other superclusters, the GA contains numerous rich Abell clusters. X-ray surveys at E>2 kev are much less affected by reddening and sky confusion at $|b| > 5°$ and thus are an efficient way of searching for x-ray luminous rich Abell clusters. We have searched for "pointlike" x-ray sources (clusters of galaxies and active galaxies) which could be associated with the GA (Jahoda and Mushotzky 1989) and have not found any down to a luminosity limit of $\sim 3 \times 10^{43}$ ergs/sec, which is considerably below the mean luminosity of Abell richness 0 clusters . We thus conclude that the GA if "real", is not similar to other superclusters, which, by construction, are composed of rich Abell clusters. This raises the interesting question of whether other superclusters are similar to the GA in being composed primarily of "field" galaxies and poor groups and has important implications for studies of large scale structure.

If the x-ray surface brightness enhancement detected in the region of the GA is due to hot gas located in the potential well of the supercluster we may estimate its core mass as: $M(gas) \leq 1.2 \times 10^{16} M_\circ (\Delta I_{GA})^{1/2} (T_{10})^{1/4} (D_{86}\Theta_{20})^{5/2}$; where Θ_{20} is the radius of the GA in units of 20 degrees and D_{86} is the distance in units of 86 Mpc and T_{10} is the temperature in units of 10 kev .

2.4 Superclusters

Persic et al. (1989b) have used the same data base to place tight upper limits on diffuse radiation from optically selected "pointlike" superclusters(e.g. they have assumed that the superclusters were of angular size $<2^0$ in size) . They find that for a sample of $<z> \sim .046$ an upper limit of $F_x < 5 \times 10^{-12}$ ergs /cm^2/sec. Using the HEAO-1 luminosity function of clusters of galaxies they estimate that the Abell clusters in the superclusters would contribute an expectation value of $\sim 4 \times 10^{-12}$ ergs /cm^2/sec. They conclude that, on average, the x-ray luminosity of diffuse emission due to hot gas associated with superclusters is $< 4 \times 10^{43} h^2_{50}$ ergs /cm^2/sec the order of the detected diffuse emission in the direction of the GA. A uniform volume of hot gas of total mass $10^{16} M_{16}$ solar masses, temperature T_{10} kev and linear size $10 h^{-1}_{50} R_{10}$ Mpc would have a luminosity of in the 2-10 kev band of $L_{gas} \sim 3 \times 10^{45} (M_{16})^2 (T_{10})^{1/2} (R_{10})^{-3} h_{50}$ ergs/sec.

The HEAO-1 data thus indicate that the non-cluster associated mass in hot gas associated with optically identified superclusters is $< 1.5 \times 10^{14}$ solar masses, on the order of the mass in hot gas associated with rich clusters. This limit already places tight constraints

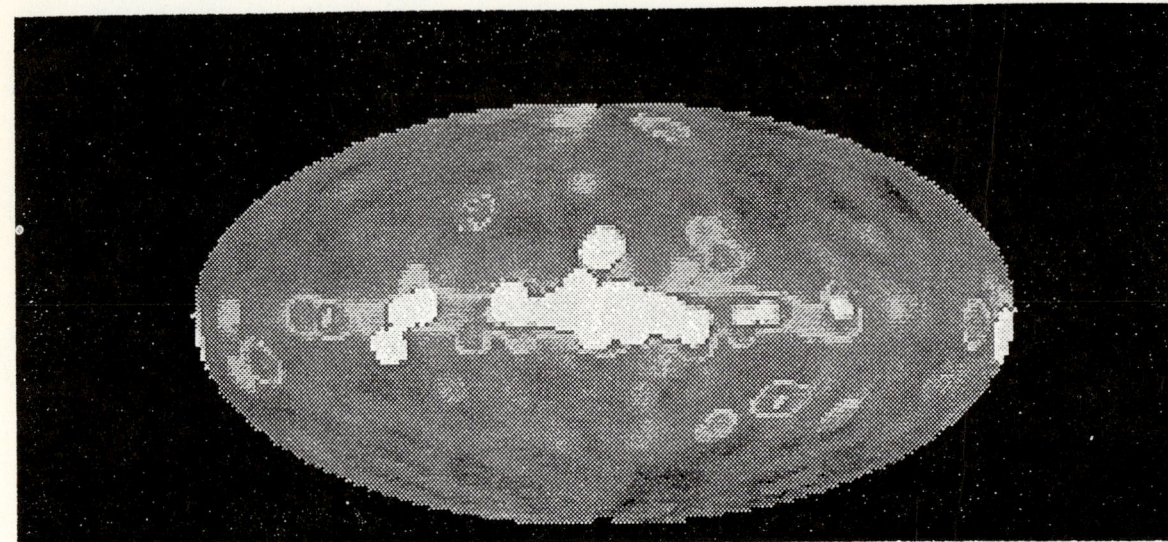

Figure 2b The 2-20 kev x-ray sky smoothed with a 6 degree gaussian. Point sources are left in the data set but the display has been clipped and stretched to emphasize weak surface brightness features. All sources brighter than 6×10^{-10} are saturated. White corresponds to bright spots

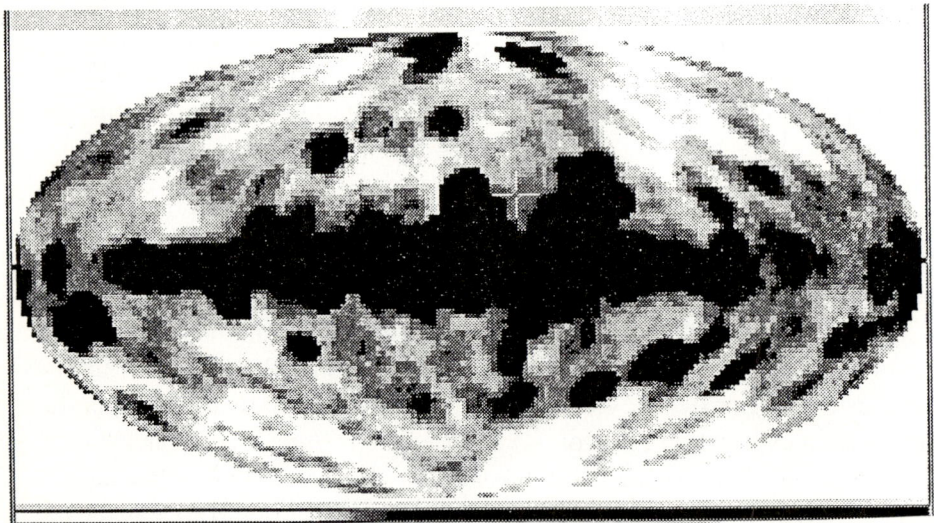

Figure 2c The same data as figure 2b but the stretch has been increased- also black and white are exchanged such that dim spots are white

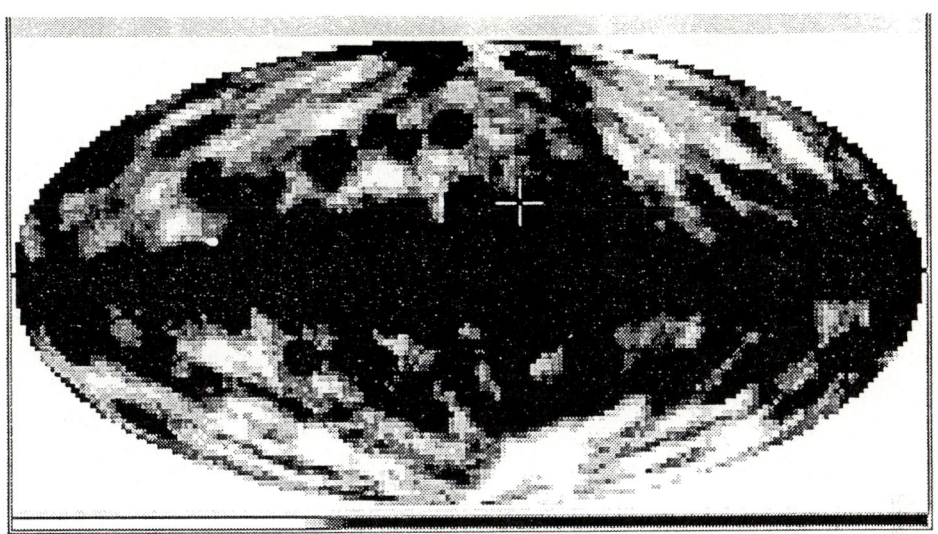

Figure 2d The same data again but now stretched such that only places in the sky with less than the median flux are displayed as white regions.

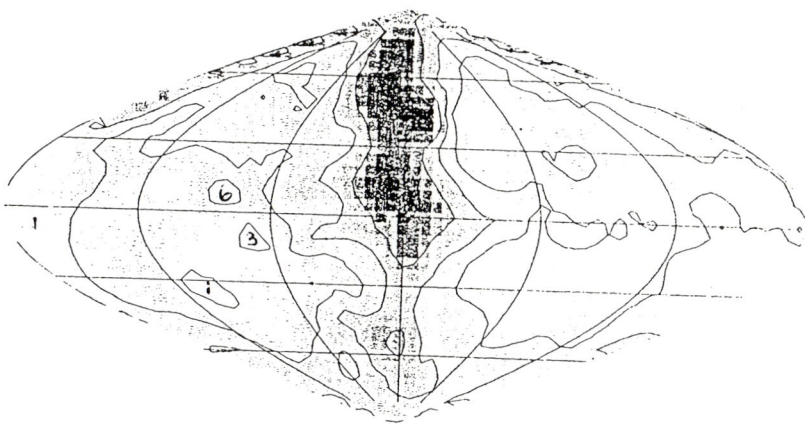

Figure 3 Projection of the data shown in figure 2a in supergalactic coordinates. The axis has been rotated so that supergalactic l=180 is at the center of the diagram. Regions marked 1 and 3 are deficits in the projected Abell cluster distribution shown in figure 5 . #6 is an enhancement in the Abell cluster distribution

(Persic, Rephaeli and Boldt 1988) on the possible baryonic mass density than can be associated with superclusters or the fraction of supercluster mass that can be in hot gas .

The all sky x-ray surface brightness data already indicate that enhancements of $(\Delta I/I)_x > 2\%$ on scales greater than 300 square degrees are rare. Then using the same formalism as for the GA and using a distance of 200 Mpc and an angular radius of 10^0 (scaling from the GA) we can already place upper bounds of $M(gas) < 8 \times 10^{15} (D_{200} \Theta_{10})^{5/2}$ solar masses for any supercluster like object of size >35 Mpc whether or not it has been previously cataloged. If we assume superclusters to have a scale of 20 Mpc (Bahcall 1988) and properties similar to that of the GA $(\Delta I/I) \sim 3\%$ of the x-ray background, $L \sim 10^{44}$ ergs /cm²/sec, $(\Delta \rho/\rho) \sim 2$) then to detect such objects at $z < 0.5$ one needs a system capable of measuring the x-ray surface brightness at the level of 1-2% on scales of $>1^\circ$. Thus an appropriately sensitive x-ray surface brightness survey might be the best way of searching for "non-Abell cluster" superclusters or any other virialized concentration of baryons which might make a significant contribution to the mass of the universe.

3 New Structures

It is apparent to the eye (figure 2b,c,d) that there are several "bright" and "dim" regions in the x-ray with relative differences of 1-2%. While visual inspection is always risky, comparison of the HEAO-1 data with the Uhuru data (Turner and Geller 1980) and the Ariel -V results (Warwick et al. 1979) shows fair overall agreement in both the location and amplitude of the surface brightness distribution. Projection of the HEAO-1 data in supergalactic coordinates (fig 3) shows that most of the large scale structure (other than in the GA region) is not directly assignable to the local supercluster.

While there are no obvious associations with the superclusters identified by Bahcall and Soneira (1987) or with the voids identified by Batuski and Burns (1988) work with A. Dekel and J. Primack shows that several of these large scale "dim" spots can be associated with voids in the distribution of clusters of galaxies (figure 5). In particular in the southern galactic hemisphere the two large scale dim spots are closely associated in projection with holes in the flux (N/r^2) distribution of clusters. These features are several hundred square degrees in "size" and represent roughly 2% diminutions in the average sky flux. The features in the north are smaller in solid angle but many of them can also be associated with features in the Abell cluster flux and with the Bootes void. Given the statistical errors in our data only 1/3 of a 2% diminution with solid angles >100 square degrees is expected in the whole sky by chance.

The volume emissivity of Abell clusters is too low (Boldt this symposium) for the effect to be due to their relative absence. If we assume that the Abell clusters are tracers of the XRB producing matter distribution then, using $(\Delta I/I)_x = (\Delta \rho/\rho) (\lambda/\lambda_H) f$ and assuming that the effective depth of the Abell catalog is $(\lambda/\lambda_H) \sim 0.1$ (e.g. $z \sim 0.1$), in order to get the total 2% diminution requires that $d\rho/\rho \sim -1$ for $f \sim 0.3$ (appropriate for the local volume emissivity of the XRB (Boldt this symposium)) similar to that claimed for the

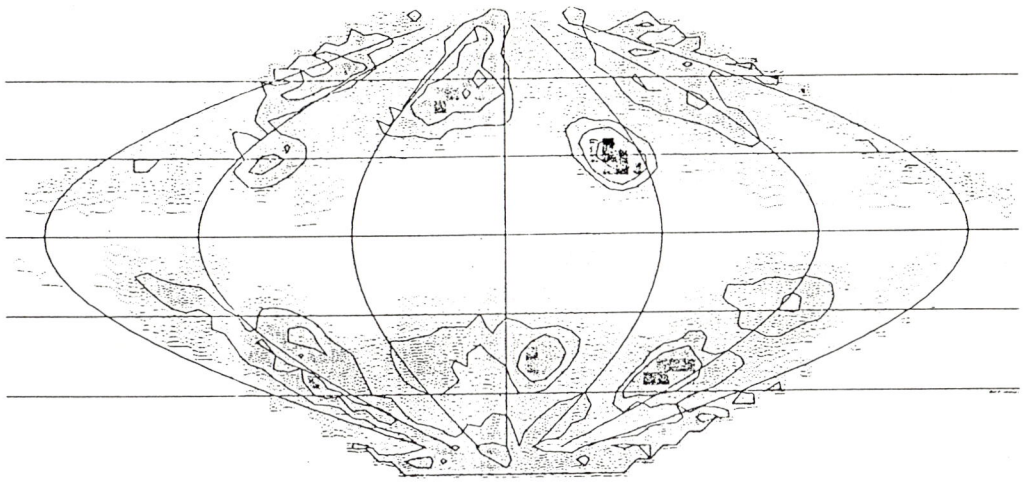

Figure 4 The surface density of D≤4 Abell clusters smoothed by a 10° boxcar in Galactic coordinates on the same scale as figure 2a .

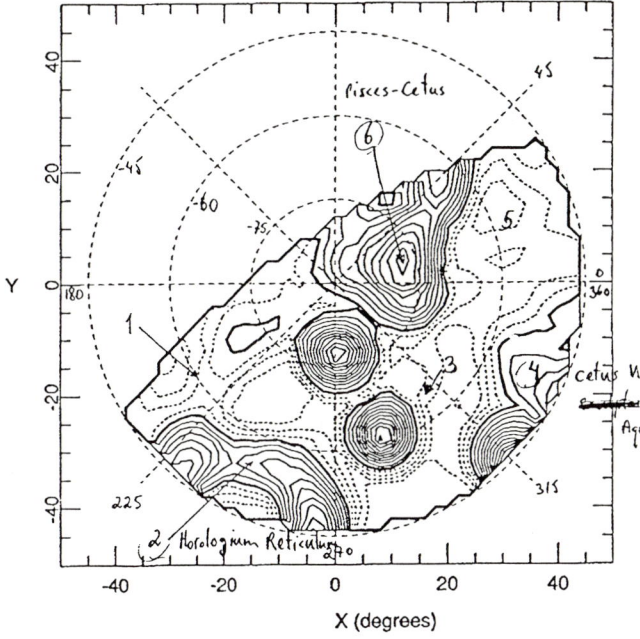

X (degrees)

Figure 5a. The flux (N/r^2) of Abell clusters in the south galactic cap. (Primack and Dekel priv comm) in supergalactic X and Y coordinates and smoothed with a 6° gaussian. The bright spots 2,4,6 are the Horologium-Reticulum Supercluster, the Cetus Wall and Pisces-Cetus supercluster. The dim spots 1 and 3 are shown in figure 3 and 5b

The ACF of the Diffuse Flux

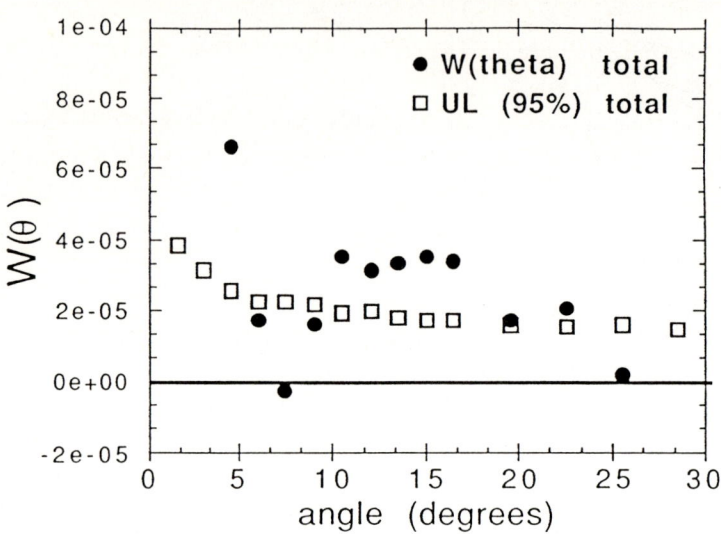

Figure 6 The autocorrelation function of the x-ray sky. The data are the dots and the 95% upper limit for the null hypothesis (i.e. that there is no signal in the ACF) is indicated by the squares.

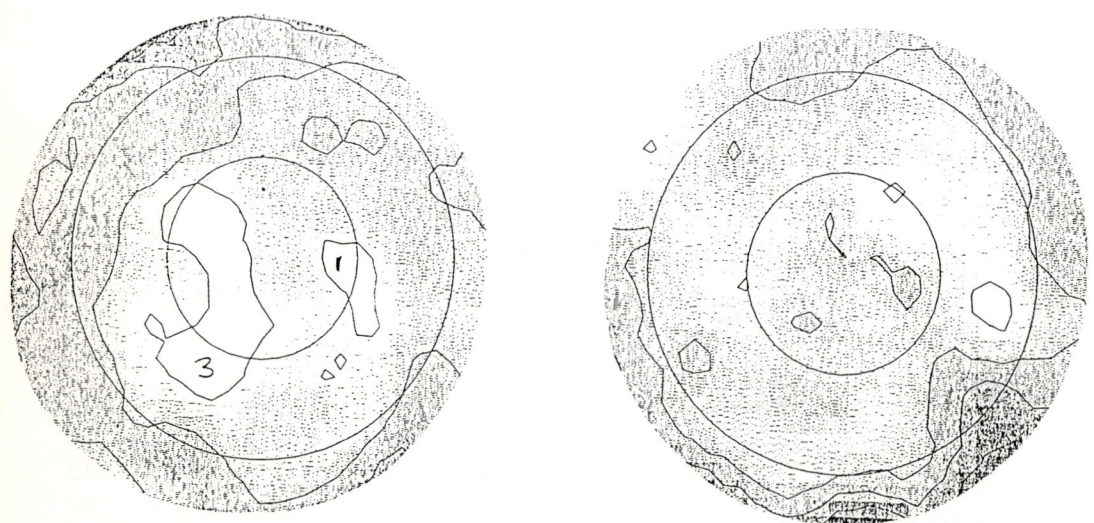

Figure 5b: The data shown in figure 2a projected at the north (right panel) and south (left panel) galactic caps, /b/=60,30 are shown as solid lines and the contours are the same as in fig 2a . The features numbered 1 and 3 in figure 5a are indicated.

Bootes void . The effective linear dimension of these "spots" is then ~ $100(z_{eff}/0.1)h_{50}^{-1}$ Mpc, similar to the Great Walls and the Bootes Void . We thus speculate that we are detecting the power on large scales claimed in the galaxy distribution data . It is not clear at present why we are primarily detecting "voids" rather than bright spots- two possibilities are 1) that the bright spots show up as x-ray sources which are explicitly excluded from the map or 2) that the peaks in the galaxy distribution are of larger amplitude than peaks in the distribution of the sources responsible for the XRB- e.g. that the sources of the XRB are less strongly clustered than Abell clusters as might be expected if they are low luminosity active galaxies. Of course then one wonders why the "voids" in the Abell cluster distribution are seen so well.

This work has just started and cannot at the present time give the detailed correlation between the tracers of large scale structure and the XRB. However, because our sensitivity is rather limited we could only have seen these features if there is a relatively large amplitude on large angular scales. The fact that several such features are "seen" in the data vindicates Rees original suggestion that the study of the XRB is the best way to find large scale structure.

The correlation of the HEAO-1 A-2 data with galaxy catalogs (Jahoda et al 1991) is reported by Lahav at this conference.

4 The Autocorrelation Function of the XRB

The clustering properties of the XRB constrain the properties of the objects that are major contributors to it. A measurement of $W(\theta)$ at some angle provides an upper limit to the contribution of a non-evolving population with a certain fixed correlation length (evolutionary models can be constructed to increase the contribution of a given population to the XRB with a fixed value of $W(\theta)$) If the correlation function of that population is a power law of slope 1.8 with a length scale of R_0 then (DeZotti et al 1990, Danese, this symposium) the auto-correlation function (ACF) due to a given population is approximately $\Gamma(\theta) \sim -.035 \ Cs(R_0/10Mpc)^{0.9}(z_{max}/0.2)^{0.1}(\theta/3°)^{-0.4}$

Where Cs is the contribution of a given population to the XRB. The best estimates of C_s from a non-evolving population is $C_s \sim 0.3$ (Boldt 1991) Thus one expects $\Gamma(\theta) \sim 8 \times 10^{-3}$ ($W(\theta) = 6 \times 10^{-5}$) at 6 degrees if $R_0 = 10$ Mpc consistent with the results of Boyle et al (1990) for QSOs. Previous results (Persic et al 1989, Carrera et al 1991) have shown that the autocorrelation function of the XRB is less than 4×10^{-4} on scales of 3-10 degrees. The HEAO-1 fluctuation data already constrain models of the sponge-like or shell-like structure of the universe (Meszaros and Meszaros 1988).

We have used the A-2 shown above to determine the ACF of the XRB. The very large number of pairs allowed by the all sky data base and the high signal to noise (2:1) of the A-2 result in a very small formal error ~ 2×10^{-5} in $W(\theta)$. The data were linearized with the best fitting csc(b) model for the galaxy and any linear time dependence for the detector efficiency removed. We report (figure 6) the first detection (99%) confidence of correlation in the XRB at scales of ~6-20 degrees with $W(\theta)$~3×10^{-5}, a factor of ~10 below the previous upper limits. The errors in $W(\theta)$ are difficult to determine since they are not independent . The 95% upper limits for the null hypothesis, that there is no power in the ACF, shown in figure 6 are determined by calculating the ACF of 1000 randomized samples of the sky, scale as the square root of the number of samples and lie in the range 2.2-1.6×10^{-5} between 6 and 20 degrees. The slope of $W(\theta)$ vs θ is also consistent with that predicted from simple arguments out to scales of ~ 20 degrees.

Using the above equation the best estimate of the flux contributed to the XRB by objects clustered with the scale length of quasars is
C_s=.32(+/-.1) $(W(\theta)/3 \times 10^{-5})^{1/2}(R_0/10Mpc)^{-0.9}(z_{max}/0.2)^{-0.1}(\theta/6°)^{0.4}$ consistent with the estimates from the local volume emissivity of Seyfert galaxies (Danese et al 1986) and the correlation of the XRB with nearby galaxies (Jahoda et al 1991). At present it is not clear how the presence of "voids" and "lumps" in the XRB affects the ACF at these angular scales. However we note that the present results are very restrictive on any non-evolving population of contributors to the XRB, constraining any population which contributes more than 60% of the XRB to have a scale length less than 5 Mpc, the scale length of galaxies. The effects of evolution are quite model dependent (DeZotti et al 1990, Carrera et al 1991) but it is rather difficult (cf DeZotti et al figure 3) for any population with a scale length of 10 Mpc or larger, such as quasars, to make up more than ~50% of the XRB.

Limits on the dipole in the x-ray background were obtained by Shafer (1983) from the HEAO-1 A-2 data using detectors with an effective beam size of ~20 square degrees. He found that the x-ray data were consistent with the same velocity and direction as the µ-wave anisotropy . The major uncertainty was due to emission from the much larger galactic component, the small number of independent sky elements, the dominance of sky fluctuation noise over Poisson noise and the lack of information about structures like the GA. Boldt (1987) gives a nice graphic representation of the problem. We are trying to see if we can improve on his results using data with better angular resolution and different approaches to modeling the galactic contribution and the effects of the large scale structures.

5 Spatial Fluctuations in the XRB

We have performed an analysis similar to that shown by Hayashida (this conference). subtracting the -1σ dim spots in the XRB from the +1σ bright spots. We find, in very good agreement with Hayashida that this difference spectrum can be well fit over the 2-20 kev band (figure 7) by a power law of energy slope 0.88+/-0.1 with $N(H)<8 \times 10^{21}$ atms/cm^2. There is a weak indication in the data of spectral flattening at E>10 kev. The HEAO-1 A-2 experiment had 3 major advantages over Ginga for measurements of the diffuse background 1) a lower internal background by about a factor of 3) ; 2) a means of directly measuring the diffuse flux without relying on modeling the background); 3) a larger solid angle so that the ratio of diffuse to internal flux is ~3:1. Of course, the Ginga data has considerably larger total number of counts.

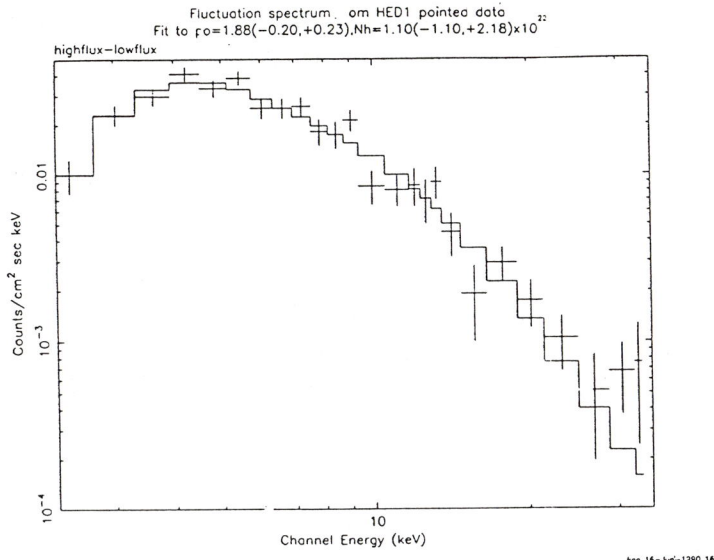

Figure 7 a: The fit of the HEAO-1 A-2 fluctuation data to a power law . The fit is acceptable out to an energy of ~30 kev.

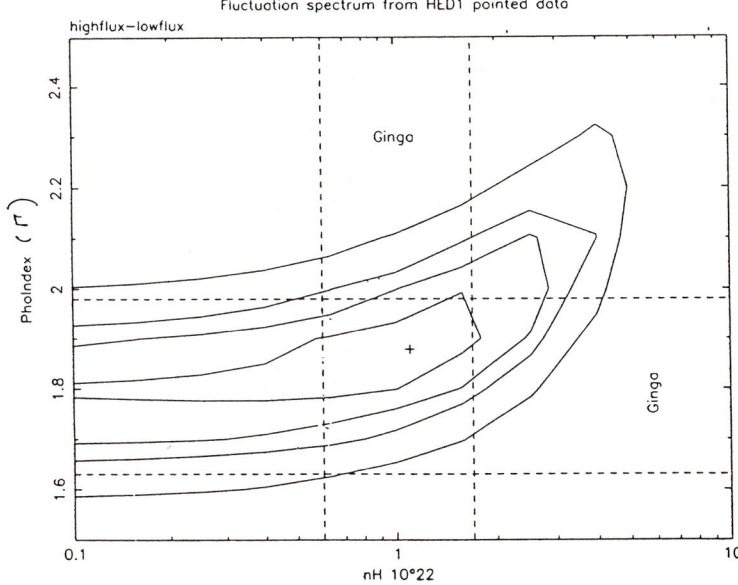

Figure 7b The 40, 68, 90,99% confidence contours for the slope and column density of the energy spectrum of the fluctuations . The best fit to the Ginga data of Warwick and Stewart 1989 is indicated.

C. Lanczycki and K. Arnaud have used the HEAO-1 "blank-sky" pointed data, that is observations of regions of the sky (Tennant 1983) in which there are no cataloged x-ray sources brighter than 1.5×10^{-11} ergs/cm^2/sec in the 2-10 kev band, to obtain considerably higher (~5000 sec/exposure) signal to noise observations. The data set is restricted to |b|>30° to avoid "contamination" from a galactic component (Koyama 1989, Jahoda and Mushotzky 1989) and has 148 observations. This is a completely independent data set from that used by Shafer (1983) and has been analyzed independently. The effective depth of this sample lies between the previous HEAO-1 fluctuation result and the Ginga results. Our preliminary results (Lanczycki, Shafer, Arnaud, Mushotzky 1991 in prep) are in good agreement with the Ginga fluctuation analysis :1) Both the slope (γ=2.3-2.7, 90% confidence) and the normalization of log N-log S= 2.2 (+0.7,-0.5)$\times 10^{-15}$ ergs/cm^2/sec agree well with previous results in the 2-10 kev band but the normalization is a factor of 3 higher then the Einstein Medium Survey results.

Distribution of the Flux Ratios in the Hard and Soft X-ray Band

20 AGN Combined IPC and MPC Analysis

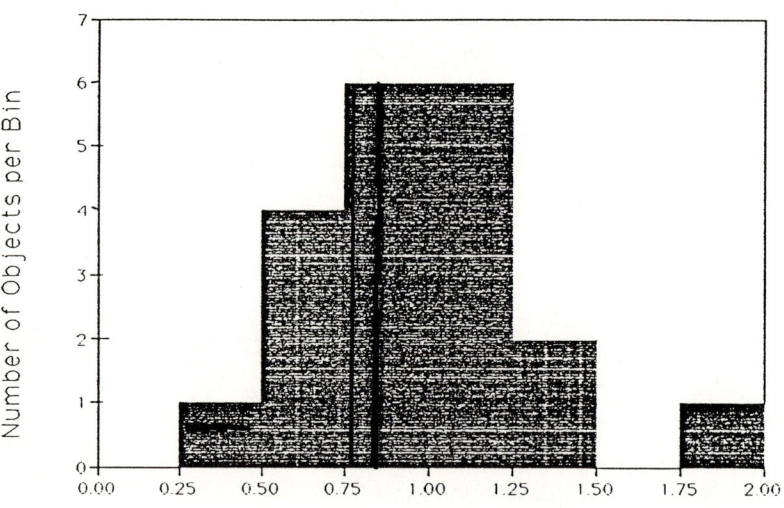

Figure 8 Ratio of the distribution of fluxes in the 2-10 kev and 0.3-3.5 kev band derived from combined spectral fitting of Einstein IPC and MPC data for 20 active galaxies. The expected ratio is ~1.0 for a photon index of 1.7

The discrepancy cannot be due to a misunderstanding of the IPC spectral response since the ratio of the fluxes derived in the 0.3-3.5 kev band to 2-10 kev band from a combined fitting of the Einstein Observatory IPC and the MPC data to 20 bright active galaxies agrees well (figure 8) with the assumption of a ~1.7 photon index power law over the entire energy band if the effects of x-ray absorption are included (Urry, Edelson and Mushotzky in preparation). A larger sample (Mushotzky 1991) shows that while there is a large spread in the soft/hard x-ray ratios (due presumably to the effects of absorption and soft x-ray excesses) the median value also agrees with the ratio predicted from a simple power law model. Because of the larger beamsize than Ginga and the known value of W(θ) for this beam it is unlikely that this data are subject to the effects of power in the auto-correlation function to distort the normalization of log N- log S.

6 Future Work

As indicated above recent studies of the x-ray surface brightness distribution of the sky have been able to make interesting statements about our galaxy, the Great Attractor, the two-point correlation of the XRB, the topology of the large scale structure of the universe and nearby superclusters. In addition a "new" unidentified component of the sky, the large scale "dim" spots, has been detected. The HEAO-1 A-2 data base is just being explored for this new component and the study of possible optical counterparts to these structures has just started. We anticipate interesting results in the future. Unfortunately, the 14 year old A-2 data are the best for this purpose until the launch of XTE in 1996.

Acknowledgements:We thank A.Dekel and J. Primack for communicating their Abell cluster results, C. Lanczycki and K. Arnaud for communication of their fluctuation results, and E. Boldt for continuing and enlightening discussions.We also thank E. Boldt and the entire HEAO-1 A-2 team for their care and attention to detail 15 years ago that has made this data set available for analysis.

References:
Bahcall, N. 1988 Ann Rev Astron + Astrophys 26,631
Bahcall, N. and Soneira R. 1984 Ap.J. 277,27
Batuski, D. and Burns, J. 1985 A.J. 90, 1413
Boldt, E.A. 1987 Physics Reports 146,216
Boldt, E.A. and Leiter, D. 1987 Ap.J. Lett 322, L1
Boyle,B. et al private communication
Carrera, F.J. et al. 1991 M.N.R.A.S. 249, 698
Danese, L. De Zotti, G. Fasano, G. and Franceschini, A. 1986 A+ A 161,1
De Zotti, G.F. et al. 1990 Ap.J. 351,22
Giacconi, R et al. 1979 Ap.J. Letters 234, L1

Giacconi, R. and Zamorani, G. 1987 Ap.J. 313,20

Goicoechea,L.J. and Martin-Mirones, J.M M.N.R.A.S. 242, 493 1990

Guilbert, P and Fabian,A. 1986 M.N.R.A.S. 220,439

Jahoda,K. 1989 B.A.A.S.20, 1086

Jahoda,K. and Mushotzky ,R. 1989 Ap.J. 346 638

Jahoda,K, Lahav,O., Mushotzky, R.F. and Boldt, E. 1991 Ap.J. Lett 378,L37

Klypin, A. A., Sazhin, M. V. Strukov, I. A., and Skulachev, D. P.Pos'ma Astron. Zh. 13, 259 1987

Koyama, K. 1988 in Physics of Neutron Stars and Black Holes Edited by Y. Tanaka pg 55 Universal Academy Press Tokyo Japan

Lahav, O. 1987 M.N.R.A.S. 225,213

Lynden-Bell, D et al. 1988 Ap.J. 326, 19

Meszaros,A and Meszaros, P. 1988 Ap.J. 325 25

Mushotzky, R.F. 1991 Proceedings of the 28 th Yamada Conference on Frontiers of High Enrgy Astrophysics Nagoya Japan 1991

Peebles, J. 1980 Physical Cosmology

Persic, M. Rephaeli,Y and Boldt E. 1988a Ap.J. Letters 327,L1

Persic, M. et al. E. 1989a Ap.J. Letters 336 L47

Persic, M. etal 1989 Ap.J.,364, 1

Rees, M. 1979 in X-ray Astronomy pg 377 edited R. Giacconi and G. Setti NATO Advance Study Institutes Series

Scaramella, et. al. 1989 Nature 338, 562

Schwartz, D and Gursky, H. 1974 X-ray Astronomy ed R. Giacconi and H Gursky

Schwartz, D. 1979 pg 453 X-ray Astronomy (COSPAR) Baity and Peterson eds Pergamon Press

Serlemitsos P 1988 J. of Applied Optics

Shafer,R. 1983 PhD Thesis University of Maryland :Spatial Fluctuations in the Diffuse X-ray Background

Smoot,G. et al 1990 in "After the First 3 Minutes" pg 95 eds S.Holt, C. Bennett,V. Trimble AIP Conference Proceedings #222

Strukov, I. A., Skulacher, D. P., Boyarsky, M. N., and Kachev, A. N. Pis'ma Astron. Zh. 13, 163 1987.

Turner, E, and Geller, M. 1980 Ap.J. 236,1

Tyson,J. 1988 A.J. 96,1

Warwick,R, and Fabian,A. 1979 Nature 280, 39

Warwick, R, Pye,J and Fabian, A. 1979, M.N.R.A.S. 190,256

White, S., Davis, M. and Frenk,C 1984 M.N.R.A.S. 209,27p

Cosmological implications of the small scale isotropy of the X–ray background

F.J. CARRERA AND X. BARCONS

Departamento de Física Moderna, Universidad de Cantabria, 39005 Santander, Spain.

1 INTRODUCTION

The importance of the study of the X–ray background (XRB) arises from the fact that it very likely comes from an early epoch of the Universe ($z \sim 2 - 10$) which is thought to coincide with that of the formation of galaxies and of the structures in the Universe that are now being revealed in optical deep surveys. Therefore, the nature of the sources whose integrated emission gives rise to the XRB and their properties and evolution is really worth investigate, because it may provide us some clues about the evolution of the Universe.

A possible approach to this problem is the fluctuation and isotropy analysis, which studies the integrated emission of sources too weak to be revealed in the surveys. As the usefulness of fluctuations analysis ($P(D)$ curve) has already been discussed elsewhere in these proceedings (see L. Danese and L. Toffolatti's contributions) we will concentrate on the characterization of the isotropy of the XRB and its angular autocorrelation function (ACF). In section 2 we discuss its relation to cosmic observables, in section 3 we comment briefly on the data used to get the ACF on arcmin scales. Implications along with comparison with other results are presented in section 4. These results are summarized in section 5 as well as the power of this and similar analyses.

2 THE ACF

The use of the ACF in the study of source clustering was presented in Barcons & Fabian (1988). Only a few points will be highlighted here.

The ACF is defined as $W(\theta) = \frac{\langle D(\hat{n}) \cdot D(\hat{n}') \rangle}{\langle I \rangle^2}$ where $\cos \theta = \hat{n} \cdot \hat{n}'$, $D(\hat{n}) = I(\hat{n}) - \langle I \rangle$, and $\hat{n}$ and $\hat{n}'$ are unit vectors along observing directions. It gathers information about how similar are on average the deflections D at each separation θ, normalized to the mean intensity $\langle I \rangle$.

The expected ACF from a source population (possibly extended) with luminosity profile $h(\vec{r})$, spatial correlation function $\xi(\vec{r}, z)$ and volume emissivity $j \equiv n(z)L(z)$ observed through a telescope/detector characterized by a 'beam' profile $G(\hat{n})$ is given by

$$W(\theta) = \frac{c}{H_0} \frac{\int_{z_{min}}^{z_{max}} dz\, (1+z)^{-2}(1+2q_0 z)^{-1/2}\frac{n(z)}{d_L^2(z)}}{\langle I \rangle^2} \int d^2q\, e^{i\vec{q}(\hat{n}-\hat{n}')}\hat{G}^2(\vec{q})\hat{\sigma}^2(\vec{q}/l(z), k_z = 0)$$

$$\times \frac{\pi}{2}[\langle L^2(z)\rangle + (2\pi)^{3/2}n(z)\langle L(z)\rangle^2 \xi(\vec{q}/l(z), k_z = 0)] \tag{1}$$

where

$$\langle I \rangle = \frac{1}{4\pi}\frac{c}{H_0} \int_{z_{min}}^{z_{max}} dz\, (1+z)^{-4}(1+2q_0 z)^{-1/2}n(z)\langle L(z)\rangle \tag{2}$$

The first term is what we call the poissonian term and depends only on the extension of the sources and the beam profile, while the second includes the spatial correlation function of the sources. Depending on the resolution of the detector and on the scales we are interested in one of them dominates. In this work we have obtained an upper limit to the ACF of the XRB as measured with a 1 arcmin beam on scales of about 3 arcmin which, at a redshift of 1 corresponds to $\sim 1\,h^{-1}$ Mpc (h= 100 km s^{-1}Mpc^{-1}). AGN can be considered as point sources and the main contribution beyond $\theta \geq 1$ arcmin will come from the clustering term.

3 THE DATA

We have used a 17192 seconds exposure image of NGC5548 taken by the ROSAT PSPC, of which only a central circle of 15 arcmin radius has been considered. To get the ACF we first must get rid of the sources present, including NGC5548, a cluster and several other putative sources. We removed any 32" × 32" pixel with ≥ 5 cts, which we estimate to correspond to a flux $S_{limit} \sim 8 \times 10^{-15}$erg cm^{-2}s^{-1} (0.5–2 keV). This amounts to about 30% of the XRB flux given by Wu *et al.* (1991) from *Einstein* IPC data, and is very similar to the value obtained by Shanks *et al.* (1991). The results that will be presented here refer only to this 70% of the XRB still unresolved ('residual'). To be sure that we have subtracted properly the two main sources, we have rebinned the data into 192" × 192" after also correcting for vignetting, and the final rebinned field does not reflect any substantial enhancement at the position of the removed sources.

In order to calculate the ACF, we have binned the subtracted and de–vignetted 16" × 16" pixels into 64" × 64" pixels and then calculated

$$W(\theta) = \frac{\sum_{\text{angle}(\hat{n},\hat{n}')=\theta\pm 32"} D(\hat{n}) \cdot D(\hat{n}')}{N \cdot \langle I_{XRB} \rangle^2}$$

N being the number of pairs fulfilling the condition in the numerator and $\langle I_{XRB} \rangle = 2.031 \times 10^{-4} \mathrm{cts}^{-1} \mathrm{arcmin}^{-2}$ (hard band) as estimated from the outer parts of the image. The result is given in Figure 1: no significative signal is found on any scale beyond ~ 2 arcmin. The shape of the PSF of the PSPC might be the origin of the slight enhancement below 2 arcmin. Taking a (very conservative) upper limit of $W_{OBS}(3') < 0.01$ (2σ) leads to very interesting constraints on the evolution of points sources contributing to this 'residual' XRB, as we will see. We will also discuss how better upper limits as that obtained by Hasinger *et al.* (1991) –ROSAT Deep Survey– strengthens our conclusions.

4 CONTRIBUTION OF POINT SOURCES TO THE XRB

We assume a source correlation function

$$\xi(r;z) = \left(\frac{1+z}{1+z_1}\right)^{-3-\epsilon}\left(\frac{r}{r_0/(1+z_1)}\right)^{-\gamma}$$

where z_1 is the reference redshift for clustering, $\gamma = 1.8$, and ϵ is the evolution parameter of clustering ($\epsilon = 0.$ for stable clustering, $\epsilon = -1.2$ for comoving clustering). We then have

$$W_{OBS} \geq f^2 r_0^\gamma\, W_{CALC}(z_{max}, j(z), \epsilon)$$

independently of the local volume emissivity (see Boldt, these proceedings). A combined power law and exponential form for its evolution is taken $j(z) \propto e^{C\tau}(1 + z)^{3+\alpha}K(z)$ where $\tau = (t - t_0)/t_0$ is the look–back time and $K(z)$ the K–correction for which we have used an energy index of 0.3 for the sources as this is the expected index of the residual background. In Figure 2 we have plotted the contour levels of $r_0 f^{2/\gamma} = 1.8, 2.6, 3.4\,\mathrm{h}^{-1}$ Mpc in the (C, α) space for $z_{min} = 1$ $z_{max} = 5$ and comoving clustering (slow evolution). We can see that a maximum clustering length of $\sim 3.4\,\mathrm{h}^{-1}$ Mpc is allowed for point sources giving rise to the XRB, in the case of stable clustering (much faster evolution) the upper limit is $\sim 5\,\mathrm{h}^{-1}$ Mpc.

5 DISCUSSION

The isotropy of the XRB in scales of a few arcmin has been tested in the energy band 0.5–2 keV, founding no significant signal at any scale $\leq 12'$ where our analysis is sensitive. A very conservative upper limit of $W(3') < 0.01$ has been obtained, which implies that the clustering evolution of point sources which integrated emission produces the residual XRB ($\sim 70\%$ of the total XRB in that band) is comoving if they cluster on scales $\sim 3.5\,\mathrm{h}^{-1}$ Mpc, higher correlation lengths ($\leq 5\,\mathrm{h}^{-1}$ Mpc) are allowed only if the clustering evolution is faster (stable clustering).

From Persic *et al.* (1989) we know that $W_{CALC}(\theta) \propto \theta^{-0.8}$ then, if $W(3') \leq 0.1$ means $r_0 f^{2/\gamma} < 5$ for AGN as shown in this work, an upper limit of $W(5') < 2 \times 10^{-3}$

(Hasinger *et al.* 1991) would mean $f^{2/\gamma} < 2.6$ (stable clustering): this limit severely constrains the kind of sources that can give rise to the XRB after removing one third of it coming from bright QSOs.

The isotropy of the XRB is a good probe of the structure and properties of the Universe beyond our present capabilities of directly 'looking into' it, as is shown by this and other recent works, and is able to help us to make a picture of the high redshift Universe.

6 REFERENCES

Barcons, X. & Fabian, A.C., 1988. *Mon. Not. R. astr. Soc.*, **230**, 189

Hasinger, G.R. *et al.*, 1991. In preparation.

Persic, M., De Zotti, G., Boldt, E.A., Marshall, F.E., Danese, L., Franceschini, A. & Palumbo, G.G.C., 1989. *Astrophys. J.*, **336**, L47

Shanks, T., Georgantopoulos, I., Stewart, G.C., Pounds, K.A., Boyle, B.J. & Griffiths, R.E., 1991. *Nature*, **353**, 315

Wu, X., Hamilton, T.T., Helfand, D.J. & Wang, Q., 1991. *Astrophys. J.*, **379**, 564

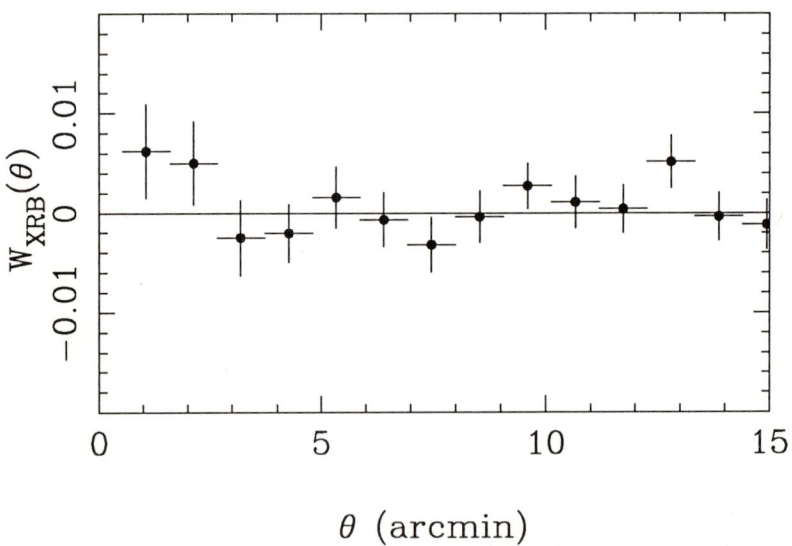

Figure 1. The autocorrelation function of the 0.5–2. keV X–ray background with 1σ error bars.

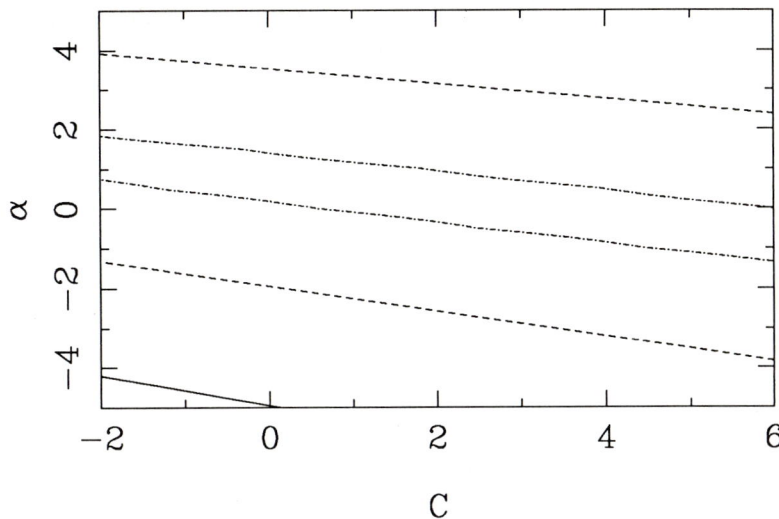

Figure 2. Contour levels of $r_0 f^{2/\gamma} = 1.8$ (solid line), 2.6 (dashed line) and $3.4\,\mathrm{h}^{-1}$ Mpc (dot-dashed line) in the (C, α) space for comoving clustering and $z_{max} = 5$.

Contribution from the Local Universe to the 2-10 keV X-Ray Background

OFER LAHAV

Institute of Astronomy, Madingley Rd., Cambridge CB3 0HA, UK.

ABSTRACT. We discuss a detection of cross-correlation of the diffuse 2-10 keV X-ray *HEAO-1* A2 data with the UGC and ESO galaxy catalogues at 'zero-lag', $\hat{W}_{xg} = (3 \pm 1) \times 10^{-3}$. The volume X-ray emissivity of the local universe is estimated to be $\rho_x = (2.8 \pm 1.0) \times 10^{38} \, h_{50} \, \mathrm{ergs \, s^{-1} \, Mpc^{-3}}$. This implies that about 15% of the 2-10 keV Background can be due to unevolved X-ray sources, distributed out to high redshift. A larger fraction can be obtained if the sources evolved, or if the cosmological constant is non-zero.

1 INTRODUCTION

The origin of the diffuse X-ray background (XRB) at 2-10 keV is still a puzzle. The COBE tight upper limit on the Comptonization parameter, $y < 4 \times 10^{-4}$, rules out a uniform hot intergalactic medium model for the XRB (e.g. Field & Perrenod 1977; Guilbert & Fabian 1986; Lahav, Loeb & McKee 1990). It is more likely that the XRB is due to unresolved point sources. There is growing evidence that at the soft band (0.1-2 keV) at least 30% of the 'Background' is due to quasars (e.g. Shanks *et al.* 1991), but known quasars cannot easily explain the 2-10 keV background since they have much steeper spectrum. Galaxies, being X-ray emitters, are another class of objects which may explain the XRB (e.g. Griffiths & Padovani 1990). Since galaxies and other extragalactic sources are clustered it is expected that the XRB will also exhibit a positive auto-correlation signal . Currently there are mainly upper limits (Persic *et al.* 1989; Carrera *et al.* 1991) and one possible detection (Mushotzky *et al.* in these Proceedings) of the XRB auto-correlation function.

Another approach is to search for a direct correlation between the fluctuations of the XRB and a catalogue of objects (e.g. galaxies) which may be themselves X-ray emitters or correlated with the X-ray emitters. The advantage here is that we know where the galaxies are in space and what are their properties. A previous search (Turner and Geller 1980) only put upper limits on correlations between the *Uhuru* XRB and the Zwicky galaxy catalogue.

Using the more sensitive 2-10 keV *HEAO-1* A2 data and the UGC & ESO galaxy catalogues we detected a signal of cross-correlation. The work discussed and interpreted here was done in collaboration with K. Jahoda, R. Mushotzky and E. Boldt (NASA/Goddard). For more technical details of the analysis see Jahoda, Lahav, Mushotzky & Boldt (1991).

2 CROSS-CORRELATION OF THE HEAO1 XRB AND UGC&ESO GALAXIES

The X-ray data used are from the all-sky *HEAO-1* A2 survey (Boldt 1987; Jahoda and Mushotzky 1989). The galaxy sample consists of all galaxies in the UGC (Nilson 1973) and ESO (Lauberts 1982) catalogues with major diameter $\geq 1'$. The UGC catalogue ($\delta > -2.5°$) and the ESO catalogue ($\delta < -17.5°$) do not overlap, and the ESO sample is about 15% deeper (e.g. Lahav *et al.* 1988). Therefore, we obtain two independent estimates of the cross-correlation. There are about 0.8 and 1.2 galaxies per square degree in UGC and ESO respectively.

To have fair comparison of the X-ray and optical data sets, we smeared the galaxy data with the beam of the A2 experiment ($3 \times 1.5°$ FWHM). We excluded sources of Piccinotti *et al.* (1982) with $L_x \geq h_{50}^{-2} 10^{44}$ erg s^{-1} (where h_{50} is the Hubble constant in units of 50 km s^{-1} Mpc^{-1}), 8 X-ray clusters with $z \leq 0.03$, the region of the Magellanic Clouds and bright Galactic sources.

We cross-correlated the data at 'zero-lag'. An estimator for $W_{xg} \equiv \frac{\langle \delta I \cdot \delta N \rangle}{\langle I \rangle \langle N \rangle}$ is

$$\hat{W}_{xg} = \frac{n_{cells}(\sum I_i N_i)}{(\sum I_i)(\sum N_i)} - 1, \tag{1}$$

where I_i is the XRB intensity and N_i is the (smeared) number of galaxies in the ith cell. We generally used 17 square degree cells at Galactic latitude $|b| \geq 30°$, but also examined the dependence of $\hat{W}_{xg}$ on Galactic and Supergalactic latitude cut-off, cell size, and galaxy catalogue. Most of the subsets of the data give $\hat{W}_{xg} \approx 0.003$.

We determined the confidence intervals using the Bootstrap method by selecting XRB/optical data-pairs at random, calculating $\hat{W}_{xg}$, and repeating the procedure enough times (1000) to build up the full probability distribution of expected values. The results for UGC and ESO are shown as the shaded histograms in Figures 1a and 1b. In the case of ESO the correlation is dominated by a small number of points with high values of both I_i and N_i located near the northern edge of the Great Attractor, the most prominent feature in the Supergalactic plane (see fig. 8 of Lynden-Bell *et al.* 1988). Indeed when the Great Attractor region is eliminated from the analysis we get $\hat{W}_{xg} \approx 0.003$, in agreement with the rest of the sky. However, when we exclude the Super-galactic plane ($|SGB| < 20°$) from the UGC analysis, the signal is not dramatically changed. We conclude $\hat{W}_{xg} = 0.003 \pm 0.001$ for the combined UGC and ESO samples (the confidence limits span the range which includes 68% of the Bootstrap values).

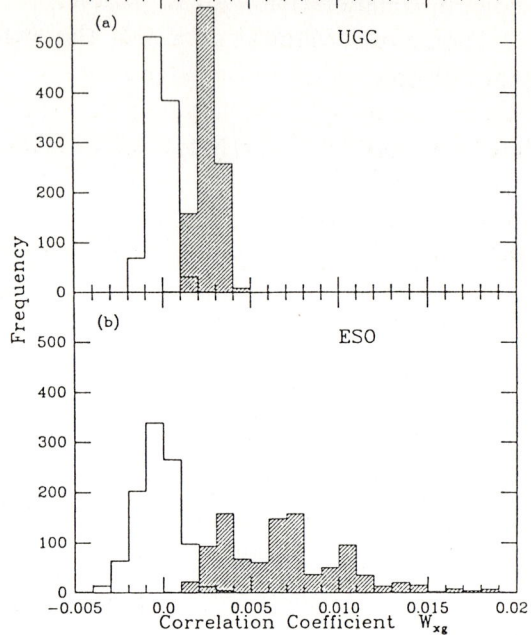

Figure 1. The Bootstrap histograms of $\hat{W}_{xg}$ for UGC (a) and ESO (b). These data are for 17 square degree cells with $|b| \geq 30°$. The shaded histograms show the full probability distribution determined with the bootstrap method. The open histograms represent the probability distribution for $\hat{W}_{xg}$ for data which are uncorrelated, and indeed they are centred on zero.

We have also splitted the optical samples into morphological types. A subset of E+S0 galaxies gives $W_{XE} \approx 1.5 \times 10^{-3}$ and 1.1×10^{-2} for UGC and ESO, while a subset of Sa+Sb gives $W_{XS} \approx 3.0 \times 10^{-3}$ and 8.9×10^{-3} for UGC and ESO respectively (these values were kindly provided by K. Jahoda). The 'fair sample' UGC may indicate that spirals contribute more to the correlation.

A possible simple relation between X-ray surface brightness I_i and galaxy surface density N_i is

$$I_i = A \, N_i + I(0). \qquad (2)$$

The term $I(0)$ represents all emission not correlated with the galaxies, whether it arises within the volume sampled by the galaxies or beyond. A direct fit of the data to eq. (2) gives $A\langle N\rangle/\langle I\rangle \approx 0.01$. Similar value is obtained by combining equations (1) and (2),

$$\frac{A\langle N\rangle}{\langle I\rangle} = \frac{\hat{W}_{xg}}{\hat{W}_{gg}}, \qquad (3)$$

where $\hat{W}_{gg} \approx 0.23$ is the 'zero lag' galaxy auto-correlation estimator.

3 THE X-RAY VOLUME EMISSIVITY OF THE LOCAL UNIVERSE

We assume that the volume emissivity ρ_x is constant over the volume sampled by the galaxies and adopt for the 'effective depth' of the catalogues $R_* \equiv \int_0^\infty \phi(r)dr = (70\pm20)h_{50}^{-1}\mathrm{Mpc}$ for UGC and $(80\pm20)h_{50}^{-1}\mathrm{Mpc}$ for ESO, where $\phi(r)$ is radial selection function of the galaxies. This gives for typical values

$$\rho_x = 4\pi A\langle N\rangle/R_* \approx 2.5 \times 10^{38}\, h_{50}\,\mathrm{erg\, s^{-1}\, Mpc^{-3}}. \tag{4}$$

To estimate the *total* local volume emissivity we added the contribution from the excluded AGNs (but not from X-ray clusters) and found $\rho_x = (2.8 \pm 1.0) \times 10^{38}\, h_{50}$ $\mathrm{erg\, s^{-1}\, Mpc^{-3}}$. For comparison with other estimates of the local emissivity see Boldt in these Proceedings. If the mean number density of bright optical galaxies is ≈ 0.003 galaxies per $(h_{50}^{-1}\mathrm{Mpc})^3$ then the emissivity per galaxy is $\sim 10^{41}\, h_{50}^{-2}\mathrm{erg\, s^{-1}}$. But it is not clear yet how many of the UGC and ESO galaxies actually contribute to the cross-correlation signal.

4 EXTRAPOLATION TO HIGH REDSHIFTS - HOW TO MAKE 100% OF THE BACKGROUND ?

Given the present epoch volume emissivity ρ_x , we can predict the $2-10$ keV intensity due to population of sources out to a redshift z_{max} for a given cosmological model

$$I(< z_{max}) = \frac{c}{4\pi H_0}\rho_x f \approx 0.13\, \frac{f}{0.5}\, I_{tot}, \tag{5}$$

where we adopted for the total observed surface brightness in the $2 - 10$ keV band $I_{tot} = 5.2 \times 10^{-8}\mathrm{erg\, s^{-1}cm^{-2}sr^{-1}}$.
The 'look-back factor' (Boldt 1987),

$$f(z_{max}, \Omega_0, \lambda_0, \alpha, Q) = H_0 \int_0^{z_{max}} (1+z)^{-\alpha}\, e^{Qt/t_0}\, \frac{dt}{dz}\, dz, \tag{6}$$

depends on the maximum redshift, the cosmological model, the spectral index of the sources α and the evolution parameter Q. To generalize it to include the cosmological constant $\lambda_0 = \Lambda/(3H_0^2)$ (currently of interest as a way to increase the age of the universe) we write

$$H_0\frac{dt}{dz} = \{\Omega_0(1+z)^3 - (\Omega_0 + \lambda_0 - 1)(1+z)^2 + \lambda_0\}^{-1/2}\, (1+z)^{-1}. \tag{7}$$

For example, if $z_{max} = 5$, $Q = 0$ and $\alpha = 0$ one gets $f(\Omega_0 = 1, \lambda_0 = 0) = 0.61$ and $f(\Omega_0 = 0.1, \lambda_0 = 0.9) = 1.12$, so the cosmological constant increases by about a factor of 2 the total X-ray emission simply due to a geometrical effect. The effect of evolution is even more dramatic; e.g. for $z_{max} = 2$ and $(\Omega_0 = 0; \lambda_0 = 0)$ one gets $f(\alpha = 0.7, Q = 5) = 3.4$ and $f(\alpha = 0, Q = 5) = 5.4$, so it is possible to get the total Background (if of course spectral and energy constraints can also be satisfied; see e.g. Daly 1991). If no evolution is assumed, $\lambda_0 = 0$ and $\alpha = 0.7$ then f ranges from 0.36 $(\Omega_0 = 1; z_{max} = 1)$ to 0.55 $(\Omega_0 = 0; z_{max} = 4)$.

5 DISCUSSION

Our estimate of the local ρ_x implies that non-evolving sources (like those correlated with present-epoch galaxies) can contribute no more than $13 \pm 5\%$ to the 2-10 keV Background. This result suggests that the sources of the XRB are either not present (and correlated with galaxies) in the nearby universe, or have undergone substantial evolution since the epoch where the bulk of the XRB was produced.

It is worth emphasizing that since we correlate 'zero-lag' cells of finite solid angle, we are unable to distinguish, from $\hat{W}_{xg}$ alone, whether this real correlation is the signature of some of the galaxies being the X-ray sources, or the signature of a physical non-zero-lag correlation scale between galaxies and other X-ray sources (e.g. clumps of hot gas or black holes). To illustrate this point, imagine two populations a and b. The cross-correlation at of counts N at coinciding ('zero-lag') cells of solid angle Ω is (cf. Peebles 1980 eq. 45.6 for the case of auto-correlation)

$$\langle (N_a - \bar{N}_a)(N_b - \bar{N}_b) \rangle_\Omega = C + \frac{\bar{N}_a \bar{N}_b}{\Omega^2} \int \int_\Omega w_{ab}(\theta) d\Omega_a d\Omega_b, \qquad (8)$$

where $w_{ab}(\theta)$ is the non-zero-lag cross-correlation function. If a and b are independent populations (say stars and galaxies) then $C = 0$, but if b is a random subset of a (say that some of the galaxies are emitting at a different wavelength) then $C = \bar{N}_b$, which accounts for the self-correlation of discrete objects (by Poisson statistic). So if $w_{ab} \propto \theta^{1-\gamma}$ then for square cells $W_{ab} = c_1 \Omega^{-1} + c_2 \Omega^{-(\gamma-1)/2}$, which decreases with Ω for $\gamma \geq 1$. In principle, by applying the 'zero-lag' cross-correlation to cells of different shapes and sizes it is possible to decouple the discreteness term from the intrinsic correlation function $w_{ab}(\theta)$. Our results (see Table 1 of Jahoda *et al.* 1991) indicate that $\hat{W}_{xg}$, in particular for UGC, is indeed a decreasing function of the solid angle Ω, but are not sufficient yet to estimate which of the two effects is stronger. (In our case eq. 8 has to be modified to include the intensity I and the beam geometry.)

It seems likely that the X-ray emitters are some of the optical galaxies and that we probe the faint-end of the luminosity function of AGN-like galaxies. It may well be that a small number of AGNs (say of the order of 100) is sufficient to produce the observed correlation. This can be verified by simple simulations and by direct cross-correlation of AGNs with UGC and ESO galaxies (T. Miyaji, OL *et al.* , in preparation).

It would be interesting to extend the cross-correlation method, including non-zero-lag, to deeper galaxy samples and XRB data at different bands. We currently study a cross-correlation of the Ginga scans at the North Galactic Pole with the UGC, IRAS, APM and Abell samples (OL, B. Hassell, F. Carrera, X. Barcons *et al.* , in preparation).

Acknowledgements. I thank my collaborators , E. Boldt, K. Jahoda and R. Mushotzky for their contribution and many stimulating discussions. I also thank NASA/GSFC and the Yukawa Institute for Theoretical Physics (Kyoto) for their hospitality.

REFERENCES

Boldt, E. (1987). *Phys. Rep.*, **146** (4), 215.

Carrera, F.J., *et al.* (1991). *M.N.R.A.S.*, **249**, 698.

Daly, R.A. (1991). *Ap. J.*, **379**, 37.

Field, G.B., and Perrenod, S.C. (1977). *Ap. J.*, **215**, 717.

Griffiths, R.G., and Padovani, P. (1990). *Ap. J.*, **360**, 483.

Guilbert, P.W. and Fabian, A.C. (1986). *M.N.R.A.S.*, **220**, 439.

Jahoda, K. and Mushotzky, R.F. (1989). *Ap. J.*, **346**, 638.

Jahoda, K., Lahav, O., Mushotzky, R.F, and Boldt, E. (1991). *Ap. J.*, **378**, L37.

Lahav, O., Rowan-Robinson, M., and Lynden-Bell, D. (1988). *M.N.R.A.S.*, **234**, 677.

Lahav, O., Loeb, A. and McKee, C.F. (1990). *Ap. J.*, **349**, L9.

Lauberts, A. (1982). *The ESO-Uppsala Survey of the ESO(B) Atlas*, European Southern Observatory.

Lynden-Bell, D., *et al.* (1988). *Ap. J.*, **326**, 19.

Nilson, P. (1973). *Uppsala General catalogue of Galaxies*, Uppsala Astr. Obs. ann., **6**.

Peebles, P.J.E. (1980). *The Large Scale Structure of The Universe*, Princeton University Press, Princeton.

Persic, M., *et al.* (1989). *Ap. J.*, **336**, L47.

Piccinotti, G., *et al.* (1982). *Ap. J.*, **253**, 485.

Shanks, T., *et al.* (1991). *Nature*, **353**, 315.

Turner, E.L. and Geller, M.J. (1980). *Ap. J.*, **236**, 1.

New Constraints on Clustering of X-ray Sources from Confusion Noise

LUIGI TOFFOLATTI[1,2] & XAVIER BARCONS[1]

[1]Dept. de Física Moderna, Universidad de Cantabria, Santander, Spain
[2]Osservatorio Astronomico, Padova, Italy

1 INTRODUCTION

The analysis of the probability distribution of fluctuations in integrated backgrounds (known as the $P(D)$ analyis; Scheuer, 1957; 1974) is a powerful way to constrain downwards in flux the source counts, i.e. the $N(> S)$ curve, thus providing clues to the number of the sources making up (or contributing to) the corresponding background. The basic reason is that the major contribution to the shape of the $P(D)$ distribution comes from flux levels $S \sim 1\sigma$ (at which there is ~ 1 source per beam) much fainter than those achieved by direct detection methods. The use of the fluctuations analysis is particularly important in the case of the extragalactic X–ray background (XRB) because the origin of a large fraction of the XRB is not yet known. *COBE* measurements of the cosmic microwave background (CMB) spectrum (Mather *et al.*, 1990) have ruled out an interpretation of the XRB due to thermal bremsstrahlung from a homogeneus, hot intergalactic medium, while known populations of discrete sources contribute less than 50% of the total intensity (see De Zotti *et al.*, 1991; Shanks *et al.*, 1991).

Given that all the sources contributing to the XRB are expected to cluster (mostly Seyfert galaxies in the 2-10 keV band, and QSOs in the 1-3 keV band), following Barcons (1991), we have included a simple clustering model in the $P(D)$ analysis. This improvement in analising the fluctuations data is very helpful for two main reasons: a) it can help to explain the inconsistency between the 2-10 keV source counts, as derived from fluctuations analysis, and those observed from the *Einstein* Medium Sensitivity Survey (MSS) (arising if a spectral conversion is assumed which ignores absorption effects); b) allows to put new constraints on the correlation length of sources contributing to the XRB (which is very important since the XRB probes the Universe at redshift $z \lesssim 10$ and its anisotropies reflect the spatial distribution of the underlying source populations when the Universe has already undergone nonlinear evolution). Here, we give first a brief account of the current status of the extragalactic 2-10 keV source counts (Section 2). Then we present the basic formalism on how to relate the confusion noise (i.e., the $P(D)$ curve) to source counts and clustering

(Section 3) and eventually, we present the results obtained comparing the $P(D)$ from *GINGA LAC* data with our current theoretical estimations (Section 4).

2 THE EXTRAGALACTIC 2-10 KEV SOURCE COUNTS

The bright end of the 2-10 keV source counts ($S > 3 \times 10^{-11}$ erg cm^{-2} sec^{-1}) has been determined from the *Ariel 5*, *Uhuru* and *HEAO-1 A2* all sky surveys. To extend downwards the source counts we have to rely on spatial fluctuations studies. The "standard" $P(D)$ analysis (i.e. without taking into account the clustering of X–ray sources) of the fluctuations data from the *HEAO-1 A2* (Shafer, 1983) and *GINGA LAC* (Warwick & Stewart, 1989; Warwick & Butcher, 1991) experiments constrain the $N(> S)$ relation to a form close to the Euclidean prediction, $N(> S) = KS^{-1.5}$, down to flux levels a factor of ~ 30 below the Piccinotti *et al.* (1982) limit. Specifically, the 2-10 keV source counts are consistent with (see Barcons & Fabian, 1991)

$$N(> S) = 3.4 \times 10^{-19} S(2 - 10)^{-1.5} deg^{-2} \qquad (1)$$

for fluxes $S \gtrsim 5 \times 10^{-13}$ erg cm^{-2}sec^{-1}. At smaller energies (0.3-3 keV), the *Einstein Observatory* provided several surveys at different flux limits: the MSS and its extension (Gioia *et al.*, 1990) and the Deep Survey (DS; Giacconi *et al.*, 1979). All of the *Einstein Observatory* surveys and fluctuations analyses (Hamilton & Helfand, 1987; Barcons & Fabian, 1990) are consistent with

$$N(> S) = 1.4 \times 10^{-19} S(0.3 - 3)^{-1.5} deg^{-2} \qquad (2)$$

down to fluxes $S \gtrsim 2 \times 10^{-14}$ erg cm^{-2}sec^{-1}. The comparison between the two determinations, taking into account that $S(2 - 10) \approx S(0.3 - 3)$, if we adopt a canonical spectral index $\alpha \sim 0.7$ (which ignores absorption effects), soon shows that there is a discrepancy of a factor $\simeq 2.5$. Several facts may contribute to it: AGN could be absorbed at low energies and high-luminosity low surface brightness clusters could be missed in the *Einstein* surveys or, conversely, the "standard" fluctuations analysis can overestimate the normalization of the $N(> S)$.

3 THE P(D) CURVE WITH CLUSTERING

The $P(D)$ distribution (where D stands for the deflection or deviation from the mean intensity) contains all the one–point information of the background as viewed with a given beam. As $P(D)$ curves are usually non–gaussian and show very extended tails towards positive values of D, their second moment is dominated by the brightest sources below the detection threshold and so it can be misleading in evaluating the background fluctuations. The use of the whole fluctuation distribution fitting it to the histogram of the observed deflections (via minimization of χ^2) is then much more appropriate. It is important to note that while in the "standard" (i.e. without clustering) treatment (see Shafer, 1983) the sources for which $N(> S) \sim 1$ per beam will dominate (the most abundant faintest ones produce only increasingly small gaussian noise), if we allow *all sources* to cluster we obtain that *all fluxes* will contribute to the broadening of the $P(D)$ curve which will be dominated by *the weakest and more numerous sources* (see Barcons, 1991). The contribution to the second moment of the distribution (σ^2) due to clustering is then $\sigma^2 = \langle I \rangle^2 \Xi$ (Barcons & Fabian, 1988) where Ξ represents the double integral of the correlation function over the beam.

As a first approach to simplify the problem we have assumed a top–hat response function, which does not alter so much the final result as the width of the $P(D)$ distribution scales as $(K_d \Omega_e)^{1/\gamma-1}$ (Condon, 1974) for simple power law forms of $N(> S)$ (Ω_e being the effective solid angle of the beam, K_d the normalization and γ the power law slope of the differential source counts). Anyway, if the beam function is not a top–hat, then off–axis bright sources can also influence the central shape of the $P(D)$ curve. To describe the spatial distribution of X-ray sources we have assumed a uniform distribution of clusters of sources with mean surface density λ clusters per unit solid angle and profile $\eta(\vec{x}) = (\eta_0/2\pi l^2)\exp(-\vec{x}^2/2l^2)$ around the cluster center. This leads to a shot noise for the 2D density field and to a gaussian two–point correlation function with correlation length $\theta = \sqrt{2}l$ and normalization $\xi(0) = (4\pi\lambda l^2)^{-1}$ (see Barcons, 1991 for the relevant formulae). It is actually expected that when the number of sources per beam is $\lesssim 2$, shot noise and gaussian clustering models must converge, since cumulants of order $\gtrsim 2$ will then be irrelevant. While the observed spatial distribution of optically selected galaxies and clusters of galaxies is known to deviate from this model (being best fitted by a single power law), on the other hand, the lack of information on the correlation function of X–ray selected sources does not allow to well constrain it. Moreover, for any adopted spatial distribution of sources we have to consider all cumulants which, excluding the shot noise model, are completely unknown for orders n > 2 (depending on the n–point correlation functions).

4 THE GINGA DATA: NEW CONSTRAINTS TO THE CLUSTERING CORRELATION LENGTH

The data used in the present $P(D)$ analysis consist of a total of 132 independent samples measured at galactic latitudes ($|b''| > 25$ deg), in regions of the sky devoid of sources. (Butcher *et al.*, 1991; Carrera *et al.*, 1991). To fit the observational $P(D)$ data we have chosen a Euclidean slope for the $N(> S)$ relation (according to all the previous determination of the X–ray source counts, at least down to fluxes of interest here) and we have allowed only sources at fluxes $S > S_{DS}$ (where $S_{DS} = 2.6 \times 10^{-14}$ erg cm^{-2}sec^{-1} at 1-3 keV) to cluster, setting $\xi(0) = 1$ and leaving θ as a free parameter. Moreover, the corresponding $P(D)$ distribution has been convolved with a gaussian (with $\sigma \lesssim 0.1$ cts sec^{-1}) representing the Poisson counting noise (whose r.m.s count rate represents only the 1-1.5% of the XRB intensity as measured by the *LAC*) and other instrumental noises.

Fig.1 summarizes our main result obtained by the minimization of χ^2 in the parameter space (K, θ), where K is the normalization of the $N(> S)$ curve and θ is the correlation length for the clustered sources. As a general result, it is clear that *including clustering in the $P(D)$ analysis gives a better fit to the data than using a uniform distribution of sources*, due to the fact that the commonly used procedure of convolving the $P(D)$ curve with a gaussian (representing the "excess" noise) reduces the "skewness" of the curve. In the case of the *GINGA LAC* data, we see that if a uniform distribution of sources is still compatible (at $< 2\sigma$ level) with the measured distribution of fluctuations, the best fit is found for a correlation length of $\theta_{min} = 0.13$ deg ($L \simeq 1.8h^{-1}$ Mpc at redshift $z = 1$) and a normalization $K_{min} \simeq 80$, in units of $(LAC$ cts sec$^{-1})^{1.5}$sr^{-1}, a factor of ~ 0.5 of the *Piccinotti* normalization. Moreover, we find the following limits (at the 95% confidence level)

$$\theta \lesssim 0.36 \; deg \; (L \lesssim 5.5h^{-1}Mpc \; at \; z = 1), \quad K \gtrsim 50 \; (LAC \; cts \; sec^{-1})^{1.5} sr^{-1} \qquad (3)$$

which shows that: a) allowing for a non uniform distribution of the bright 2-10 keV sources, the *MSS* normalization of the $N(> S)$ relation is still compatible with the value found at higher energies, but only at $\sim 2\sigma$ level; b) the sources mainly

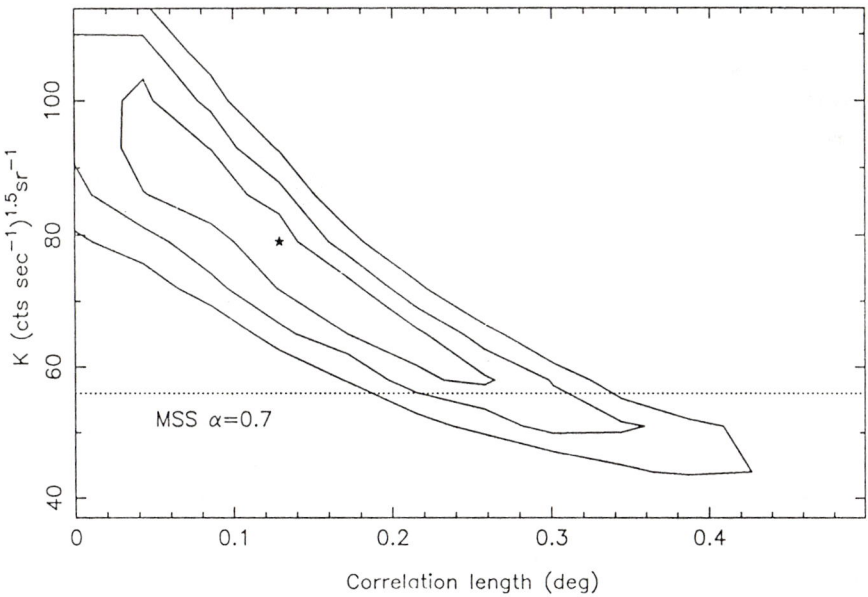

Figure 1. Constant χ^2 contours in the parameter space (K,θ) adopting a Euclidean slope for the $N(> S)$ relation. The continuos lines indicate the 1,2 and 3σ probability levels and the star the minimun of the fit. The dotted line correspond to the *MSS* normalization transformed to the *GINGA LAC* 4-12 keV band. The conversion factor here used is 1 LAC count sec^{-1} (4-12 keV) = 2.3×10^{-12} erg cm^{-2}sec^{-1} at 2-10 keV (see Hayashida *et al.* 1989) with a beam effective amplitude $\Omega_e = 5.83 \times 10^{-4}$ sr.

contributing to the measured fluctuations are mostly galaxies, or sources which shows the same clustering properties of galaxies (faint AGN?). It is interesting to note that the present limit is very close to the value recently found by Boyle (1991) (a constant comoving correlation scale of $\sim 6h^{-1}$ Mpc for 400 optically selected QSOs out to $z = 2.5$) and to the clustering scale of optically selected galaxies. Anyway, a further analysis of other data taken with a different beam amplitude (i.e. *HEAO-1 A2* all sky survey) seems necessary to confirm our current result.

Acknowledgements We thank F.J. Carrera for scientific help and all the members of the *GINGA* collaboration for making available the data. L.T. acknowledges a MEC grant from the Spanish Ministry of Education and Science. Partial financial support for this work was provided by the CICYT.

References

Barcons, X., 1991. *"The Infrared and Sub–mm Sky after COBE"*, Les Houches (France), March 1991. Eds. Signore, M. & Dupraz, C., in the press.

Barcons, X. & Fabian, A.C., 1988. *Mon. Not. R. astr. Soc.*, **230**, 189.

Barcons, X. & Fabian, A.C., 1990. *Mon. Not. R. astr. Soc.*, **243**, 366.

Barcons, X. & Fabian, A.C., 1991. *Adv. Space Res.*, **11-No.8**, (8)55.

Boyle, B.J., 1991. Proc. Texas–CERN Symposium, in press.

Butcher *et al.*, 1991. In preparation.

Carrera, F.J., Barcons, X., Butcher, J.A., Fabian, A.C., Stewart, G.C., Warwick, R.S., Hayashida, K. & Kii, T., 1991. *Mon. Not. R. astr. Soc.*, **249**, 698.

Condon, J.J., 1974. *Astrophys. J.*, **188**, 279.

De Zotti, G., Martín-Mirones, J.M., Franceschini, A. & Danese, L., 1991. Preprint.

Giacconi, R., *et al.*, 1979. *Astrophys. J. Lett.*, **234**, L1.

Gioia, I., Maccacaro, T., Schild, R.E., Wolter, A., Stocke, J.T., Morris, S.L. & Henry, S.P., 1990. *Astrophys. J. Suppl.*, **72**, 567.

Hamilton, T.T. & Helfand, D.J., 1987. *Astrophys. J.*, **318**, 93.

Hayashida, K., *et al.*, 1989. *Publ. Astr. Soc. Japan*, **41**, 373.

Mather, J.C., *et al.*, 1990. *Astrophys. J.*, **354**, L37.

Piccinotti, G., Mushotzky, R.F., Boldt, E.A., Holt, S.S., Marshall, F.E., Serlemitsos, P.J. & Shafer, R.A., 1982. *Astrophys. J.*, **253**, 485.

Scheuer, P.A.G., 1957. *Proc. Camb. Phil. Soc.*, **53**, 764

Scheuer, P.A.G., 1974. *Mon. Not. R. astr. Soc.*, **166**, 329

Shafer, R.A., 1983. Ph.D. Thesis, University of Maryland, USA.

Shanks, T., Georgantopoulos, I., Stewart, G.C., Pounds, K.A., Boyle, B.J. & Griffiths, R.E., 1991. *Nature*, **353**, 315.

Warwick, R.S. & Stewart, G.C., 1989. *"Two Topics in X–ray Astronomy"*, Bologna (Italy), 13–20 September 1989. ESA SP-296, p.727 (November 1989).

Warwick, R.S. & Butcher, J.A., 1991. Preprint.

Chapter IV

The contribution of

known classes of sources

The Extragalactic Volume X-ray Emissivity

ELIHU BOLDT

Laboratory for High Energy Astrophysics, NASA / Goddard Space Flight Center, Greenbelt, Maryland 200771

ABSTRACT

The extragalactic X-ray volume emissivity characteristic of the present epoch is evaluated in several complementary ways and compared with the universal value representing that averaged over all epochs, as directly given by the observed CXB (Cosmic X-ray Background). The total energy flux of the CXB sets a well-defined *lower* limit to this average comoving value at $q_o \equiv (9 \rightarrow 11) \times 10^{39} h_{50}$ ergs s^{-1}Mpc^{-3} for $\Omega=0$ and $\Omega=1$ respectively. Using the distribution of gravitational mass as a valid tracer of X-radiation, it is found that relating the peculiar velocity of the local group of galaxies to the total dipole moment of the all-sky distribution of X-ray flux provides an absolute *upper* bound to the present-epoch volume emissivity in the 2-10keV band at $0.06q_o$. Even after making currently plausible corrections for the limited bandwidth involved here this upper limit still falls short of the universal value. The X-ray/optical relation for normal galaxies is used to determine that their local volume X-ray emissivity (2-10keV) is $\sim 0.003q_o$. Comprehensive X-ray luminosity functions recently obtained for present-epoch sources imply $0.021q_o$ for the local AGN volume emissivity (2-10keV) and $0.014q_o$ for rich clusters of galaxies. The newly established correlation between fluctuations in the surface brightness of the X-ray sky and foreground galaxy counts implies that all galaxy correlated X-ray emission, other than that from rich clusters, corresponds to a local volume emissivity (2-10keV) that amounts to $(0.028\pm0.010)q_o$. With currently acceptable luminosity evolution and bandwidth corrections, AGNs could well account, energetically, for at least half of the universal average volume X-ray emissivity. Recent studies suggest, however, that a possible substantial *residual* CXB arising at redshifts beyond known objects (i.e., z>5) might produce a significant effect on the IGM (intergalactic medium).

1. INTRODUCTION

The CXB (cosmic X-ray background) is particularly fascinating because it gives us a remarkably precise "total answer" about the extragalactic X-ray sky. Of foremost importance is that it has a definite characteristic energy (W_o=40keV) in the sense that the measured background radiation (Marshall et al. 1980; Rothschild et al. 1983; Gruber 1991)may be represented by a simple expression which, specified by W_o (Boldt 1987, 1989), accurately describes the surface brightness spectral density over the broad band $E = 3 \to 100$keV (see Fig. 1), viz:

$$dI/dE = 7.8\ E^{-0.29}\ \exp(-E/W_o)\ s^{-1}\ cm^{-2}\ sr^{-1}. \tag{1}$$

This equation (1) may be integrated in closed form over all photon energies to yield the following well defined omnidirectional total energy flux ($4\pi I$) for the extragalactic X-ray sky:

$$4\pi\ I = c\ t_o \int \omega\ (n\ L)\ d\tau\ =\ 2.8 \times 10^{-6}\ ergs\ s^{-1}\ cm^{-2} \tag{2}$$

where $\omega \equiv (1+z)^{-1}$, z is the redshift at a lookback time τ [in units of the age (t_o) of the universe] and n is the comoving number density of sources of mean X-ray luminosity (L) at z. Analogous to the 3^oK microwave background, the flux given by (2) represents a significant *distinct* component of the radiation field of the universe[1], albeit of different origin(s). The well-defined integral property provides us with a valuable constraint that must be strictly satisfied by any proposed model for how this component arises.

Since the X-ray sky surface brightness at energies >100keV tends to somewhat exceed that given by the spectrum in (1) the actual omnidirectional flux observed over extremely large bandwidths is somewhat higher than given by (2); with a $2mc^2$ (1MeV) bandwidth the discrepancy is approximately 20%. For the standard 2-10keV limited spectral band much used for comparison with known discrete source populations the omnidirectional flux is 24% of the value given by (2). In this discussion, we deliberately choose to exploit the full

---

[1] Although the local energy density of this X-ray component is only $\sim 10^{-4}$ that of the microwave background it is about a hundred times more than the thermal energy density that would pertain to an IGM temperature $\approx 10^5 (\Omega_{IGM}\ /\ 0.01)^{-1}$ degrees, where Ω_{IGM} is the IGM density relative to the closure value.

broadband integral (2) in addition to the more traditional 2-10keV band since it presents us with a direct well-defined measure of the *complete* X-ray volume emissivity (luminosity density) to be associated with the CXB. In particular, we obtain the following general expression for the expectation value of X-ray volume emissivity $<q_x>$, as averaged over all lookback times:

$$<q_x> \equiv \int q_x \, d\tau = \int (n \, L) \, d\tau = [4\pi \, I / [(<\omega>)ct_0]] \qquad \text{for } \tau = 0 \rightarrow 1 \qquad (3a)$$

where $4\pi I$ is given by (2), $q_x = (n \, L)$,

$$(<\omega>) \equiv \int \omega \, (n \, L)d\tau \; / \; \int (n \, L)d\tau \qquad (3b)$$

and the scale length $ct_0 = 1.23 \times 10^{28} \, h^{-1}$cm (for $\Omega=1$) $\rightarrow 1.85 \times 10^{28} h^{-1}$cm (for $\Omega=0$), with $h \equiv H_0/(50 \text{km s}^{-1} \text{ Mpc}^{-1})$. We may express $(<\omega>)$ in terms of the mean travel time $<\tau>$ for extragalactic X-ray sky photons to reach us after their emission (i.e., mean lookback time); for the case[2] $\Omega=0$ (3b) becomes particularly simple and revealing in this regard, viz:

$$(<\omega>) \equiv (1- <\tau>) \qquad \text{for } \Omega=0 \qquad (4a)$$

where τ(for $\Omega=0$) $= z/(1+z)$ and $<\tau> \equiv [\int \tau \, (n \, L) \, d\tau] / [\int (n \, L) \, d\tau]$. Since discrete observable sources appear to be limited in redshift space to $z < 5$ (Schneider, Schmidt and Gunn 1989), corresponding to $\tau < 5/6$ (for $\Omega =0$), I prefer to think of the summation of X-radiation from this directly known region of the universe as the cosmic X-ray *foreground* and to consider the associated average volume emissivity as "local". It is this foreground that we are concerned with when we address the topic of luminosity functions for the various constituent source populations that we are familiar with and their possible evolution. For an invariant comoving emissivity (i.e., no evolution in nL) over $\tau=0 \rightarrow 5/6$, $<\tau> = 5/12$. From the aforementioned range permitted for $<\tau>$ in regard to a *foreground* we infer via (4a) the corresponding limits[3] on $(<\omega>)$, viz:

$$\underline{5/6} > (<\tau>) > 5/12 \quad \rightarrow \quad \underline{1/6} < (<\omega>) < 7/12(=0.58), \quad \text{for } \Omega=0 \qquad (4b)$$

---

[2] Ω is the ratio of *total* gravitational mass density to the closure value.

[3] For the case of $\Omega =1$, eq.(3b) yields the limits $1/6 < (<\omega>) < 0.64$, and the associated characteristic energies $[W_0/(<\omega>)] = 63$keV $\rightarrow 240$keV.

and their associated characteristic energies $[W_o/(<\omega>)] = 69\text{keV} \rightarrow 240\text{keV}$.

We can use the observed energy flux (2) associated with the extragalactic X-ray sky for a *direct* determination of the present-epoch volume emissivity (q_o) if we make the simple hypothesis that the *comoving* value is epoch invariant, as is the case for the density of conserved subrelativistic matter (e.g., see Boldt 1987). Assuming this epoch invariance to be universal (i.e., even beyond foreground sources), we obtain

$$(4\pi\ I)/ct_o = \int q_x\ \omega\ d\tau\ = q_o\ \int \omega\ d\tau \tag{5a}$$

$$q_o = (4\pi\ I)H_o\ /\ [c \int (1+z)^{-3}(1+\Omega z)^{-1/2}dz]\ =\ (9 \rightarrow 11)\text{x}10^{39}h\ \text{ergs s}^{-1}\text{Mpc}^{-3} \tag{5b}$$

where[4] the lower and upper limits correspond to $\Omega=0$ and $\Omega=1$ respectively.   Since q_o given by (5) happens also to be the ω-weighted average for q_x, it constitutes a *lower* limit to the time-averaged $<q_x>$ defined by (3a), provided that dq/dt ≥ 0. Furthermore, we note that this value of q_o (5b), covering all energy X-rays, is a clear order of magnitude *higher* than our recently determined upper limit to the actual local level for the 2-10keV limited band (see § 6).

2. BASIC GAUGES

What can we say about any "true" background, that coming from beyond the portion of redshift space populated by known sources? We note that the redshifts to be considered would be limited by the characteristic photon energy value to be associated with the dominant (as yet unspecified) sources of this background radiation. As emphasized by Fabian (1990), this characteristic limiting photon energy is likely to be related in some fundamental way to the rest mass of the electron. For highly compact sources, electron-positron pair effects limit the characteristic energy of the emerging radiation to $\leq mc^2$ (Leiter and Boldt 1982; Zdiarski 1988; Fabian, Done and Ghisellini 1988); if this energy is redshifted to $\leq 40\text{keV}$ in the present epoch these objects could be at redshifts well beyond known sources and thereby qualify as candidates for producing the background.   This is to be contrasted with AGNs exhibiting radiative transfer effects associated with reflection from cold matter, where the reflected spectrum has a break at about 120keV due to Compton electron recoil (Lightman and White 1988; Fabian et al. 1990; Rogers and Field 1991). For such objects a roll-off at 40keV in the

---

[4] We have here made use of the identity: $t_o\ \omega\ d\tau = (H_o)^{-1}(1+z)^{-3}(1+\Omega z)^{-1/2}dz$.

spectrum as observed at z=0 would imply source redshifts at z≈2 (Fabian et al. 1990); they would thereby fall into the classification that I have designated for foreground objects. Considering the composite X-ray sky (foreground plus background) we note that $W_0/\omega <$ mc^2 implies $\omega > 0.08$ (i.e., z < 12).

The numerical value for the overall mean emissivity corresponding to (3) may be expressed as the product of the cosmic time averaged comoving number density ($<n>$) and luminosity ($<L>$) of the sources involved, viz:

$$(< q_x >) = (<n>)\,(<L>) = \int n\,L\,d\tau = 4.5 \times 10^{39}\ \mu\,h\,/\,(<\omega>)\ \text{ergs s}^{-1}\ \text{Mpc}^{-3} \tag{6}$$

where $(<n>) = \int n\,d\tau$, $\mu = 1$(for $\Omega=0$), $\mu = 3/2$(for $\Omega=1$), and $(<\omega>)^{-1} \approx 2$ (for $<\tau> \approx 0.6$) $\rightarrow 6$ (for *all* emission at z =5). To establish a scale for this discussion of q_x we consider the bolometric luminosity density $(q_b)_0$ arising from present-epoch field galaxies over the full range of their luminosities. The best-fit Schechter luminosity function (Schechter 1976; Felten 1977; Efstathiou, Ellis, and Peterson 1988) for these galaxies yields:

$$(q_b)_0 = 9.7 \times 10^7 h L_0\ \text{Mpc}^{-3} = 3.7 \times 10^{41} h\ \text{ergs s}^{-1}\ \text{Mpc}^{-3}. \tag{7}$$

The characteristic luminosity for the Schechter function is $L^* = 4.7 \times 10^{10} h^{-2} L_0 = 1.8 \times 10^{44}$ h^{-2} ergs s^{-1} (Efstathiou, Ellis, and Peterson 1988), slightly larger than that for our galaxy. Taking L^* as our unit galaxy luminosity, the corresponding equivalent number density n^* of such standardized objects would be

$$n^* \equiv (q_b)_0\,/\,L^* = 2.1 \times 10^{-3}\ h^3\ \text{Mpc}^{-3}. \tag{8}$$

The average specific X-ray luminosity to be associated with galactic sources of the CXB is then obtained from (6) and (8), as follows:

$$<L> \equiv (< q_x >)\,/\,n^* = 2.1 \times 10^{42}\ (<\omega>)^{-1}\ \mu\ h^{-2}\ \text{ergs s}^{-1}. \tag{9}$$

Since $(<\omega>)^{-1}\,\mu > 2$, this overall average X-ray luminosity (9) is clearly greater than that for normal galaxies, the vast majority of which have $L < 10^{42} h^{-2}$ ergs/s (Fabbiano 1989; Fabbiano, Kim, and Trinchieri, 1991; Canizaras and Blizzard 1991).

In order to facilitate comparisons of several particularly interesting present-epoch X-ray source populations with the fiducial number density (8) and the average specific X-ray luminosity for CXB sources (9), the number density and mean luminosity for each class are exhibited together in Table 1. The luminosities refer to the 2-10keV band; observations made with the 0.2-3.5keV band of the HEAO-2 Einstein Observatory are converted to the higher one used here under the assumption that the energy spectral index is $\alpha \approx 0.7$. The underlying luminosity functions employed for normal E/SO galaxies and spirals are discussed in §5.

Table 1. Present-epoch X-ray Source Populations

Population	Mean Luminosity $\log (< \underline{L} >)$	Number Density $\log (< \underline{n} >)$
Normal Spiral Galaxies ($\underline{L} = 10^{38} - 10^{43}$) (Fabbiano, Kim & Trinchieri 1991)	39.3	-2.2
Normal E & SO Galaxies ($\underline{L} = 10^{38} - 10^{43}$) (Fabbiano, Kim & Trinchieri 1991)	40.1	-2.9
Compact Groups (Hickson et al. 1989)	<41.5	<-3.0
Starburst Galaxies (Rephaeli et al. 1991)	42.0	-4.6
Bright AGNs ($\underline{L} > 10^{42}$) (Della Ceca & Maccacaro 1991)	43.0	-4.7
Rich Clusters ($\underline{L} > 2 \times 10^{43}$) (Edge et al. 1990)	44.1	-6.0

$\underline{L} \equiv [L\,(2\text{-}10\text{keV})]\,/\,(h^{-2}\text{ergs s}^{-1})$ $\underline{n} \equiv n\,/\,(h^3 \text{Mpc}^{-3})$

3. CLUSTERS OF GALAXIES

The most X-ray luminous extragalactic present-epoch sources are rich clusters of galaxies. Using a flux limited sample (excluding Virgo) of 45 such objects (where the redshifts involved are at z<0.15) Edge et al. (1990) have determined a best-fit Schechter type luminosity function for which the characteristic luminosity is $L^*(cluster) = 8.4 \times 10^{44} h^{-2} ergs$ s^{-1}(2-10keV). The volume emissivity corresponding to this luminosity function is q(clusters) $= 1.4 \times 10^{38} h$ ergs s^{-1} Mpc^{-3}, with an estimated statistical precision (σ) of about 50%. Considering those 10 clusters listed by Edge et al. (1990) which constitute a sample (excluding Virgo) that is volume limited to z < 0.03 for $L > 6.6 \times 10^{43} h^{-2} ergs s^{-1} h^{-2} ergs s^{-1}$ we establish a clear-cut lower limit on the volume emissivity (i.e., uncorrected for galactic obscuration or the reduced volume accessible for detecting lower luminosities) for a more local region. This straight-forward addition of luminosities for z < 0.03 yields a lower limit $q = 2.4 \times 10^{38}$ h ergs s^{-1} Mpc^{-3}, about double that which would be expected from the luminosity function characterizing the larger sample associated with the more extended region of space.

For cluster X-ray luminosities greater than $2 \times 10^{43} h^{-2} ergs s^{-1}$ the expected mean luminosity is $1.3 \times 10^{44} h^{-2} ergs s^{-1}$ whereas the actual average of the 10 clusters in this category at z<0.03 is $3.8 \times 10^{44} h^{-2} ergs s^{-1}$, three times larger than anticipated. Although limited by the statistics of small numbers, such an effect could be indicative of the sort of evolution for clusters within relatively recent epochs noted by Edge et al. (1990) and Gioia et al. (1990). In the context of the present discussion this tells us that we can not in fact establish a truly well-defined *local* volume emissivity for clusters of galaxies; the number of objects over a suitably local region is too few.

4. MASS DENSITY AS A TRACER OF X-RAY EMISSIVITY

What is the most local X-ray volume emissivity of astrophysical interest? Perhaps it should pertain to the smallest region for which a physically meaningful global measure is available. I here suggest that it be operationally defined for the smallest sphere which contains essentially all the anisotropically distributed mass giving rise to the gravitational acceleration responsible for the observed peculiar motion of the Local Group (LG) of galaxies. Considering a sample of galaxies with recession velocities complete up to about 7000km/s (i.e., $\leq$140 h^{-1} Mpc) it is found that most of the resulting optical dipole moment used to represent the mass dipole moment of this underlying matter is produced within a radius corresponding to 4000 km/s

(Lahav, Rowan-Robinson and Lynden-Bell 1988; Lynden-Bell, Lahav and Burstein 1989). Known large structural configurations of galaxies beyond this sample are probably too far away to make much of a perturbation to the dipole moment (Lahav 1991a). Those galaxies (mostly spiral) observed with IRAS (Strauss and Davis 1988) and X-ray bright AGNs detected with HEAO-1 (Miyaji and Boldt 1990; Miyaji, Jahoda and Boldt 1991) also yield dipole moments that apparently saturate at redshifts comparable to that for the optical sample.

We note that the sample of rich Abell clusters (R≥1) is essentially devoid of objects with recession velocities less than about 4000 km/s and that for an average number density of ≈ $10^{-6} h^3 Mpc^{-3}$ (Bahcall 1988) we would only expect a couple; in fact three X-ray clusters are known in this volume, viz: Virgo, Centaurus and A1060, with a composite 2-10keV luminosity of $1.0 \times 10^{44} h^{-2} erg/s$ (Lahav, Edge, Fabian, and Putney 1989; Edge et al. 1990). Hence, our proposed "local" volume is not appropriate for providing a reliable estimate of that portion of the present-epoch X-ray volume emissivity arising from bound systems as large and sparse as rich clusters. The number of X-ray bright AGNs with $L_x > 10^{43}$ $h^{-2} ergs/s$ expected in this volume is an order of magnitude larger than clusters (Piccinotti et al. 1982) and would already yield an X-ray output of about $4 \times 10^{44} h^{-2} ergs/s$; AGNs of lower luminosity are expected to increase their number by another order of magnitude (Padovani, Burg and Edelson 1990; Persic et al 1989). In essence then we are dealing with a region of space in which the effects of a statistically poor sample of X-ray clusters may be readily excluded and where the remaining output in X-rays involves the superposed radiation from all the individual galaxies present.

Apart from the three nearest clusters in the sample (see preceding), the 53 X-ray brightest clusters ((Lahav, Edge, Fabian, and Putney 1989) make a negligible contribution to the all-sky X-ray dipole moment. Excluding these three, the entire X-ray sky yields a dipole moment whose apex is close to that of the LG peculiar velocity (Boldt 1989; Boldt 1990) and whose magnitude (D_x) is

$$D_x = (2.0 \pm 0.4) \times 10^{-9} \text{ ergs cm}^{-2} \text{ s}^{-1} \text{ (2-10keV)}. \tag{10}$$

X-radiation as a tracer of gravitational attractors (and repellers) has been described in detail by Goicoechea and Martin-Mirones (1990). However, under the assumption that there is relatively small motion if any between the proper frames of the X-ray and microwave backgrounds, about half of the all-sky X-ray dipole moment (10) is likely to be associated with the Compton-Getting effect (Shafer 1983; Shafer and Fabian 1983). The Compton-Getting dipole arises from the solar velocity relative to the proper frame of the background, a direction which is also fairly close to that of the LG peculiar velocity. Hence, the overall X-ray sky dipole represents an upper limit to the magnitude of the dipole to be associated with all the underlying gravitational matter. Assuming that variations in X-ray emissivity (q) and mass density (ρ) track each other such that, over an appropriate region, $D_q/D_\rho = b\,(q/\rho)$, we obtain (Boldt 1990):

$$q = (H_0/v)\,\Omega^{0.6}\,(D_q/b) \tag{11}$$

where D_q is the X-ray dipole moment caused by the spatial variation of emissivity relative to the mean value (q), D_ρ is the mass dipole moment due to the corresponding spatial deviations of the density from the mean value (ρ) within the same radius horizon, $\Omega \equiv [\rho/\rho(\text{critical})]$ is the mass density parameter, v is the LG velocity (relative to the proper frame of the microwave background), b is the bias parameter (Lahav et al. 1989) relating the contrast of X-ray emissivity with that of the underlying mass density and H_0 is the Hubble constant. Noting that $D_q < D_x$ and v = 600km/s (Lahav, Rowan-Robinson and Lynden-Bell 1988) in evaluating (11) yields the constraint

$$q < 5.9 \times 10^{38}(2.7\Omega^{0.6}/b)\,h\,\text{ergs s}^{-1}\,\text{Mpc}^{-3}. \tag{12}$$

The appropriate value of b to be used for (12) has been obtained from an X-ray flux-limited sample of 30 AGNs detected with the HEAO 1 A-2 experiment (Piccinotti, et al. 1982). Relating the X-ray flux dipole moment for this all-sky sample to the mass dipole moment responsible for the peculiar motion of the LG (Miyaji and Boldt 1990) yields a value consistent with $b=2.7\Omega^{0.6}$ obtained by Lynden-Bell, Lahav and Burstein (1989) for their optically selected sample of galaxies. Using a lower flux threshold Miyaji, Jahoda and Boldt (1991) have increased the HEAO 1 A-2 number of X-ray selected AGNs to 56 and used this augmented sample to check the robustness of the X-ray result; in this instance the analysis yields $b = (2.6\pm0.5)\Omega^{0.6}$, indicating that the initial conclusion is indeed a stable result. Assuming that this local emissivity arises entirely from galaxies, the upper limit on the average specific X-ray luminosity corresponding to (12) is

$$<L> \equiv q/n^* \; < \; 3.5 \times 10^{41}(2.7\Omega^{0.6}/b)\,h^{-2}\,\text{ergs s}^{-1} \quad (2\text{-}10\text{keV}). \tag{13}$$

This upper limit (13) is about two orders of magnitude larger than the X-ray luminosity of our galaxy (Boldt 1974; Fabbiano 1989).

5. FOREGROUND GALAXIES

Information provided by the newly released comprehensive X-ray catalog of galaxies observed with the HEAO-2 Einstein Observatory (Fabbiano, Kim, and Trinchieri 1991) can be used to make a straight-forward calculation of the local X-ray volume emissivity arising from normal galaxies. In particular, the best-fit relation between X-ray luminosity L_X ergs s^{-1}(0.2-3.5keV) and optical luminosity L_B (in L_O units) is:

$$\text{Log } (L_X) = 1.0 \text{ Log } (L_B) + 29.3 \qquad \text{for spiral galaxies,} \qquad (14)$$

$$\text{Log } (L_X) = 1.8 \text{ Log } (L_B) + 21.5 \qquad \text{for elliptical (E) and SO galaxies.} \qquad (15)$$

where $\text{Log } L_B \equiv 0.4(5.48-M_B) \equiv \text{Log } L - \text{Log } L_O$. For our present purpose it is convenient to recast expressions (14) and (15) as

$$L_X / L = \varepsilon \qquad \text{for spiral galaxies,} \qquad (16)$$

$$L_X / L = \varepsilon \, (\eta \, L/L^*)^{0.8} \qquad \text{for elliptical (E) and SO galaxies.} \qquad (17)$$

where $L^* = 4.7 \times 10^{10} h^{-2} L_O$ is the characteristic luminosity of the Schechter luminosity function for galaxies (Efstathiou, Ellis, and Peterson 1988), $\varepsilon = 5.1 \times 10^{-5}$ and $\eta = 8.3$. If we assume an effective energy spectral index $\alpha \approx 0.7$ for the 0.2-10keV spectrum of galaxies, then the L_X(0.2-3.5keV) that appears in (16) and (17) is numerically equal to L_X over the 2-10keV band to be addressed here. Considering the luminosity function $(\varphi^*/L^*) \, (L/L^*)^{-\beta} \exp(-L/L^*)$ for all field galaxies and noting that, at $L \approx L^*$, 67% are spirals and 19% E/SO galaxies (Franceschini et al. 1988), we may use (16) and (17) to evaluate the total volume emissivity arising from such "normal" galaxies as

$$q(\text{galaxies}) = \varepsilon \, \varphi^* \, L^* \, [\, 0.67 \, \Gamma(2-\beta) + 0.19\eta^{0.8} \, \Gamma(2.8-\beta) \,] \qquad (18)$$

where $\Gamma(x)$ is the gamma function, $\varphi^* = 2.0 \times 10^{-3} \, h^3 \text{Mpc}^{-3}$ and $\beta = 1.07$ (Efstathiou, Ellis, and Peterson 1988). Evaluating (18) with the appropriate values for the various parameters presented here yields q(galaxies) = 8x10^{-5} φ* L* = 3x10^{37}h ergs s^{-1} Mpc^{-3} (2-10keV).   (19)

In order to obtain an upper limit on the local emissivity arising from sources other than normal galaxies we need to subtract the value given by (19) from that given by (12). Taking b=2.6Ω$^{0.6}$ for evaluating (12), this subtraction yields

q < 5.8x10^{38}h ergs s^{-1} Mpc^{-3} (2-10keV).   (20)

The limit provided by (20) is to be compared with the volume emissivity corresponding to the power-law luminosity function for L$_{43}$ ≥1 directly obtained from the HEAO 1 A-2 AGN sample (Piccinotti et al. 1982), viz:

q = (nL)$_o$ =2.0x10^{38} (L$_{43}$)$^{-.75}$ h ergs s^{-1} Mpc^{-3} (21)

where L$_{43}$ ≡ [L(2-10keV)]/[10^{43}h^{-2}ergs/s] . By extending this power law (21) to lower luminosities and equating it to the upper limit (20) we learn that such an extrapolation yields an overestimation when considering threshold luminosities L$_{43}$ ≤ 0.24, and that the luminosity function itself must therefore flatten in this regime. The power-law X-ray luminosity function directly observed for AGNs with L$_{43}$ >1 corresponds to <L$_{43}$> = 2.3. This average X-ray luminosity is two orders of magnitude greater than the upper limit on the mean specific X-ray luminosity (13) of all present-epoch field galaxies. The associated volume emissivity (21) of this AGN subsample defined by L$_{43}$ > 1 already *exceeds* a third of the total [as represented by the upper limit given by (20)]. And the AGN number density involved here is only n ≈ 9x10^{-6} h^{3}Mpc^{-3}, an order of magnitude less than that for all Seyfert galaxies (Persic et al. 1989; Padovani, Burg and Edelson 1990).

6. X-RAY SURFACE BRIGHTNESS CORRELATION WITH GALAXY COUNTS

A relatively direct approach for obtaining the present-epoch emissivity and a more restrictive upper limit on the mean X-ray luminosity of local extragalactic sources involves measuring the correlation between CXB surface brightness fluctuations and galaxy counts for present-epoch objects of number density (n) comparable to n*. Under the assumption that the number density of these cataloged galaxies constitutes a linear tracer of the X-ray volume emissivity (i.e., dq/q = dn/n) the measurement of such a correlation would provide <q> directly. Using this procedure with Uhuru CXB data and the Zwicky catalog, Turner and

Geller (1980) could only put an upper bound on CXB/galaxy correlations and concluded that less than half of the CXB originates from sources substantially represented among bright galaxies. With the assumption that n < n_0(all sources), a more sensitive correlation analysis could also be used to establish a stringent upper limit on the mean luminosity of all contributing sources, viz: ($\langle L \rangle$) $\equiv$ q/n_0 < q/n.

Jahoda et al. (1991) have recently succeeded in detecting a signal in the cross-correlation of the CXB (2-10keV) surface brightness as measured with the HEAO-1 A2 observatory and the galaxy surface density derived from diameter limited samples (corresponding to objects at z<0.03) from the UGC and ESO catalogs (Lahav 1991b). Pixels excluded from this analysis involve all those contaminated by any A2 cataloged discrete source(s) (Piccinotti et al. 1982) other than low luminosity AGNs (L <10^{44} h^{-2} ergs/s). In effect, this excludes all A2 cataloged sources within our galaxy, all A2 cataloged AGNs that are too distant (z>.03) to be correlated with the galaxy sample and all nearby rich clusters (Edge et al. 1990) which are likely to distort the linear correlation being sought. Under these conditions, the resulting correlation implies the following local volume emissivity (2-10keV):

$$q = (2.5\pm1.0)\times10^{38} \ h \ ergs \ s^{-1} \ Mpc^{-3}. \tag{22}$$

This value (22) is to be understood as that portion of the total local emissivity which 1) is linearly related to the number density of galaxies, and 2) excludes the contribution from high X-ray luminosity AGNs (L >10^{44} h^{-2} ergs/s). Using the AGN luminosity function of Piccinotti et al (1982) for L = (1$\rightarrow$10)x 10^{44} h^{-2} ergs/s, we infer the increase of 0.3 x10^{38} h ergs s^{-1} Mpc^{-3} needed to obtain the corrected total emissivity from (22), as follows:

$$q(corrected) = (2.8\pm1.0)\times10^{38} \ h \ ergs \ s^{-1} \ Mpc^{-3}. \tag{23}$$

7. WHAT DOES THE LOCAL VOLUME EMISSIVITY TELL US?

Assuming that the sources associated with the volume emissivity given by (23) have an energy spectral index $\alpha \approx 0.7$, that their comoving volume emissivity is invariant with redshift and that the metric is flat (Ω=1), we obtain that the corresponding flux is (12±4)% of the 2-10keV CXB [$4\pi I_{CXB}$=6.6x10^{-7}cm^{-2}s^{-1}]. By construction, the emissivity given by (23) does not include the contribution of rich clusters. In general the hot intracluster gas for poor clusters (<50 galaxies/cluster) corresponds to kT<3keV (Kriss, Cioffi, and Canizares, 1983; Price, Burns, Duric and Newberry 1991); hence, the thermal emission from such clusters is not expected to make a significant contribution to the hard X-ray emission considered here. Assuming that (23) arises from galaxies alone we can use our previous calculation of the contribution of normal galaxies (19) to obtain the net volume emissivity arising from special galaxies (e.g., AGNs, starburst), viz:

$$q(special) = (2.5\pm1.0)\times10^{38} \text{ h ergs s}^{-1} \text{ Mpc}^{-3}. \tag{24}$$

The starburst galaxy 2-10keV luminosity density corresponding to HEAO-1 observations (Rephaeli et al. 1991) amounts to only about a tenth of this estimate for q(special). In fact, comparing this volume emissivity (24) arising from extraordinary galaxies with that given by (21) at $L_{43} > 1$ for X-ray bright AGNs, we find that their apparent equality (to within statistics) implies that the total emissivity arising from *all* special galaxies of lower X-ray luminosity (Elvis, Soltan and Keel 1984; Griffiths and Padovani 1990) must be less than 3.5 $\times10^{38}$ h ergs s^{-1} Mpc^{-3} (at ~3σ confidence). In the absence of evolution, the corresponding upper limit on their composite contribution to the observed CXB would be 15%.

Since the effective number density of galaxies in the samples used by Jahoda et al. (1991) is n $\approx 1.2\times10^{-3}h^3$Mpc^{-3}, less than that for all galaxies, we can use (23) to obtain an upper limit on the average X-ray luminosity of all present-epoch galaxies as

$$L(2\text{-}10keV) < q(corrected) / n = 2.3\times10^{41} \text{ h}^{-2} \text{ ergs/s.} \tag{25}$$

This limit (25) for the average X-ray luminosity is a hundred times less than the mean luminosity $<L_{43}> = 2.3$ characteristic of the bright AGNs at $L_{43} > 1$ to be associated with the emissivity given by (21), for which the number density of sources is only $\approx10^{-5}h^3$Mpc^{-3}.

What about the possible enhanced contributions to the local volume emissivity arising from compact groups of galaxies? This is of interest because of apparent IR enhancements observed with IRAS (Hickson et al.1989) which suggest that the interactions between galaxies in such groups may trigger star formation and/or promote nuclear activity. However, Hickson et al. (1989) have set an upper limit to the mean X-ray luminosity (2-10keV) associated with compact groups of galaxies at $3\times10^{41}h^{-2}$ergs s^{-1} and noted that their number density is only about 1% that of all bright IRAS galaxies. This implies that their contribution to the local emissivity must be small compared with that given by (23).

It is particularly revealing to compare the estimates of local volume emissivity discussed in §6 with the lower-limit value (5) for the universal time-averaged X-ray volume emissivity directly implied by the integral CXB energy flux (2). To do this we must first correct the local volume emissivity given by (23) for the 2-10keV limited bandwidth involved. For $\alpha \approx$ 0.7, the value corresponding to a bandwidth of mc^2 is

$$q_0(\Delta E = mc^2) = (19 \pm 7) \times 10^{38} h \text{ ergs s}^{-1} \text{Mpc}^{-3}. \tag{26}$$

The contribution from rich clusters (see § 3) to be added to this value (26) is negligible. That this local level (26) is only about 20% of the level given by (5b) implies that, in order to account for the observed CXB energy flux, there must be substantial evolution in the full-band X-ray volume emissivity .

8. EVOLUTION OF FOREGROUND SOURCES

Based on a sample of over 420 X-ray selected AGNs at z<2 extracted from the Einstein Observatory (HEAO-2) Extended Medium Sensitivity Survey (EMSS), Maccacaro et al.(1991) have determined a best-fit to the evolution in luminosity required and the corresponding baseline local luminosity function over the range $10^{42} \rightarrow 10^{46}$ergs/s, albeit for a lower energy band (0.3-3.5keV) than the standard 2-10keV band we have addressed here. For unabsorbed AGN spectra characterized by an energy spectral index $\alpha \approx 0.7$, however, we have the fortunate circumstance that L(2-10keV) $\approx$ L(0.3-3.5keV). The de-evolved "local" luminosity function for AGNs derived from the EMSS, based on the best-fit luminosity evolution $L_x(0) = e^{-Ct} L_x(z)$ with C=4.18, shows a substantial flattening at $L_{43}<0.5$. A good fit to this luminosity function over $L_{43} = 10^{-1} \rightarrow 10^3$ with a broken power-law (matched at L_{43}=2.5) yields the following local volume emissivity corresponding to all luminosities:

$$q(0.3\text{-}3.5\text{keV}) = (1.5 \pm 0.1) \times 10^{38} \text{ h ergs s}^{-1} \text{ Mpc}^{-3} \tag{27a}$$

where the stated precision corresponds to a 1σ statistical uncertainty alone. However, as pointed out by Della Ceca and Maccacaro (1991), a closer look at the EMSS data indicates that the exponential form for luminosity evolution requires that this evolution be luminosity dependent, contrary to the assumption underlying (27a). Considering an alternate functional form for AGN evolution in which $L_x(z) = \omega^{-C} L_x(0)$ Della Ceca and Maccacaro (1991) find that the assumption of pure luminosity evolution becomes acceptable; fitting this to the EMSS data for AGNs up to z$\approx$2 (i.e., ω< 1/3) they obtain C= 2.56$\pm$0.17. The de-evolved "local" luminosity function for AGNs based on this form of luminosity evolution may be adequately fit with a broken power-law (matched at $L_{43}=3.6$) that yields the following local volume emissivity corresponding to all luminosities:

$$q(0.3\text{-}3.5\text{keV}) = 2.1\text{x}10^{38} \text{ h ergs s}^{-1} \text{ Mpc}^{-3}. \tag{27b}$$

However, as emphasized by Maccacaro et al. (1991), the apparent paucity of AGNs at $L_{43} < 1$ could in part be due to the intrinsic absorption preferentially present in low luminosity objects. In particular, we note that the volume emissivity given by (27b) is in good agreement with the HEAO-1 result (21) obtained at the luminosity threshold $L_{43} > 1$ for the 2-10keV band directly. If we do assume that (27b) is correct for the 2-10keV band, that the associated sources have a spectral index $\alpha \approx 0.7$, that the comoving volume emissivity is invariant with redshift and that the metric is flat ($\Omega=1$), then we obtain that the corresponding omnidirectional flux is only 9% of the 2-10keV CXB. Taking (27b) as the AGN contribution to (24) we obtain that the upper limit (at 3σ) to the composite contribution of all *other* special galaxies (e.g., starburst), without evolution, would be 15%, the same limiting level discussed in §7.

Although somewhat incomplete and not fully identified, the sample of X-ray sources detected with the HEAO-1 A1 experiment provides us with a potentially extensive sample of AGNs measured at an effective energy of about 5keV (Wood et al. 1984), significantly higher than that available with the HEAO-2 Einstein Observatory. Using a catalog of HEAO-1 A3 based identifications of A1 sources now available from R. Remillard at MIT, Marshall (1991) has performed a preliminary analysis of the sample which indicates that the evolution of the X-ray luminosity function of AGNs is stronger than inferred at lower energies from the HEAO-2 EMSS. A first estimate of the local luminosity function for the A1 AGNs in this preliminary study suggests that the volume emissivity (2-10keV) would be larger than implied by (27a), by up to a factor of about two. However, such a higher value would be closer to that given by (27b) and comparable to the estimates of the local volume emissivity (23, 24) obtained from the correlation of fluctuations in the surface brightness of the X-ray sky (2-10keV) with foreground galaxy counts.

9.- WHAT'S LEFT?

Assuming that the foreground sources responsible for q(0) as given by (24) represent a population that has evolved such that $q = B_X [q(0)] \omega^{-C}$, we can construct the ratio of the average emissivity expected from this evolved population to the actual average emissivity ($<q_X>$), given by (6), corresponding to the observed broadband CXB flux, as follows:

$$(\int q \, d\tau)/(<q_X>) = [\int B_X \, q(0) \, \omega^{-C} \, d\tau] /(<q_X>)$$

$$= (0.056 \pm 0.022) \, (B_X/\mu) \, [(<\omega_X>)/(<\omega_f>)] \, (\int \omega^{1-C} d\tau) \tag{28}$$

where q(0) is the 2-10keV volume emissivity at τ=0 (i.e., ω=1), B_X is a well-defined bandwidth correction factor[5] which takes into account that the emissivity given by q(0) pertains to a limited portion of the X-ray band, where $<\omega_X>$ obtains from the redshift distribution of *all* extragalactic sources of X-radiation and $<\omega_f>$ from "foreground" sources alone, viz:

$$(<\omega_f>) \equiv (\int \omega^{1-C} \, d\tau) / (\int \omega^{-C} d\tau). \tag{29}$$

Using the value C=2.56 obtained by Maccacaro et al. (1991) from the EMSS within the framework of an open geometry (i.e., Ω=0) the ratio expressed by (28) becomes (with μ =1)

$$[\int B_X \, q(0) \, \omega^{-2.56} \, d\tau] /(<q_X>) = (0.056 \pm 0.022) \, B_X \, Y_\tau \, [(<\omega_X>)/(<\omega_f>)] \tag{30}$$

where the evolutionary factor Y_τ ($\equiv \int \omega^{-1.56} d\tau$) and τ are exhibited (Table 2) as a function of the maximum redshift (z), with $\omega = (1-\tau)$ for $\Omega = 0$.

Table 2. Evolutionary Factor (for Ω=0): $Y_\tau = \int (1-\tau)^{-1.56} d\tau$

z	1	2	3	4	5
Y_τ	0.85	1.52	2.10	2.61	3.09
τ	0.50	0.67	0.75	0.80	0.83

[5] With $\alpha \approx 0.7$: B_X=5.2 for ΔE=0.5mc^2, B_X=6.8 for ΔE=mc^2, B_X= 8.8 for ΔE=2mc^2. For sources having the spectral shape given by (1): B_X= 4.2.

Using Y_τ obtained from Table 2 for $z = 4$ [the largest redshift considered by Maccacaro et al. (1991)], $B_x = 8.8$ and $(<\omega_x>) = (<\omega_f>)$, we thereby infer from (30) that

$$[\int B_x \, q(0) \, \omega^{-2.56} \, d\tau] / (<q_x>) = 1.3 \pm 0.5. \tag{31}$$

Hence, within this evolutionary scheme, the present epoch volume emissivity as given by (24) for special galaxies (i.e., AGN, starburst etc) is already compatible with the universal average arising from *all* sources. However, if the present epoch AGN volume emissivity is indeed as low as the lower EMSS value (27a) and we assume that $z=3$ [i.e., the redshift limit for *continuous* quasar evolution (Osmer 1982) and also for identified X-ray sources], $B_x \leq 6.8$ (for $\Delta E \leq mc^2$) and that $(<\omega_x>) < (<\omega_f>)$, then the contribution of the corresponding evolved population to the universal average emissivity would be less than half[6]. Under such circumstances, we leave open the possibility that most of the X-ray sky surface brightness could originate at redshifts $z>3$. Producing such a dominant component of the observed flux has been discussed (Leiter and Boldt 1982; Boldt and Leiter 1987; Zdziarski 1988; Daly 1991) in terms of spectral evolution whereby precursor AGNs at $z>3$ radiating mainly in the X-ray band have an appropriate spectrum characterized by an energy given by $W_0/\omega \equiv (1+z)W_0 > 160$keV.

If most of the observed CXB comes from as yet unidentified sources at redshifts $z>5$, beyond that of already known objects, then the resulting ratio (r) of X- radiation density at $z \approx 5$ to that of the IGM thermal energy density could be appreciable, viz:

$$r \approx 10 \, h^{-2} (T_5)^{-1} (\Omega_{IGM})^{-1} [(1+z)/6] \tag{32}$$

where $T = T_5 \times 10^5 \, ^\circ K$ and Ω_{IGM} is the IGM mass density relative to the closure value. The corresponding ratio (R) of the rate of X-ray Compton heating the IGM to that of cooling is

---

[6] This limit pertains to both of the evolutionary schemes considered by Maccacaro et al. (1991).

$$R = (4\pi I)(W_0/mc^2)(1+z)^5 \, g \, \sigma_0 \, (kT/t_{cool})^{-1}$$

$$\approx 2 \, h^{-1} \, (T_5)^{-1} \, [(1+z)/6]^{4-\varepsilon} \, (t_{cool}/t_{exp}) \tag{33}$$

where g (≈ 0.4) is a factor based on the Klein-Nishina deviations from the Thomson cross-section (σ_0), the X-ray spectral shape and electron recoil kinematics, where $\varepsilon \approx 0$ (for $\Omega \ll 1$) and $\varepsilon = 1/2$ (for $\Omega = 1$), and where $t_{exp} = 0.5(H_0)^{-1}(1+z)^{-1-\varepsilon}$ is the characteristic cooling time due to adiabatic universal expansion alone. As pointed out by Collin-Souffrin (1991), Compton heating of the IGM by such an X-ray background could, under certain scenarios, provide much of the required input for yielding a temperature at $z \approx 4$-5 that is sufficient to maintain a degree of ionization compatible with the Gunn-Peterson test for a low density IGM ($\Omega_{IGM} \leq 0.01$).

Della Ceca and Maccacaro (1991) and Maccacaro et al. (1991) inferred the X-ray luminosity evolution for AGNs from EMSS data for sources up to $z \approx 2$ (i.e., $\omega < 1/3$) within the framework of an open geometry ($\Omega = 0$). If the actual geometry in fact corresponds to closure ($\Omega = 1$), however, the luminosities used in their analysis have in effect been overestimated by a factor that increases approximately as $(1+z)$. Since they did not consider this important case we do so here under the assumption that C =2 for $\Omega = 1$, precisely the same as the evolution prescribed for AGN bolometric luminosity that emerges within the framework of a flat metric when one uses the comprehensive model for the underlying galactic fuel supply described by Murphy, Cohn and Durisen (1991). Assuming that the foreground sources responsible for q(0), given by (24), represent a population that has evolved so that $q = B_x [q(0)] \, \omega^{-2}$ we can again construct the ratio of the average emissivity expected from this population to the actual average emissivity ($<q_x>$) corresponding to the CXB energy flux. Within the framework of a flat metric (i.e., $\Omega = 1$) with $\mu = 3/2$ the ratio expressed by (28) becomes

$$[\int B_x \, q(0) \, \omega^{-2} \, d\tau] / (<q_x>) = (0.056 \pm 0.022) \, B_x \, (2Y_\tau /3) \, [(<\omega_x>)/(<\omega_f>)] \tag{34}$$

where the evolutionary factor $Y_\tau (\equiv \int \omega^{-1} \, d\tau)$ and τ are presented (Table 3) as a function of the maximum redshift (z), with $\omega = (1-\tau)^{2/3}$ for $\Omega = 1$.

Table 3. Evolutionary Factor (for $\Omega=1$): $Y_\tau = \int (1-\tau)^{-2/3} d\tau$

z	1	2	3	4	5
Y_τ	0.88	1.27	1.50	1.66	1.78
τ	0.65	0.81	0.88	0.91	0.93

Considering $z \lesssim 5$, $B_x \lesssim 6.8$ (for a bandwidth $\leq mc^2$) and $(<\omega_x>) \lesssim (<\omega_f>)$, we thereby infer from (34) that

$$[\int B_x\, q(0)\, \omega^{-2} d\tau] / (<q_x>) \lesssim 0.5 \pm 0.2. \tag{35}$$

Hence, for a flat metric, the average *evolved* volume emissivity based on the present epoch luminosity density arising from special galaxies (e.g., AGN, starburst), as given by (24), may well be significantly smaller than the universal average implied by the CXB.

10. SOME OUTSTANDING ISSUES

Although our understanding of the foreground appears to be converging we still can not exclude the intriguing possibility of a substantial "true" background (e.g., due to an as-yet unidentified population of predominantly X-radiating sources at large redshifts). What are some of the next steps to be taken?   Clearly, the EMSS AGN data should be reanalyzed within the context of a flat metric in order to see if indeed the inferred luminosity evolution of foreground AGNs is significantly less than obtained assuming $\Omega=0$, as done by Maccacaro et al. (1991). Do these foreground AGNs exhibit broadband spectral evolution? We should investigate a sizable sample of foreground AGNs for possible *spectral* evolution with redshift, but this is not likely to be accomplished soon. If there is some AGN X-ray spectral evolution, then we need to know how this affects the determination of X-ray luminosity evolution inferred from data over a severely limited energy interval ($\Delta E << W_0 = 40 keV$), such as the EMSS. In order to get a precise value for the local emissivity a careful determination of the "break" in the present-epoch AGN luminosity function at $L_{43} \approx 1$ should be made in the well behaved 2-10keV band directly, thereby avoiding the complicating spectral effects (e.g., absorption, soft components) present at lower energies. The work started by Marshall (1991) on the recently identified AGNs in the HEAO-1 A1 catalog of X-ray sources at ~5keV (Wood et al. 1984) appears to be a promising step in this direction. Observations of the broadband ($\Delta E > 100$keV) spectra of several present-epoch AGNs with GRO would certainly provide needed improvement in precision for the sort of bandwidth correction estimates that we have attempted here. Is the inhomogeneity of the present-epoch universe (see Appendix) misleading us in any drastic way in our determination of the "local" emissivity? Finally, we should continue to search for clever ways to ascertain if in fact there is much energy density in X-rays at large redshifts (z>3), such as considering possible observable effects of the CXB on the IGM (Collin-Souffrin 1991).

In the preparation of this review I have benefited from valuable discussions with S. Collin-Souffrin, R. Daly, K. Jahoda, D. Leiter, H. Marshall and P. J. E. Peebles.   I am especially grateful to O. Lahav for introducing me to the concept of using the all-sky dipole moment of observed electromagnetic radiation as a measure of local gravity and for his vital role in determining the CXB correlation with galaxy counts.

APPENDIX

In order to minimize systematic effects, potentially significant (but weak) large-scale anisotropies of the X-ray sky (2-10keV) have been investigated by examining variations of the surface brightness observed with HEAO-1 A2 (Boldt 1987) along a great-circle belt defined by the ecliptic equator (±24 degrees in width). Except for ecliptic longitudes close to the galactic and supergalactic equators, the X-ray sky along this band is remarkably isotropic. The maximum anisotropy which is clearly extragalactic is that associated with the direction defined by the intersection with the supergalactic plane, an effect that was anticipated many years ago (Boldt 1967). The fore-aft asymmetry in surface brightness corresponding to this direction is about 1.5%, much of which is probably due to the Compton-Getting dipole effect arising from our motion relative to the proper frame of the background. Based on the dimensions of the supergalaxy (local supercluster) discussed by deVaucouleurs (1976) and Oort (1983), we conclude from this composite asymmetry that the X-ray volume emissivity within the supergalactic disk is no more than one order of magnitude greater than the average value (23) characterizing the present epoch. Searches for enhanced X-ray volume emissivity in other superclusters by Tawara et al. (1989), Persic et al. (1990), Day et al. (1991) and Mushotzky (1991) have yielded comparable results.   S. Boughn and R. Daly (private communication) have started a study to obtain a more precise limit for the enhancement in X-ray volume emissivity to be associated with the local supercluster by evaluating the correlation of detailed supergalactic structure with the X-ray surface brightness variations observed with the HEAO-1 A2 all-sky survey. The CXB surface brightness variation studied by Jahoda et al. (1991) suggest that the supergalactic X-ray enhancement may only be substantial near the center.

References

Boldt, E. 1967, "The Local Supergalaxy as the Structured Aspect of a Universal Background of X-rays" NASA/GSFC X-611-67-486

Boldt, E. 1974, in *High Energy Particles and Quanta in Astrophysics*, ed. F. McDonald and C. Fichtel (Cambridge: MIT Press) p.386

Boldt, E. and Leiter, D. 1987, ApJ, 322, L1

Boldt, E. 1987, Phys. Reports, 146, 215

Boldt, E. 1989, in *X-ray Astronomy, 2. AGN and the X-ray Background.* ESA SP-296, Noordwijk, p. 797

Boldt, E. 1990, in *Observatories in Earth Orbit and Beyond*, Proceedings of IAU Colloquium 123, ed. Y. Kondo (Dordrecht: Kluwer) p.451

Canizares, C. and Blizzard, P. 1991, ApJ, 382, 79

Collin-Souffrin, S. 1991, Astron.Astrophys., 243,5; also this workshop

Daly, R. 1991, ApJ, 379,37

Day, C., Fabian, A., Edge, and Raychaudhury, S. 1991, MNRAS, 252, 394

Della Ceca, R. and Maccacaro, T. 1991, Proc. Astron. Soc. Pac., in press; Bologna Astrophysics Preprint: BAP 06-1991-020-OAB

deVaucouleurs, G. 1976, ApJ, 205, 13

Edge, A., Stewart, G., Fabian, A. and Arnaud, K. 1990, MNRAS, 245, 559

Efstathiou, G., Ellis, R. and Peterson, B. 1988, MNRAS, 232, 431

Elvis, M., Soltan, A., and and Keel, W. 1984, ApJ, 283, 479

Fabian, A., Done, C. and Ghisellini, G. 1988, MNRAS, 232, 21P

Fabian, A., George,I., Miyoshi, S., and Rees, M. 1990, MNRAS, 242, 14P

Fabian, A. 1990, presented at *Texas/ESO-CERN Symposium on Relativistic Astrophysics,* Brighton, UK, preprint

Fabbiano, G. 1989, Ann. Rev. Astron. Ap., 27,87

Fabbiano, G., Kim, D., and Trinchieri, G. 1991 ApJ Suppl., submitted

Felten, J. 1977, Astron. J., 82, 861

Franceschini, A., Danese, L., De Zotti, G., and Toffolatti, L. 1988, MNRAS, 233, 157

Gioia, I., Henry, J., Maccacaro, T., Morris, S., Stocke, J. and Walter, A. 1990, ApJ, 356, L35

Goicoechea, L. and Martin-Mirones, J. 1990, MNRAS, 244, 493

Griffiths, R. and Padovani, P. 1990, ApJ, 360, 483

Gruber, D. 1988, personal communication

Gruber, D. 1991, this workshop

Hickson, P., Menon, T. Palumbo, G. and Persic, M. 1989, ApJ, 341, 679

Jahoda, J., Lahav, O., Mushotzky, R. and Boldt, E. 1991, ApJ, 389, L37

Kriss, G., Cioffi, D., and Canizares, C. 1983, ApJ, 272, 439

Lahav, O., Rowan-Robinson, M. and Lynden-Bell, D. 1988, MNRAS, 234, 677

Lahav, O., Edge,,A., Fabian, A. and Putney, A. 1989, MNRAS, 238, 881

Lahav, O. 1991a, in *After the First Three Minutes*, AIP Conference Proceedings 222
    (S. Holt, C. Bennett and V. Trimble, editors) p. 421
Lahav, O. 1991b, this workshop
Leiter,D. and Boldt, E. 1982, ApJ, 160, 1
Lightman, A. and White, T. 1988, ApJ, 335, 57
Lynden-Bell, D, Lahav, O., and Burstein, D. 1989, MNRAS, 241, 325
Maccacaro, T., Della Ceca, R., Gioia, I., Morris, S., Stocke, J. and Wolter, A.
    1991, ApJ, 374, 117
Marshall, H. 1991, Proc. Astron. Soc. Pac., in press; UC/Berkeley EUV Preprint 469
Marshall, F., Boldt, E., Holt, S., Miller, R., Mushotzky, R., Rose, R., Rothschild, R. and
    Serlemitsos, P. 1980, ApJ, 235, 4
Miyaji, T. and Boldt, E. 1990, ApJ, 353, L3
Miyaji,T., Jahoda, K. and Boldt, E. 1991, in *After the First Three Minutes*, AIP
    Conference Proceedings 222 (S. Holt, C. Bennett and V. Trimble, editors) p.431
Murphy,B., Cohn, H. and Durisen., R. 1991, ApJ, 370, 60.
Mushotzky, R. 1991, this workshop
Osmer, P. S. 1982, ApJ, 253, 28
Oort, J. 1983, Ann. Rev. Astron. Astrophys., 21, 373
Padovani, P., Burg, R. and Edelson, R.1990, ApJ,353,438
Persic, M., De Zotti, G., Boldt, E., Marshall, F., Danese, L., Franceschini, A. and
    Palumbo, G. 1989, ApJ, 344,125
Persic, M., Jahoda, K., Rephaeli, Y., Boldt, E., Marshall, F., Mushotzky, R.,
    and Rawley, G. 1990, ApJ, 364, 1
Piccinotti, G., Mushotzky, R. Boldt, E., Holt, S., Marshall, F., Serlemitsos, P., and
    Shafer, R. 1982, ApJ, 253, 485
Price, R., Burns, J., Duric, N. and Newberry, M. 1991, AJ, 102, 14
Rephaeli, Y., Gruber, D., Persic, M. and MacDonald, D. 1991, ApJ Letters, 380, L59
Rogers, R. and Field, G. 1991, ApJ, 370, L57; ApJ, 378, L17
Rothschild, R., Mushotzky, R., Baity, W., Gruber, D., Matteson, J. and Peterson, L.
    1983,ApJ, 269, 423
Schechter, P. 1976, ApJ, 203, 297
Schneider, D., Schmidt, M. and Gunn, J. 1989, Astron. J., 98, 1951
Shafer, R. 1983, PhD thesis dissertation, University of Maryland at College Park,
    NASA TM 85029
Shafer, R. and Fabian, A. 1983, in IAU Symposium 104, Early Evolution of the Universe
    and its Present Structure, ed.G. O. Abell and G. Chincarini (Dordrecht:Reidel), p.333
Strauss, M. and Davis, M. 1988, in *Large Scale Motions in the Universe*, Proceedings of the
    Vatican Study Week, eds. G. Coyne and Vera Rubin, Princeton: Princeton University
    Press
Tawara, Y., Kawada, M., Takano, S., Koyama, K., and Awaki, H. 1989, in *X-ray
    Astronomy, 2. AGN and the X-ray Background.* ESA SP-296, Noordwijk, p.1065
Turner, E. and Geller, M. 1980, ApJ, 236, 1
Wood, K., Meekins, J., Yentis, D., Smathers, H., NcNutt, D., Bleach, R., Byram,E.,
    Chubb,T., Friedman, H., and Meidav M. 1984, ApJ Suppl., 56, 507
Zdiarski, A. 1988, MNRAS, 233, 739

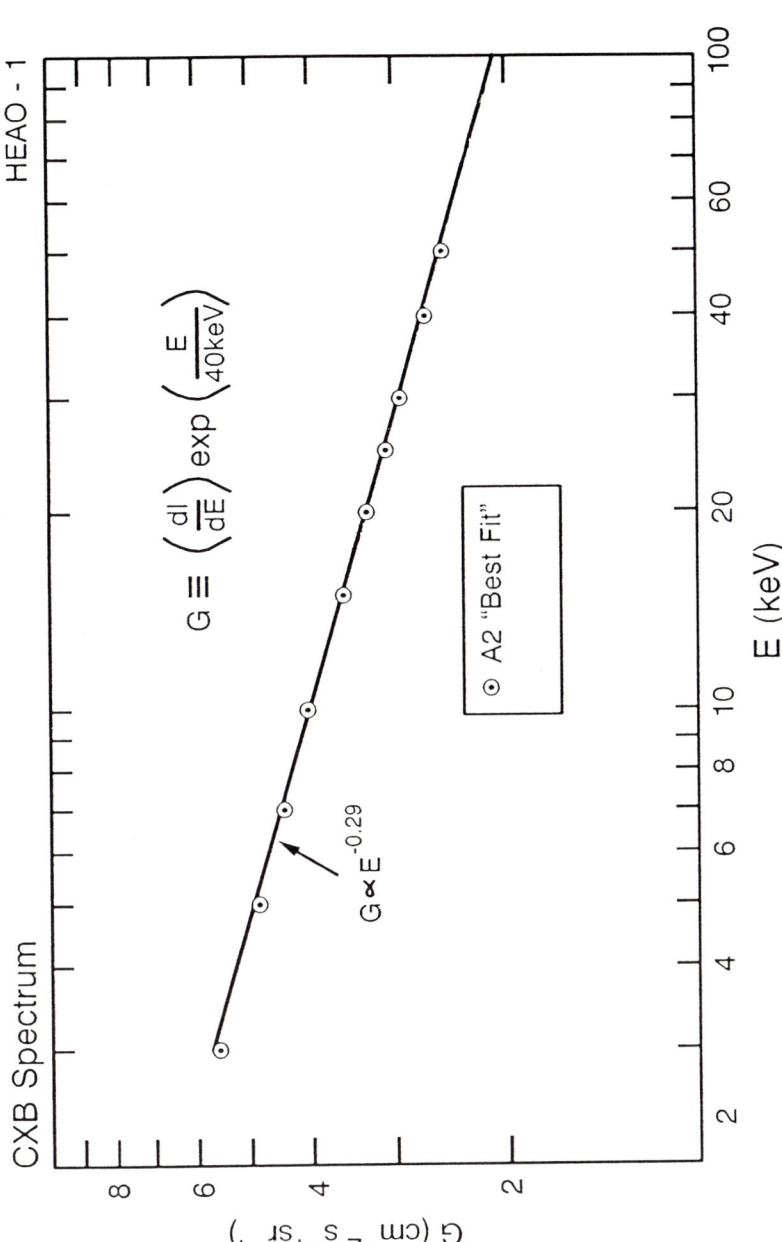

Figure 1. G [the observed CXB spectrum multiplied by exp(E/W_0)], with $W_0 = 40$keV, as a function of photon energy (E). The circled points shown represent the best-fit thermal model for measurements made with gas proportional counters of the HEAO-1 A2 experiment (Marshall et al. 1980). The power-law straight line indicated correponds to equation (1). CXB measurements at E≥20keV carried out with alkalide halide scintillators of the HEAO-1 A4 experiment are consistent with this line up to ~90keV (Gruber 1991).

Deep X-ray Source Counts

THOMAS HAMILTON

Columbia University

1. Introduction

The nature of the source of the X-ray background (XRB) has remained obscure since the background's discovery almost thirty years ago (Giacconi et al. 1962). While this problem has been approached from many directions as illustrated by the contents of this volume, the most straightforward and most expensive approach has been to build X-ray telescopes capable of studying the background with high spatial resolution and thereby attempt to resolve as much of the background as possible into individual sources which could be identified with objects visible in other wavelengths.

Inasmuch as this paper is a discussion of the most sensitive X-ray surveys to date and their importance to understanding the sources of the XRB, it is important to emphasize what is meant by XRB. In normal astronomical parlance, the XRB refers to the large isotropic flux of X-rays from presumably extragalactic sources. This flux has the spectrum of a thermal bremsstrallung at 40 keV. No satellite has been launched capable of high resolution imaging in that waveband. Hence, in most of this discussion XRB is understood to refer to that portion of the high galactic latitude, isotropic flux which is hard enough travel through the interstellar medium (and a proportional counter window), i.e. >.16 keV, and soft enough to efficiently focussed with existing grazing incidence optics, i.e. <3.5 keV. As we shall see below, the source of this band may be different from, and possibly less interesting than, the source of the XRB at 40 keV.

Well before the Einstein Observatory discovered large numbers of extragalactic X-ray sources, astronomers considered the aggregate flux of AGN to be one of the most likely sources of the XRB. Indeed, shortly after the first discovery of X-ray emission from an AGN, 3C 273 (Bowyer et al. 1970), Setti and Woltjer (1970) proposed that the integrated emission from AGN might make up the XRB. The HEAO A-2 sky survey by Piccinotti et al. (1982) included 29 objects to a limiting flux of 3.1×10^{-11} ergs cm^{-2} s^{-1} in the 2-10 keV band. Although the precise measurement of the diffuse X-ray background spectrum provided by this same experiment (Marshall et al. 1980) demonstrated that a simple superposition of AGN spectra was incompatible with the energy distribution of the background radiation (see Boldt et al. 1987 for a review), it was clear from a modest extrapolation of the X-ray AGN log N-log S relation that these sources must contribute a substantial fraction of the background. Normal galaxies and clusters of galaxies were also recognized as populations of discrete X-ray emitters which, at least at energies of less that 5 keV, provided a non negligible addition (~20%) to the background (e.g.,Setti 1985; Schmidt and Green 1986).

Although X-ray catalogs have existed from the earliest days of rocket-borne X-ray detectors, we shall hear consider surveys produced since the employment of focussing X-ray optics on long duration satellite missions. The first such mission was HEAO-2 or the EINSTEIN Observatory

2. The EINSTEIN Observatory

The launch of the Einstein Observatory in 1978 provided an opportunity to directly image the X-ray background and identify individual sources contributing to the integrated flux. In all deep surveys under discussion the basic program is the same: use the X-ray observations to determine the location of significant X-ray emitters and then optically observe the emitters position to identify an optical counterpart and measure its properties. In general the X-ray sources of interest are near the threshold of detectability of the X-ray telescope and thus have emitted such a small number of detected photons that the X-ray data itself contains little information other than the source's existence. In general, it is possible to extract considerably more information from the optical counterpart. In particular, one can generally determine what type of object it represents and measure a redshift. The history of deep X-ray surveys is therefore the a series of catalogs of X-ray fluxes and positions with corresponding, and generally more detailed, optical identifications.

A rather complete characterization of the log N-log S relation for X-ray sources >10^{-13} ergs cm^{-2} s^{-1} is found in the Extended Medium Sensitivity Survey (EMSS; Gioia et al. 1990; of the 761 sources identified optically, 541 are extragalactic objects: 13 relatively normal galaxies, 99 clusters of galaxies, 34 BL Lacs, and 395 other AGN. Including unidentified sources whose X-ray to optical flux ratios are indicative of an extragalactic origin, the log N log S for these sources implies a surface density of 4.5 discrete X-ray emitters deg^{-2} at a threshold of 10^{-13} ergs cm^{-2} s^{-1} in the 0.3-3.5 keV band.

Fainter source populations were probed through an program of long integrations (up to 10^5 s) using both the imaging proportional counter (IPC) and the high resolution imager (HRI) on Einstein.

The first deep survey published after the launch of the Einstein observatory was the survey of Giacconi et al. (1979) Written only a few months after the launch of the spacecraft, the survey was impaired by the limited knowledge of the Einstein Imaging Proportional counter available at that time. Indeed the survey reported approximately 20 sources in each of two 1 deg^2 fields. In order to interpret their results the authors selected a subset of these as a presumably extragalactic "complete sample" and used these as the basis of their analysis. Analyzed in this way, the survey observed the remarkable fact that the at the limit of the Einstein IPC sensitivity (~2 x 10e^{-14} ergs/cm^2 s over 1-3 keV) the N(S) function for faint X-ray sources followed a Euclidean extrapolation of the the value measured at much greater fluxes. That is the log N log S slope was -1.5 over several orders of magnitude. Most of the G79 sources remain optically unidentified. Investigations are continuing however, especially of their radio properties (see below).

Griffiths et al. (1983, G83) examined a field in Pavo which had the longest Einstein IPC integration time and several long HRI pointings. G83 reported a lower surface density of sources that the shallower G79 survey, however G83 was much more successful in garnering optical identifications. So the identified G83 sample represented a surface density comparable to the selected G79 sample, leading to generally similar scientific conclusions.

Nevertheless, the discrepancy in total source counts led Hamilton and Helfand (1991, HH91) to reexamine all of the deep survey fields in light of the knowledge gained in ten years of analyzing Einstein data. A similar project proceeded independently at SAO and is reported in Primini et al. (1991; P91). In order to understand how measuring the faint source content of an X-ray field could become controversial and merit three independent efforts we must consider the technical details of the functioning of the Einstein IPC

3. Technical issues

Problems in the analysis of faint X-ray surveys with both Einstein and ROSAT stem from the small number of photons from each source. Because the X-ray background and particle counter background combined are of the order of 3×10^{-4} events s^{-1} per beam, even after 10s of thousands of seconds of integration a typical beam has on the order of 10 total events. This not only produces the obvious problems of small number statistics, but also the more insidious problem of calibration. That is, to measure the systematic properties, such as the flat field, of the detector to, say, 3% precision requires a thousand background events per beam, a number which, given the duty cycle of the instrument, will take months to accumulate. Figure 1 shows a flat field for the Einstein Observatory IPC on orbit. This filed was constructed by examining all IPC data taken during the entire Einstein mission. Fields in which diffuse emission appears to be present or fields in which the mean count rate is more than twice the mission average are excluded. All coaligned observations are then coadded and searched for sources. The region of the counter near a detected source is also excluded from the database. Finally a map is created by adding the counts and exposure times for each pixel in the IPC detector. The point here is that only with the experience gained by examining all of the data produced during the mission can we gain the statistical precision to correct for systematic non-uniformities. As is clearly visible in Figure 1, the counter has considerable linear structure. These striations are parallel to the wires which make up the instruments cathode planes. Figure 2 illustrates the hazards which confronted the earliest Einstein surveys. The raw data (in event number) are clearly suggestive of a linear structure running parallel to the wire mesh, yet G79's automated source search algorithm interpreted the striation as six separate sources.

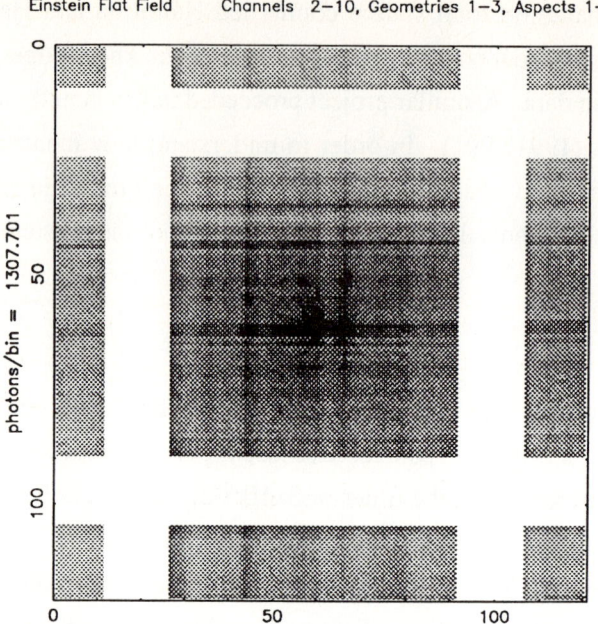

Figure 1 The quantum efficiency map (flat field) for the EINSTEIN IPC.

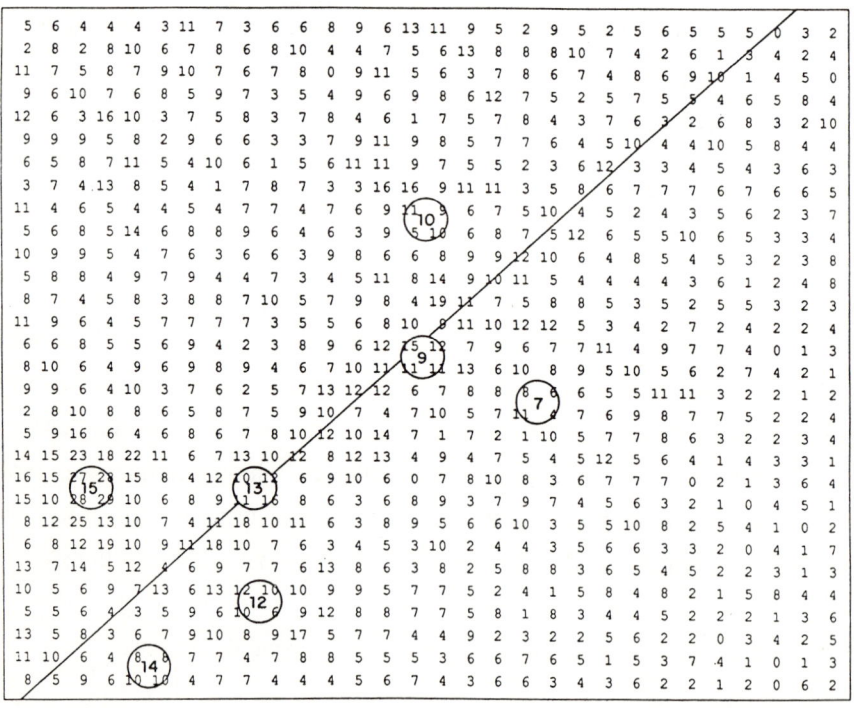

Figure 2 Raw photon counts from Eridanus survey showing spurious sources.(32" bins)

The HHW91 reanalysis was simultaneous with the independent work of P91. The technical approaches were radically different. In general HHW91 adopted a philosophy of severe editing to eliminate all possible contamination. For example, all data taken when the satellite was illuminated by the sun was excluded from analysis. This reduced the size of the available database by slightly more than half, but eliminated all possibility of contamination of the images by solar X-rays scattering of the residual upper atmosphere. P91 adopted the opposite approach. To eliminate the influence of variations in X-ray flux on the scale of tens of arc minutes, such as might result from variations in the soft galactic component of the X-ray background, we compared each potential source position with a background generated by a local annulus. P91 performed a global background subtraction, subtracting the average of the deep survey fields from each one individually and looking for residuals. In general, P91's methods were more sensitive while HHW91's were more reliable. Remarkably enough, a careful source by source comparison of the two survey reveals almost identical results. While there are technical differences between the results of G79 and HHW91, there is no substantive controversy. All of the surveys performed since the first months of the Einstein mission (G83, HHW91, and P91) agree to within reported errors on the shape of the log N-log S for sources down to 2.5 x 10^{-14} ergs cm^{-2} s^{-1}. The curve from HHW91 is shown in Figure 3

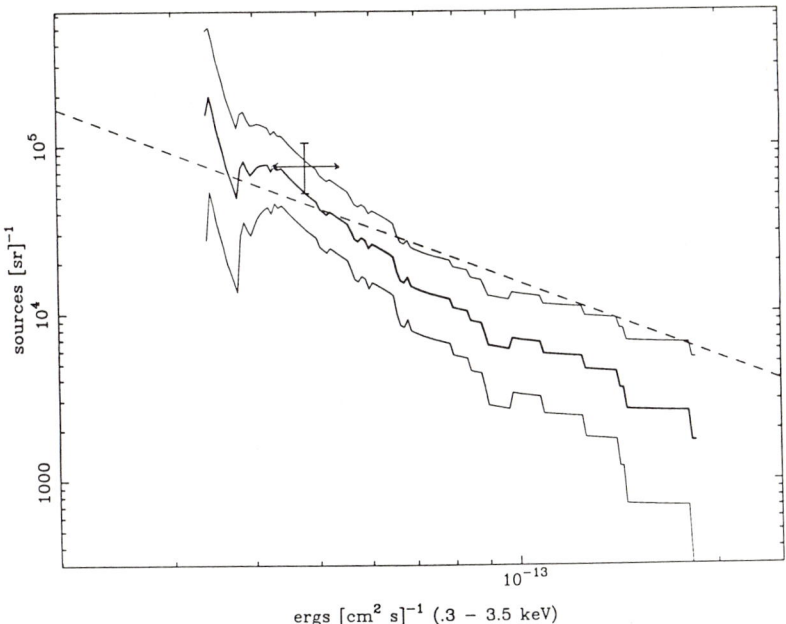

Figure 3 Integral Log n-Log S from HHW91.

4. ROSAT Surveys

While various groups analyzed and reanalyzed the Einstein data, the Max Planck Institute with its American and British collaborators built and, in 1990, launched ROSAT satellite. ROSAT's mirrors and instruments are similar to EINSTEIN's but a number of design and operation features make ROSAT's performance superior in a number of ways. For example, the ROSAT PSPC, the counterpart to EINSTEIN's IPC, has a considerably lower internal background than the IPC. Also ROSAT is operated with a continuous dither of 3 arc minutes amplitude and a 400 second period. That is, the physical pointing of the satellite is constantly wandering about the nominal pointing of the observation. All X-ray satellites with time resolved event counting detectors wander from their nominal pointings to some extent. It is cheaper to measure the wander with star trackers and correct for it later when the data is analyzed than to build a high fidelity attitude control system like the one on HST. However, EINSTEIN was held as steady as possible to minimize dependence on the post facto correction. ROSAT follows an opposite strategy of deliberately wandering in a controlled way. The virtue of this method is that irregularities in the counter sensitivity such as those which plagued analysis of Einstein deep survey data are smoothed out over a scale which is large compared to the scale of both the wire separation and the beam. This is an enormous advantage for faithful detection of extremely faint point sources as well as the study of spatial structure on a scale of arc-minutes. Of course the irregularities in the counter face have not been eliminated from the data, this clever strategy has simply mapped them from space into time. A non-varying point source observed with the PSPC thus appears to have substantial flickering with particular power at the harmonics of 400 seconds. Thus, at an X-ray background conference like this, I can applaud the wise decision of ROSAT's designers with confidence. At a conference devoted to X-ray variability I would be more circumspect.

The first deep extragalactic survey reported from ROSAT was a 49,000 second exposure of the North Ecliptic Pole (Hasinger, Schmidt, and Trumper, 1991, see Hasinger, this volume, for a more recent report). This paper reported a striking turnover in the log N-log S at a flux level just below the limit of the Einstein deep survey. Such a turnover was predicted by analysis of the fluctuations in the deepest EINSTEIN pointings (Hamilton and Helfand, 1987, Barcons and Fabian 1990) and is consistent with extrapolations of the luminosity functions of extragalactic X-ray sources (Schmidt, 1989). Quantitative understanding of the slope below the turnover is still limited although continuously expanding as more data are analyzed.

Shanks et al. have carried out a program of optical spectroscopy of sources observed within a 30,000 second ROSAT exposure and discovered an impressive number of QSOs. (Shanks et al. and Griffiths, this volume) Indeed, with a sensitivity limit of 1×10^{-14} ergs cm^{-2} s^{-1}, Shanks et al. detected QSOs sufficient to account for 35% of the XRB at 1 keV. However, they also conclude that the QSO contribution to the XRB peaks at 1 keV.

Shanks et al. were also able to constrain the slope of the log N-log S relationship for sources fainter than their detection limit by measuring the intensity fluctuations in their field. They estimate that the log N-log S continues with a slope of - 1.2. If that slope is extrapolated until the source population's integrated flux makes up the 90% of the XRB, a population of about 4000 sources per square degree is required. This is in agreement with the Einstein results of Hamilton and Helfand (1987), although the ROSAT measurement is considerably more sensitive. For a more recent and more sensitive analysis of ROSAT fluctuation see Danese (this volume).

5. The GINGA Paradox

For reasons which are not understood, fluctuation analysis of GINGA data have indicated that the source population in the GINGA band is considerably higher than would be expected by extrapolating the spectra of EMSS sources to GINGA's harder bandpass (Warwick and Stewart 1989). While some of this effect may be a result of the EMSS's inability to detect diffuse sources which nevertheless contribute to the GINGA fluctuations (Helfand and Hamilton, 1991), it seems reasonable to believe that GINGA is detecting hard sources which are not easily detected in the softer bands to which EINSTEIN and ROSAT are sensitive. However, it is plausible that inasmuch as the sources must evolve cosmologically, they represent a range in z, and some of them would be redshifted into an easily imaged band. Fortunately for the purposes of moderately deep surveys, ROSAT spent much of the summer of 1991 in a reduced pointing mode, resulting a large number serendipitous surveys. These or other ROSAT deep fields may yield examples of hard X-ray sources which would resolve the discrepancy between the GINGA fluctuation results and the EMSS.

Recent theoretical work has investigated the possibility that much of the XRB above a few keV is emitted by highly absorbed sources or sources in which a significant fraction of the flux has been reflected (Warwick and Stewart, 1989, Fabian, 1990, Rogers this volume). Although extragalactic sources with the requisite spectrum have not yet been detected in surveys, evidence for their existence may lie hidden in the vast archive of X-ray data accumulated by Einstein and being expanded by ROSAT.

6. X-ray - Radio Correlations

At Columbia David Helfand and I have combined information from various wavelengths in an attempt to locate examples of this putative group of hard sources. We combined three 20 cm VLA bandwidth synthesis mode observations of the Draco deep survey. With a total of 31 hours of observing time we were able, using standard reduction techniques, to make a map with 30 microJansky rms and compile a list of 115 radio sources over a 1800 arcmin2 area, complete to 150 microJansky. We also compiled a list of 22 positive 2 sigma fluctuations in the EINSTEIN data. Of these 22, 6 are greater than 3.5 sigma and correspond to known sources; 6 or 7 are statistical artifacts; the remaining 10 are then probably associated with real enhancements in the X-ray sky. Comparing the two lists we find 9 positional coincidences where 4 coincidences would be expected randomly. Two of the coincidences correspond to known sources, leaving 7 unknown objects. All were optically identified and 5 of the 7 were found to be more or less normal galaxies with redshifts around .3. Naturally the statistical significance of such numbers is poor so we repeated the experiment using the Eridanus deep survey field. In Eridanus we expected 3 random coincidences and found 14. (We do not yet have optical identifications). It is unclear how this result is related to the QSO abundance reported in Shanks et al. at comparable flux levels. Progress along similar lines is reported by Worrall (this volume)

Analyzing the correlation of radio and X-ray positions yielded the surprising result that the correlation of X-ray source positions, as defined by the above procedure, with radio source positions is substantially higher that the autocorrelation function for the radio positions. That is X-ray emitting radio sources are found preferentially in areas of high radio source density. This raises the possibility that the X-ray sources are not really the galaxies themselves but are some other object which is possible lensed by or physically associated with a group containing some radio bright galaxies.

An analogous technique has been applied to the entire EINSTEIN IPC database to search for radio loud objects which would have been missed by more straightforward surveys. Unlike objects associated with fluctuations in deep survey fields, these sources are not particularly faint, most of them are found in relatively short EINSTEIN observations. The fluxes are comparable to the sources found in the EMSS Working with Robert Becker of UC Davis, we have compiled a catalog of 88,000 EINSTEIN "sources" or which about half are statistical artifacts. We have compared this catalog with the positions of all 53,500 sources >30 mJy at 6 cm (in the northern sky). This flux is greater than 98% of the EMSS sources. Expecting 300 identifications through chance, we find 1,100. Eliminating 355 which were the targets of observations or in the EMSS, we then searched for sources with no counterpart >150 mJy at 80 cm. In this way we compiled a list of 286 flat spectrum radio sources associated with X-ray emission. We have observed all with the VLA to get good positions and six, so far, with the Lick 3 meter. Of those 6, 4 are galaxies, including 2 with strong emission lines, and 2 are QSOs, including one with $z = 3.87$. Multiwavelength techniques of this sort are plagued with more complicated selection effects than straightforward deep surveys, but they do provide a way to discover and explore hitherto undetected populations while we await the advent of high resolution imaging telescopes capable of probing the harder bands invisible to EINSTEIN and ROSAT.

7. Conclusion

The 13 years since the launch of the EINSTEIN observatory have seen a decidedly sporadic march towards the resolution into point sources of a greater portion of the XRB. In particular, the accepted value of the integrated flux of the XRB in the 1 keV band has moved upwards (McCammon and Sanders 1990, Wu et al. 1991) at the same time that ROSAT has enabled more faint sources to be imaged. Consideration of the discrepancy between the GINGA fluctuations and the EMSS, as well as the discrepancy between the spectra of known extragalactic sources and the observed spectrum of the XRB, has led many investigators to question whether the sources of the 1 keV XRB are identical to the sources of the harder X-rays which make up most of the background.

Future exploration of the 1 keV background will be slowed by the apparent rapid turnover of the log N-log S slope. However, the promise held out by the GINGA fluctuations of a new population of hard, relatively bright, X-ray sources at energies of a few keV creates hope that ASTRO-D and AXAF have plenty of new, hard extragalactic objects to study.

8. Bibliography

Barcons, X. & Fabian, A. 1990, M.N.R.A.S, 243, 366

Bowyer, C.S. et al. 1970, ApJ, 161, L1

Fabian, A. et al. 1990, M.N.R.A.S., 242, 14P

Giacconi, R. et al. 1962, Phys. Rev. Lett.,9, 439

Giacconi, R. et al. 1979, ApJ, 234, L1 (G79)

Gioia et al. 1990, ApJS, 72, 567

Griffiths, R.E. et al. 1983, ApJ, 269, 375 (G83)

Hamilton, T. & Helfand, D. 1987, ApJ, 318, 93

Hamilton T., Helfand, D. & Wu, X. 1991 ApJ, 379, 576 (HHW91)

Hasinger, G., Schmidt, M. & Trumper, J. 1991, J. Ast. & Ap., 246, L2

Helfand, D. & Hamilton, T. 1992, Proc. of Conference at MPE Garching Nov. 1991

McCammon, D. & Sanders, W. 1990, Ann Rev Ast & Ap., 28, 657

Piccinotti, G. et al. 1982, ApJ, 253, 485

Primini F. et al. 1991, ApJ, 374, 440 (P91)

Schmidt, M. & Green, R. 1986, ApJ, 305, 68

Schmidt, M. et al. 1989,23rd ESLAB Symp. (eds Hunt, J. & Battrick, B.), 853

Setti, G. & Woltjer, L. 1970, Ap&SS, 9, 85

Shanks, T. et al. 1991, Nature, 353, 315

Warwick, R. & Stewart, G. 1989, 23rd ESLAB Symp. (eds Hunt, J. & Battrick, B.) 727

Wu, X. et al.1991, ApJ, 378, 564

The AGN Contribution to the X-ray Background

D.A. SCHWARTZ

Smithsonian Astrophysical Observatory

ABSTRACT
This review considers the contribution of Active Galactic Nuclei (AGN, including Seyfert galaxies and quasars without distinction) to the extragalactic, diffuse X-ray background. The conclusion will be that AGN do contribute the bulk of this background. AGN were originally considered as the substantial contributor of the background. Subsequently problems arose due to the "spectral paradox" and to the low level of measured fluctuations. We argue for the resolution of these problems as follows: Observations of the X-ray spectra of AGN, in particular by the Ginga satellite, reveal a component which shows increased hardening above 10 keV. This may provide the signature of the $E^{-0.4}$ power law shape of the background, if it can be shown to be appropriately redshifted to the 3 to 12 kev band. AGN X-ray luminosity functions whose evolution is constrained by the continuity equation can achieve both the reproduction of the measured spectral shape, and predict sufficiently low fluctuations that the observed isotropy constraints are not violated.

1 INTRODUCTION – THE EXTRAGALACTIC X-RAY BACKGROUND

1.1 Extragalactic Nature of the Background
The diffuse X-ray background was discovered during the first experiment which observed cosmic X-rays (Giacconi et al. 1962). This is not surprising in view of the fact that we now know the diffuse background dominates the total flux from the X-ray sky above 5 keV (Boldt 1978; 1987). Many early rocket and balloon flights showed that the radiation was roughly isotropic (e.g., Giacconi et al. 1962; Seward et al. 1967; Bleeker et al. 1968; Matsuoka et al. 1968; Davison and Thomas 1971), and therefore an extragalactic origin was indicated.

Schwartz (1969; 1970) used OSO-III satellite data in the 7.7 to 110 keV range to measure this isotropy to limits of a few percent over the entire sky, on scales of 20 x 20 degrees. Subsequently, smaller limits to anisotropies have been measured using experiments on board the UHURU (Schwartz et al. 1976; Fabian 1975; Schwartz 1980), Ariel V (Warwick and Pye 1978), and HEAO-1 (Shafer 1983) satellites, covering the 2 to 10 keV range. These measurements progressively extended the angular scales down to 5×5 degrees, 10×0.7 degrees, and 3×3 degrees, respectively.

Two classes of models were invoked for the origin of the background: a truly diffuse process occurring throughout metagalactic space, vs. the summation of individual discrete sources of a class sufficiently weak and numerous that they were not all resolved individually. The two diffuse mechanisms considered were inverse Compton scattering of relativistic electrons on the microwave blackbody photons (Felten and Morrison 1966), and optically thin thermal bremsstrahlung from a hot plasma, which would have a density of the same order as that necessary for closing the universe (Cowsik and Kobetich 1972).

1.2 Discrete Source vs. Diffuse Origin.

This review will start from the premise that the discrete source origin is clearly preferred. We will not establish this position here, but outline the reasoning as follows.

In the inverse Compton scenario, the spectral density of electrons was estimated from a calculation of the density and spectrum of electrons which were inferred to be in radio galaxies, considering their subsequent escape into metagalactic space (Felten and Morrison 1966). The models fell short by a factor of about 100 from producing the entire intensity. Other than the intensity problem, there is no direct evidence contradicting this model, rather the observations of discrete sources seem to have supplanted any need for it.

Cowsik and Kobetich (1972) argued that inverse Compton emission could not produce sharp spectral breaks, and suggested a thermal bremsstrahlung origin of the background. Detailed theories were discussed by Field and Perrenod (1977). These authors emphasized the large energy content which would be in the gas, due to its long radiation lifetime. To reduce the total energy content of the hot gas, and to reduce the problems of its initial heating, various scenarios of a clumpy medium were invoked. More recently, Rogers and Field (1991a) considered a general formulation involving clumped hot gas, and ruled out any possible degree of clumping giving sufficient emission measure to produce the background. A definitive upper limit of the Sunyaev-Zeldovich parameter $y \propto \int n_e kT \, dl$ by the Far Infrared Absolute Spectrometer on the COBE satellite (Mather et al. 1990), provides a direct upper limit of a few percent to the contribution which such plasma might make to the diffuse X-ray Background. We note that such a contribution cannot be excluded, and would constitute a significant fraction of the baryon density of the universe.

1.3 Definitions

In defining AGN we will use "tunnel vision" from an X-ray observational viewpoint: AGN are X-ray point sources for which extragalactic redshifts are obtained by observation of broad emission lines. We thus exclude BL Lac objects, but include radio loud quasars. The latter may have somewhat different X- ray properties (cf. Zamorani et al. 1981, Wilkes and Elvis 1987; Worrall 1989), but represent $\lesssim 10\%$ of all AGN. By contributing the "bulk of the X-ray background," we make the specific, arbitrary definition that the observed energy flux, integrated from 2 to 10 keV, which is contributed by AGN is at least $10^{0.5}$ times greater than that due to the sum of all other diffuse and discrete sources, i.e., that they contribute a minimum of 75.96%.

As stated, by "cosmic X-ray background" we are referring to the energy range above 2 keV, and up to at least a few hundred keV, in which range the isotropy proves the extragalactic origin, and in which the bulk of the energy, EI(E), is contained. However, in discussing discrete sources, we must bear in mind that the most recent and extensive observations are from the Einstein and ROSAT telescope missions, covering the 0.1 to 3 keV region. Large extrapolations must be made to convert from the observed band of those discrete sources to the 2 to 10 keV band of the extragalactic diffuse background and of the source observations made by the mechanically collimated sky survey missions such as Uhuru, Ariel-V, and HEAO-1 A-2 (cf. Piccinotti et al. 1982).

2 AGN AS THE SOURCE OF THE DIFFUSE X-RAY BACKGROUND

Many authors estimated the fraction of the diffuse X-ray background which could be produced by various classes of object (cf., Schwartz and Gursky 1974). Schwartz (1979) first showed that not only could no known class of object make up the background without evolution in cosmic time, but also that the background could not arise from a new, undetected class of object which did not evolve and which was of a homogeneous density throughout the universe. Since AGN were known to have strong cosmic evolution, they were the obvious prime candidate within the scenarios of a superposition of discrete sources. Setti and Woltjer (1979) had already pointed out that objects of the typical luminosity of 3C273, and which increased in number at larger redshifts as quasars were known to do, could produce the diffuse background intensity.

With the first results of the Einstein satellite, at least three independent lines of evidence pointed to a significant AGN contribution. It must remembered that the Einstein observatory, like other imaging telescopes, operated below the 3 – 10 keV energy range in which we observe the cosmic diffuse background. Thus it is necessary to convert fluxes and extrapolate source properties from the soft to hard X-ray bands, in order to use the tremendous wealth of data made available. Typically this comparison was made in terms of the flux densities at 2 keV.

One line of evidence was due to the fact that an estimated 25% of the background was resolved out into discrete sources at the flux level 2×10^{-14} ergs cm^{-2} s^{-1} of the deep survey (Giacconi et al. 1979). About half of these were identified with quasars. Since it was unreasonable that there would be no more sources just below the deep survey sensitivity, the 25% was interpreted as a lower limit. If the extrapolation to lower flux levels were according to the 3/2 law, $N(> S) \propto S^{-3/2}$, then at an additional factor of 16 lower flux the remainder of the background would be supplied.

Another estimate was made based on Einstein observations of known quasar targets (Tananbaum et al. 1979). By estimating the mean X-ray to optical flux of these, and applying this ratio to the known counts of quasars vs. optical magnitude, estimates were made for producing 42 to 150% of the background (Avni et al. 1980; Marshall et al. 1983).

The third was based on discovery of X-ray selected serendipitous sources appearing outside the target of the IPC observations. This led to substantial numbers of identifications of previously unknown AGNs (Maccacaro et al. 1983; Gioia et al. 1984). This AGN sample directly showed, for the first time, that the X-ray luminosity function did evolve in cosmic time in the appropriate sense of being larger at higher redshifts (Maccacaro et al. 1984). The latter reference made a formal estimate that 47 to >100% of the background could be produced.

3 WHY AGN WERE DISCOUNTED AS THE ORIGIN OF THE X-RAY BACKGROUND.

Continued study of AGN as the source of the diffuse X-ray background revealed some significant difficulties. These included the facts that the AGN X-ray spectra seemed inconsistent with that of the background, that measured fluctuations were less than predictions from an extrapolation of the AGN Log N vs. Log S curve, and that more detailed studies only seemed to predict about half of the measured background intensity.

The most precise measurement of the spectral shape of the X-ray background is from the HEAO-1 A-2 experiment. Marshall et al. (1980) showed that the background accurately fit an optically thin thermal bremsstrahlung model with a temperature kT=40 keV, over the 3 to 60 keV range. Because of this measurement, many authors tried to construct detailed thermal models to explain the background (cf. Guilbert and Fabian 1986). In addition to the problems with total energy, mentioned above, the apparent fit to a single isothermal model by Marshall et al. must be regarded as highly suspicious, since we must be observing the integration over redshift of a plasma which would have a range of temperatures.

The question of whether a bremsstrahlung mechanism could make sense physically has recently been considered more quantitatively by Boldt (1987), and Giacconi and Zamorani (1987). The general line of argument was to start with the measured thermal bremsstrahlung shape, subtract a spectrum due to discrete sources with a particular measured or assumed slope, and attempt to interpret the residual

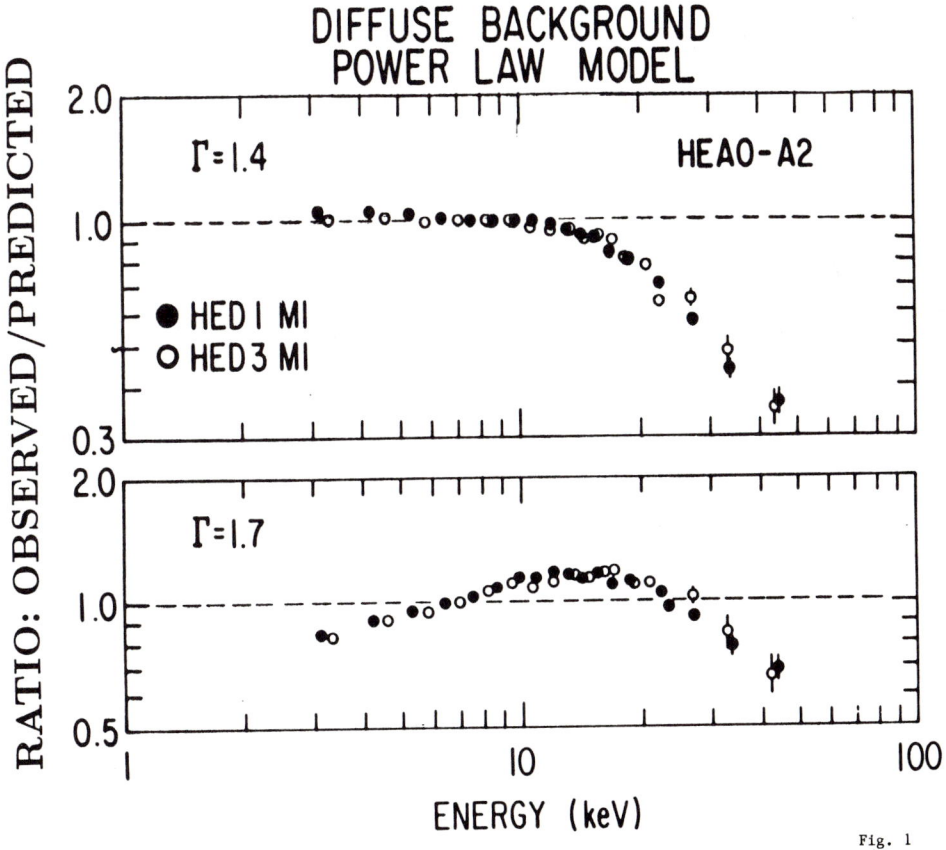

Fig. 1

spectrum. The approach of Boldt (1987) was to fit the residual spectrum to the form $I(E) = KE^{-\alpha}exp(-E/W)$, where W would be the bremsstrahlung temperature, and $E^{-\alpha}$ would represent the Gaunt factor. He showed that even with the minimum 20% contribution of foreground AGN and clusters of galaxies, that the fit would require the unphysically small value of $\alpha < .2$ (cf. Dermer 1986).

Let us therefore consider the alternate power law shape fit to the diffuse background in the 3 to 12 kev range. Figure 1, from Marshall et al. (1980) shows an accurate fit of the background, in this limited energy range, to a power law spectral shape with energy index just steeper than 0.4. The same figure shows the impossibility of representing the diffuse background using a 0.7 power law index. The importance of that number is that Mushotzky (1984) found it to be the ubiquitous index measured for AGN by the same HEAO-1 A-2 instrument.

This inconsistency gives rise to what has been called the "spectral paradox" (Boldt 1987). The paradox is not merely that the spectra are inconsistent, but rather rests on the fact that many lines of evidence argue that AGN do produce a substantial, even if not the dominant, portion of the background. For example, reanalysis of the Einstein deep surveys by Primini et al. (1991) show a minimum of 20% of the diffuse background is resolved into sources at a flux 2×10^{-14} ergs s^{-1} cm^{-2}. Griffiths et al. (1991) estimated $50\pm20\%$ of the background would arise from the quasars identified in the ROSAT surveys. At this meeting, Giacconi and Burg (1991) reported that 35 to 45 % of the ROSAT diffuse background was resolved into discrete sources at the level of 5×10^{-15} ergs s^{-1} cm^{-2}. Analysis of the Extended Medium Sensitivity Survey (EMSS) using the Einstein serendipitous sources, predicts $40\pm4\%$ of the background to arise from the X-ray selected AGN (Maccacaro et al. 1991), using an ad hoc numerical extrapolation for the evolution.

Constraints due to the fluctuations of the 1 to 3 keV background measured by Einstein, have been used to set strict upper limits of <50% to the fraction of background contributed by point sources of currently known populations (Hamilton and Helfand 1987; Barcons and Fabian 1990). From the limit to angular correlations of these fluctuations, QSOs were restricted to a limit of 30% of the background (Barcons and Fabian 1988). The strength of these conclusions rests on measuring smaller fluctuations than predicted. Since rms fluctuations are a positive definite quantity, allowance for instrumental effects would increase the discrepancy between predicted and measured fluctuations.

Some of the most recent direct estimates of the fraction of the diffuse background contributed by AGN are settling around 30 to 50%, with an internal accuracy which seems to preclude contributing the bulk of the background. Schmidt and Green (1986) considered the optical quasar counts, derived a model of how these counts would evolve with redshift, and used measured X-ray to optical flux ratios to limit quasars to 13% and quasars plus Seyferts to 42% of the background at 2 keV. Maccacaro et al. (1991) used the 427 AGN found in the Einstein EMSS to derive a luminosity function in the range 10^{42} to 10^{46} ergs s^{-1}. They furthermore used the functional form $(1+z)^C$ to extrapolate their function to different redshifts. For the best fit to C they derived $40\pm4\%$ of the cosmic background.

4 WHY AGN ARE THE DOMINANT CONTRIBUTOR TO THE X-RAY BACKGROUND.

4.1 Possible Resolution of the Spectral Paradox.

There are at least 3 categories of approach to resolving the spectral paradox: invoking evolution of the spectral form with cosmic redshift; recognition that the average of power law spectra of different indices is not equal to the power law spectrum of the average index; and the possibility that the AGN spectra are not really exact power laws over 3 or more decades, but merely that a power law gives a good approximate fit over the given energy range covered by any given experiment.

A series of papers by Boldt and Leiter (1984; 1987; Leiter and Boldt 1982) constructed a model in which the prototype AGN systems emitted Eddington limited Comptonized thermal emission from a disk during their early life, and that this integrated to produce the diffuse X-ray background. After the black hole spins up, the AGN changes state to give a non-thermal $e^{\pm}$ plasma emitting X-rays with a mean 0.65 power law index. More recently it was recognized (Boldt 1987; 1989) that the thermal bremsstrahlung state is constrained by the general argument on the viability of this mechanism after foreground sources are subtracted.

Another class of spectral evolution models simply considered a continuous ad hoc change in the spectral index according to a law such as $\alpha = 0.7exp(-B\tau(z))$ (Morisawa and Takahara 1989). Those authors showed how such spectra could be integrated, using the form $(1 + z)^C$ for the evolution of AGN luminosity, to reproduce the diffuse spectrum. Canizares and White (1989) used Einstein data to investigate the mean spectrum of various class of AGN, vs. redshift. Their results were not sensitive enough to test for such a small change in index as was expected, however they did establish that quasars at redshifts > 1 did not have a power law index as flat as 0.65. Wilkes and Elvis (1987) and Turner and Pounds (1989) have interpreted their measurements of steeper spectra in terms of a soft X-ray excess.

The fact that the Einstein spectral indices for both quasars (Wilkes and Elvis 1987) and Seyfert galaxies (Kruper et al 1990) is systematically steeper than the HEAO-1 A-2 result for the 2 to 20 keV range, is relevant to the other two categories of resolution of the spectral paradox. Surprisingly, the fact that $< E^{-\alpha} > \neq E^{<-\alpha>}$, where the brackets represent an average over a population of power law sources with different indices, has not been discussed in the literature in this context. Obviously, at the very lowest energies the average spectrum will be dominated by the few sources of steepest spectral index, while at the highest energies the few sources with flattest spectral index will dominate. These latter sources, when redshifted, could produce the 0.4 index observed for the background.

Brecher and Morrison (1969) made such a calculation, in a related context. They considered production of the diffuse background by relativistic electrons leaking out of radio galaxies. For a mean spectral index m of the radio flux density, with a Gaussian dispersion μ, they showed that the resulting x-ray spectral flux density has the Log-Gaussian form:

$$I(E) \propto E^{-m} exp\{\frac{1}{2}[\mu ln(E/E_B)]^2\}.$$

A Log-Gaussian, in the form of $f(E) = KE^{-a+b\log E}$, was an example invoked ad hoc by Schwartz and Tucker (1988; denoted ST88) to support their suggestion that AGN X-ray spectra are not described by a single, exact power law over many decades of energy. ST88 noted that possible mechanisms were the dispersion of spectral indices as just discussed, Compton reflection, or the superposition of power law and thermal bremsstrahlung components. We were motivated by the fact, referenced above, that the Einstein spectra of quasars and Seyfert galaxies were considerably steeper that the A-2 spectra. Our model of a continuously changing spectral index had only the two free parameters, a and b, whereas models in which AGN spectra were multiple exact power laws must have at least three free parameters. These would be equivalent to two spectral indices and an energy at which the components are of equal flux density. (Both models have an additional parameter for the normalization.) The Log-Gaussian expression led us to suggest that the flatter spectrum extrapolated for the range above 20 keV would be consistent with that measured for the diffuse background, after being redshifted. This provided our solution to the spectral paradox.

In his Ph.D. thesis, Qian (1989) fit the spectra of 33 bright quasars observed by Einstein to alternate forms with gradual spectral change (Schwartz, Qian and Tucker 1989). This work showed that when the foreground galactic column density was fixed at the value measured by Elvis et al. (1989), then the probability that an exact power law fit the data better than the Log-Gaussian, or equivalently the probability that b=0 was allowed, was less than 0.0001. When Wilkes and Elvis (1987) had previously fit the same data sets, they found that negative intrinsic column densities were required, which they interpreted as a requirement for an additional soft component. It is still possible that an extraneous soft X-ray component leads to the illusion of a spectrum which steepens continuously at lower energy, but the failure of Wilkes et al. (1989) to correlate their excess with any physical process of the AGN makes such a suggestion speculative.

In section 5.2 we will show how this spectral form can be used to accurately reproduce the diffuse background spectral shape.

4.2 The Signature of AGN X-ray Spectra.

Probably all the above mechanisms are relevant to resolving the spectral paradox; however, from recent Ginga observations of individual AGN we know that the suggestion of ST88 is essentially correct.

Observations with the GINGA satellite showed that almost every AGN has an iron emission feature, with equivalent width of 50 to 300 eV, consistent with resonant K-shell lines at the energy 6.4 keV of cold iron. Many individual AGN show further evidence of absorption troughs and spectral flattening at energies above the iron line, but generally with only marginal statistical precision. This iron line is interpreted as fluorescent emission from cold gas, (i.e., $T < 10^6$ K), irradiated by the primary non-thermal spectrum.

Pounds et al. (1990) studied the composite spectrum from 12 Ginga observations of 8 different AGN. The inferred incident spectrum when best fit to a power law gives an index 0.7, but leads to systematic excess residuals around 6.4 keV and above $\sim$15 keV and a systematic deficit around 8 to 12 keV. To fully account for these features many authors have discussed the "Compton reflection paradigm." In this scenario there is a single, primary, non-thermal power law spectrum. This undergoes Compton reflection, i.e., fluorescence and scattering, from cold gas (cf. Bai 1980; Titarchuck 1987; Lightman and White 1988; Guilbert and Rees 1988; George et al. 1989). The observer sees a superposition of the primary and reflected spectra, with variable relative weighting depending on the viewing angle, absorption within the AGN, possible anisotropic effects, and possible geometric configurations of an accretion disk.

The albedo due to the reflected flux density is calculated as a fraction of the primary non-thermal spectrum. At low energies, below 10 keV, the incident spectrum is greatly attenuated due to photoelectric absorption. At energies $>$ 100 keV the albedo decreases due to the decrease of the Klein-Nishina cross section compared to the Thompson cross section. This allows the incident X-rays to penetrate deeper, and have lower probability of escape. One of the most appealing features of this scenario is that since the rollover mechanism is regulated only by atomic physics, it is reasonable to expect it to operate in all AGN, at all redshifts.

When Pounds et al. (1990) fit their composite spectrum to a model consisting of the non-thermal primary plus Compton reflection, with both components attenuated by a warm absorber, they essentially eliminated all residuals. This fit predicts a continual flattening up to at least 40 keV, and therefore confirms the two predictions of ST88, that AGN X-ray spectra are not exact power laws, and that flattening of AGN spectra toward higher X-ray energies allows the redshifted AGN spectrum to mimic the 3 to 12 keV diffuse background spectrum.

Various other authors have calculated specific scenarios in which the observed AGN spectra, with the extrapolated Compton reflection component, reproduce the diffuse X-ray background spectral shape (George and Fabian 1991; Fabian et al. 1990; Morisawa et al. 1990; Rogers and Field 1991b). Note that independent of the theoretical modeling of the origin of the X-ray spectrum, the Ginga observations reveal a measured hardening of the spectrum which is sufficient to reproduce the diffuse X-ray background shape.

5 EVOLUTIONARY LUMINOSITY FUNCTION

5.1 Derivation of an Evolutionary Luminosity Function.
An accurate luminosity function would allow prediction of the AGN contribution to the X-ray background, and to source counts vs. flux. Avni (1987, private communication), coined the phrase "evolutionary luminosity function," or ELF to describe the fact that the volume density of AGN emitting at luminosity L at any redshift z, would be in general a specific function of the independent variables L and z. In principal this function would be determined by detailed observations of large numbers of AGN at each small range of z, covering the complete range of L. The Einstein EMSS has provided the largest published sample (Maccacaro et al. 1991), but still falls short of directly determining the functional form independent of a model of evolution.

Other constraints can be applied to restrict the ELF. Cavaliere, Morrison, and Wood (1971; denoted CMW) noted that the ELF must satisfy a continuity equation. They expressed this as

$$\frac{\partial N}{\partial t} + \frac{\partial}{\partial L}[\frac{dL}{dt}N(L,t(z))] = S(L,t(z)).$$

where S is the source function (number of AGN created or destroyed at the luminosity L), while t=t(z) is the cosmic time and is related to the redshift z by a specific Friedmann model. CMW, and subsequent works by Cavaliere et al. (1983, 1985) assumed that the quasars undergo a phase of increasing luminosity up to some maximum redshift z_{max}, after which S can be set to zero. They made the heuristic argument that the time rate of change of luminosity of each AGN can only depend on its luminosity, since it is hard to imagine how the central core around a massive black hole could be aware of absolute cosmic time. This led to a parameterization of the rate of change of luminosity as

$$dL/dt = -AL^p$$

where A is a constant and p>1. If the initial luminosity function at z_{max} is a power law in luminosity, then after a time t the luminosity function will break sharply above a luminosity with an initial lifetime t.

Tucker and Schwartz (1986; denoted TS86), following a suggestion of Cavaliere et al 1983, added an important degree of realism to the formalism by allowing A to be a random variable instead of an exact constant. TS86 took a probability distribution for A of the form $\rho(A) = KA^x$ for $A < A_{max}$. This causes steepening by a discontinuity of $\Delta\gamma = (p-1)/(x+1)$ at approximately the luminosity $L_{break}(z) = L_{break}(0)[g(z)]^{-1/(p-1)}$, where

$$g(z) = [(1+z)^{-(1+q_0)} - (1+z_{max})^{-(1+q_0)}]/(1+q_0).$$

The parameters K, x, and A_{max} are eliminated in terms of observables by requiring the probability density function to be integrated to unity probability, and by the observed values of $\Delta\gamma$ and $L_{break}(0)$.

Estimates of the current epoch hard X-ray luminosity function for AGN had been based on the Piccinotti et al. (1981) sample from the HEAO-1 satellite. Elvis, Soltan, and Keel (1984) extended this to lower luminosities based on an extrapolation of the correlation of $H\alpha$ emission to X-ray luminosity. TS86 took the resulting representation of the X-ray luminosity function at the current epoch to fix all parameters except the luminosity derivative index p. They showed that for a few different luminosity functions which bracketed the data, and independently of whether q_0 was 0 or 1/2, that a value of p could be chosen in the range 1.25 to 2 which would cause AGN to produce the entire 2 to 10 keV integrated flux. In that work we specifically allowed uncertainty by a factor of 2 in the normalizations.

5.2 Calculation of the X-ray Background Spectrum

If the differential luminosity function is N(L,z) AGN per unit L and per unit z, then the observed diffuse background intensity I at an observed energy E_0 is

$$I(E_0) = \frac{c}{4\pi H_0} \int_0^{z_{max}} \int_{L_1}^{L_{max}} \frac{f(E_0(1+z))LN(L,z)}{(1+z)^2(1+2q_0z)^{1/2}} dLdz$$

where $f(E)$ is the normalized mean spectral shape function at emitted energy $E = E_0(1+z)$. As a representation of the normalized spectral shapes we use the Log-Gaussian form:

$$f(E) \propto E^{-a+b\log E}$$

with a and b free parameters. Qian (1989) and Schwartz, Qian and Tucker (1989) have shown that the set of 33 bright quasars studied by Wilkes and Elvis (1987)

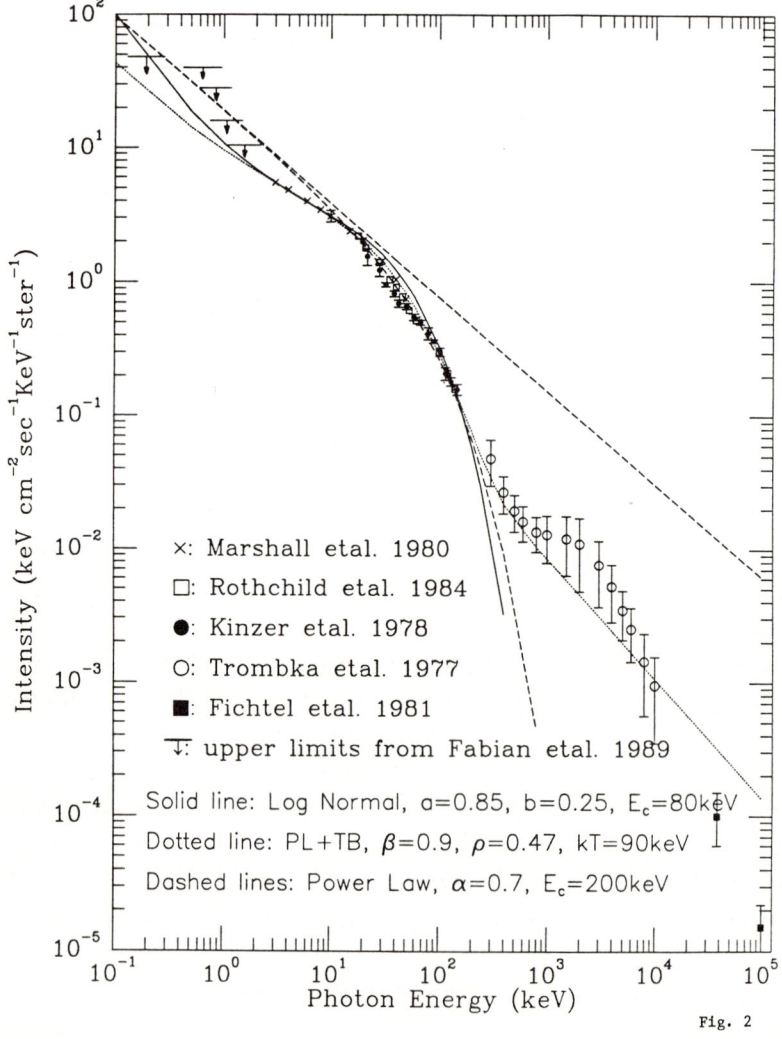

×: Marshall etal. 1980

□: Rothchild etal. 1984

●: Kinzer etal. 1978

O: Trombka etal. 1977

■: Fichtel etal. 1981

⊤: upper limits from Fabian etal. 1989

Solid line: Log Normal, a=0.85, b=0.25, E_c=80keV

Dotted line: PL+TB, β=0.9, ρ=0.47, kT=90keV

Dashed lines: Power Law, α=0.7, E_c=200keV

Intensity (keV cm^{-2}sec^{-1}keV^{-1}ster^{-1})

Photon Energy (keV)

Fig. 2

are better fit to this form, with a = 0.85, and b = 0.25, than to an exact power law with b = 0. Alternately, a power law plus Thermal Bremsstrahlung (PL+TB):

$$f(E) \propto E^{-\beta} + \rho^2 g(E, kT) exp\{-E/kT\},$$

with β and ρ as free parameters and kT arbitrarily fixed at 125 keV, fits the Einstein quasar spectra with $\beta = 1.1$ and $\rho^2 = 0.4$. We also use the constraint that $< f(E) >$ gives a slope of 0.65 ± 0.07, when fit to a power law in the $2 - 10$ keV range. Either of these cases may be considered a numerical approximation to a mean AGN spectrum consisting of a power law plus Compton reflection component.

Figure 2 shows than either of these two functions can produce a highly precise fit to the 3 - 20 keV X-ray background spectrum, as measured by Marshall et al. (1980). We regard this agreement of the integration of the extrapolated AGN X-ray spectrum with the measured diffuse X-ray background as crucial evidence that AGN are the source of the bulk of the X-ray background.

5.3 Calculation of the Log N vs. Log S Curve.

From the ELF, we may calculate the differential number of sources expected in any flux interval S to $S + dS$ as

$$dN = \int_{z=0}^{z max} V(z) N(4\pi D^2(z)S, z) dL dz,$$

with $dL = \frac{L}{S} dS$.

Figure 3 shows the predicted Log N vs. Log S curves. In figure 4 we show the predicted slope of this curve as a function of flux. In the flux range 10^{-13} to 5×10^{-12} the curve does approximate a power law, with a slope of -1.4. Calculations of the expected fluctuations based on a power law assumption, or of the extrapolated Log N vs. Log S numbers assuming a power law, may therefore be valid in this range. However, we note that the assumption of a power law is simply not valid in any lower flux range. In particular, at 3×10^{-14}, the slope is predicted to become flatter than 1.0. Therefore, the quantity S*dN/dS can be integrated to arbitrarily small S without diverging.

5.4 Prediction of an Fe Line in the Diffuse X-ray Background.

The ubiquity of Fe line emission from AGN may be a necessary corollary to the fact that AGN are the dominant contributor to the diffuse X-ray background. Schwartz (1990; 1992) has discussed observation of this line in the diffuse background, considering that it is smeared out over a wide range of energies due to the different redshifts of the emitting galaxies. Observations of the Fe line emission in AGN

AGN: LOG N vs. LOG S

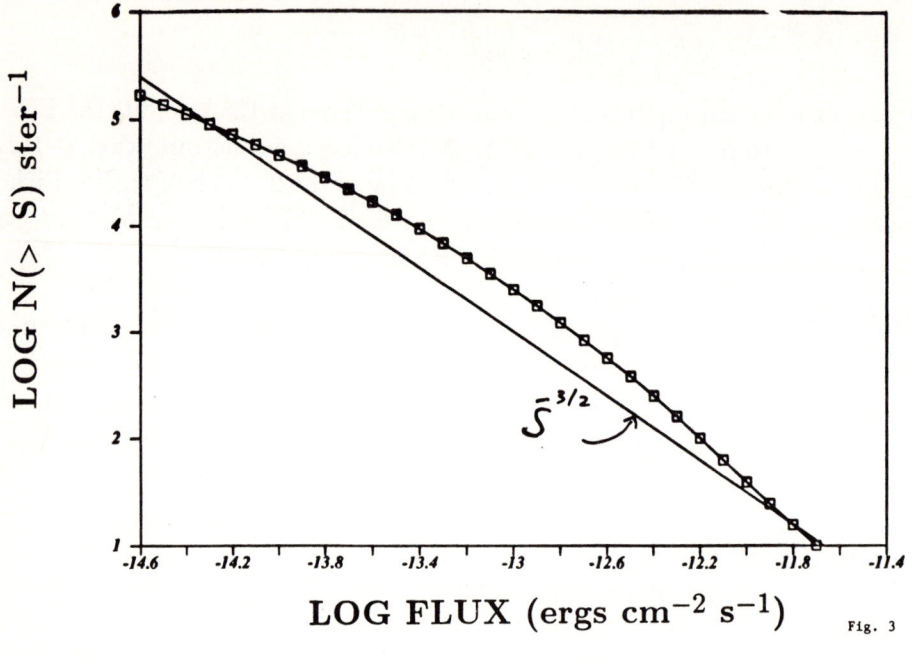

Fig. 3

d N(>S) /dS VS. S

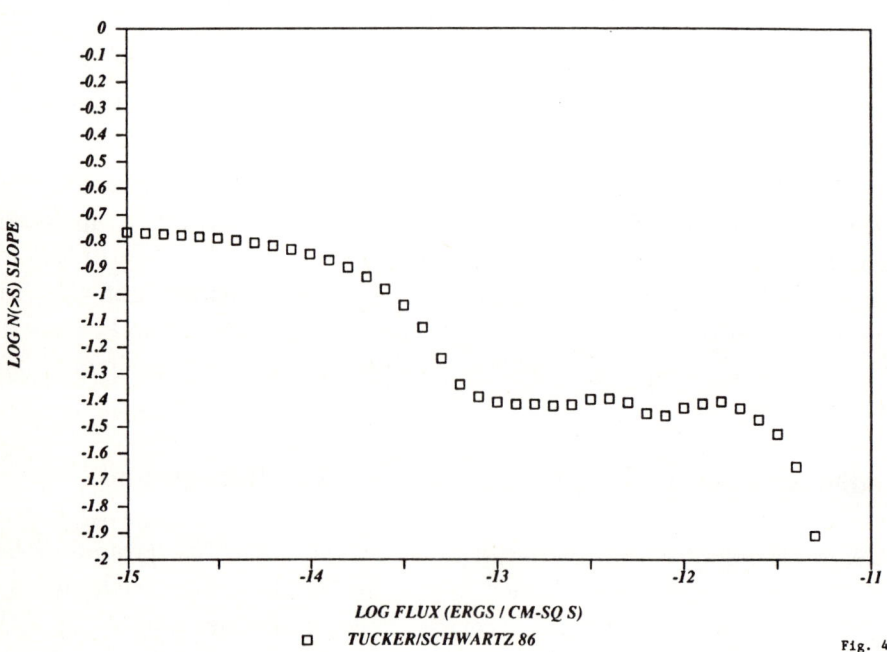

□ *TUCKER/SCHWARTZ 86*

Fig. 4

Fabian, and Ross 1990). George and Fabian (1991) have presented the following simple intuitive picture justifying 150 eV as an appropriate typical equivalent width: the equivalent absorption width by the inferred density of 10^{24} H atoms cm^{-2} is about 1 keV, the fluorescent yield of iron is about 30%, and about half the Fe line photons will be emitted in the hemisphere toward the observer.

Figure 5 shows the predicted, fractional residual continuum due to Fe line emission from the luminosity function model a of TS86. This model is distinguished by having current epoch power law slopes of 3.25 at luminosities above 3×10^{43} ergs s^{-1}. We used a deceleration parameter $q_0 = 0.5$, and a creation redshift $z_{max} = 3$. This ELF model was $dL/dt \propto L^{1.5}$. Such a model is essentially a "worst case" for detecting the line, as models with stronger evolution will show much more concentrated emission at $6.4/(1 + z_{max})$, while models with weaker evolution will show a sharper discontinuity at the precise rest frame energy.

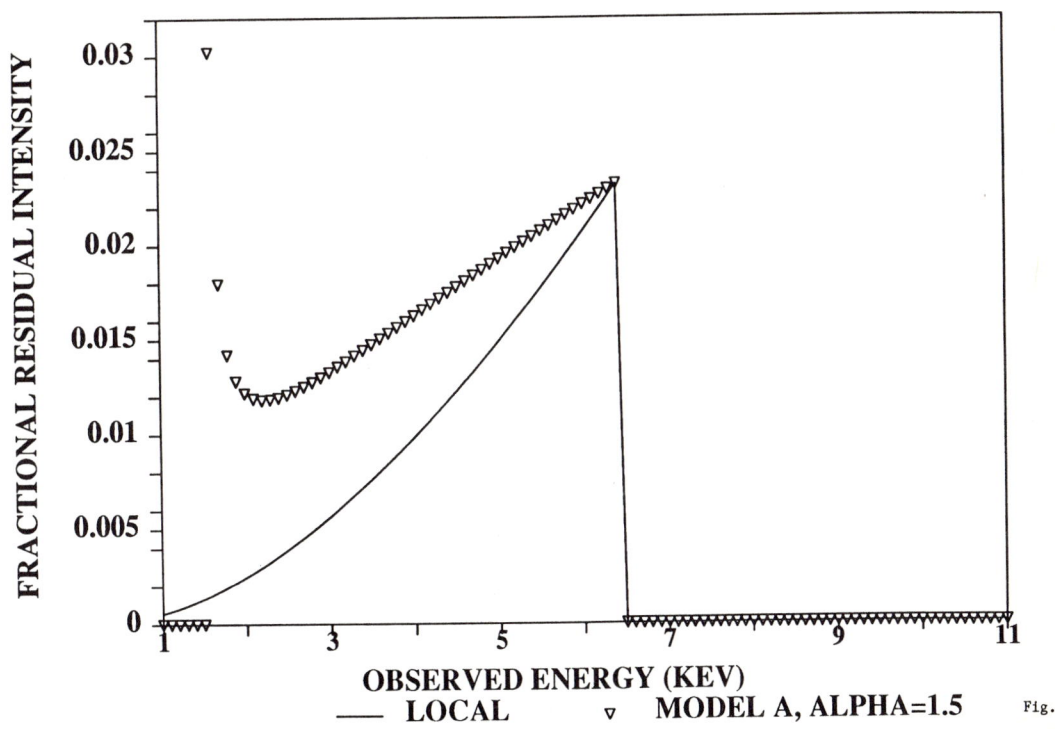

Fig. 5

5.5 X-ray Background Fluctuations

Hamilton and Helfand (1987) worked directly with the measured IPC pixel distributions, predicting the width of the distribution for various slopes of the assumed power law form for N(>S). This technique had the advantage of dealing directly with the instrumental effects via Monte Carlo modeling, but the disadvantage that the results are not expressed in physical units to which other models may be confronted. Because the measured fluctuations were less than predicted, they concluded that AGN could not produce more than 45% of the diffuse background.

We can now see from the Log N vs. Log S function reviewed here that their conclusion is an artifact of the unphysical assumption of a power law form for N(>S). In particular, since they only considered $N(> S) \propto S^{-x}$ with $x > 1$, their integral of S dN/dS diverged at low S before N(>S) became greater than their lower limit of 5000 sources deg^{-2}. However, the relation shown in Figure 3 can be integrated below 10^{-17} ergs cm^{-2} s^{-1}, giving 97% of the background intensity at a source density of 6000 deg^{-2}. Although these numbers are quite sensitive to the true AGN ELF, we have clearly removed one of the important previous obstacles for the AGN origin of the bulk of the diffuse background. However, more careful application of our ELF to the fluctuation formalism, (e.g., Dautcourt 1977; Barcons and Fabian 1988) should be studied.

Another problem with the AGN fluctuations is the lack of measured correlations on small angular scales (cf. Barcons and Fabian 1989). These have been predicted by Carrerra et al. (1991), based on optical quasar-quasar correlations. X-ray upper limits led them to conclude quasars did not supply more than 50% of the X-ray background. We suggest that our ELF, rather than pure luminosity evolution, be used in the formalism to test the robustness of the conclusions.

Fluctuation measurements also provide the normalization information to indicate that the absolute emissivity of the AGN is sufficient to produce the background. The fluctuation results of the Uhuru, Ariel V, and HEAO-1 A-2 experiments are consistent with the A-2 AGN discrete source counts normalization. As shown by TS86, this normalization can produce the entire diffuse X-ray background. However, conversion of the Einstein 0.3 to 3.5 keV flux to the 2 to 10 keV flux leads to a discrepancy in the sense of the EMSS being a factor of 2 to 3 deficient in source counts. Such normalization led Maccacaro et al. (1991) to conclude that AGN produce no more than 44% of the diffuse background. Recent fluctuation measurements by Ginga in the hard X-ray band (Warwick and Stewart 1989; Hayashida 1989; Warwick and Butcher 1991) confirm the normalization of Schaefer. The suggestion is that intrinsic absorption of AGN X-rays systematically decrease the flux measured in the IPC band. The nature of this absorption must be like a partial covering fraction, which attenuates the flux observed by the soft X-ray sky surveys, since it is improbable that 1/2 to 2/3 of AGN are totally missed by the EMSS.

6 CONCLUSIONS

The present review has established the following facts:

1. AGN spectra are observed to flatten with energy, above $\sim$10 keV.

2. This flattening can be understood together with the Fe emission line via the Compton reflection mechanism.

3. Such flattening spectra, when weighted by a physically motivated evolutionary luminosity function, can accurately reproduce the shape of the cosmic X-ray background.

4. Normalizing the AGN luminosity function to that of hard X-ray emitting Seyfert galaxies or, equivalently, to the source count normalization determined by hard X-ray fluctuation measurements, can produce the entire 2 to 10 keV diffuse integrated intensity.

5. The ELF used to derive conclusions 3 and 4 predicts background fluctuations consistent with those measured in the Einstein IPC band.

Based on these facts we conclude that AGN are the dominant contributor to the 2 – 200 keV cosmic diffuse background. To confirm or deny this conclusion we suggest the following observations:

Fit existing (HEAO-1, Einstein) and future (ROSAT, AXAF) surveys of X-ray discovered AGN to find the true form and normalization of the ELF $N(L_x, z)$. Then directly integrate this observed function to compare to the intensity and fluctuations.

Perform more extensive spectral measurements of AGN, covering at least the 0.2 to 10 and the 2 to 100 keV bands, to look for dependence on L or z, and to establish conversion factors between different flux bands.

Search for the smeared out Fe line in the diffuse background spectrum.

Observe correlations in the diffuse X-ray background fluctuations due to the quasar-

quasar correlation function.

We specifically predict that the ROSAT Log N(>S) vs. Log S slope will become flatter than -1 at a few times 10^{-1} ergs cm^{-2} s^{-1}, and that neither ROSAT nor AXAF will entirely resolve out the background, with about 20% remaining unresolved in the 1 to 3 keV band.

ACKNOWLEDGMENTS

I have benefited from collaboration with W. Tucker and Y. Qian in many aspects of this work, as referenced above. I thank X. Barcons for his organization and hospitality, and for the stringent deadlines without which this paper would not have been completed. This work has been sponsored, in part, by NASA contract NAS8-36123.

REFERENCES

Avni, Y., Soltan, A., Tananbaum, H., and Zamorani, G. 1980, Ap.J., 238, 800.
Bai, T. 1980, Ap. J., 239, 328.
Barcons, X. and Fabian, A. C. 1988, MNRAS, 230, 189.
Barcons, X. and Fabian, A. C. 1989, MNRAS, 237, 199.
Barcons, X. and Fabian, A. C. 1990, MNRAS, 243, 366.
Bleeker, J.A.M., et al. 1968, Can.J. of Phys., 46, S461.
Boldt, E. and Leiter, D. 1984, Ap.J., 276, 427.
Boldt, E. and Leiter, D. 1987, Ap. J. (Letters), 322, L1.
Boldt, E.A. 1978, NASA TM78106.
Boldt, E.A. 1987, Physics Reports, 146, 215.
Boldt, E.A. 1989 in *X-ray Astronomy: Pros 23rd ESLAB Symposium*, eds., Hunt and Battrick (ESA SP-296), 797.
Brecher, K. and Morrison, P. 1969, Phys. Rev. Letters, 23, 802.
Canizares, C. R. and White, J. L. 1989, Ap. J. 339, 27.
Carrera, F. J., et al. 1991, MNRAS, 249, 698.
Cavaliere, A., Morrison, P., and Wood, K. 1971, Ap. J., 170, 223.
Cavaliere, A., Giallongo, E., Messina, A., and Vagnetti, F. 1983, Ap.J., 269, 57.
Cavaliere, A., Giallongo, E., and Vagnetti, F. 1985, Ap.J., 296, 402.
Cowsik, R. and Kobetich, E.J. 1972, Ap.J., 177, 585.
Dautcourt, G. 1977, Astron. Nach., 298, 141.
Davison, P.J.N., and Thomas, R.M. 1971, Nature (Phys.Sci), 233, 27.
Dermer, C. D. 1986, Ap. J., 307, 47.
Elvis, M., Lockman, F. J., and Wilkes, B. J. 1989, A. J., 97, 777.
Elvis, M., Soltan, A., and Keel, W. C. 1984, Ap. J., 283, 479.
Fabian, A. C. 1975, MNRAS, 172, 149.

Fabian, A.C., George, I.M., Miyoshi, S. and Rees, M.J. 1990, MNRAS, 242, 14P.

Field, G. B., and Perrenod, S. 1977, Ap. J., 215, 717.

Felten, J. E., and Morrison, P. 1966, Ap.J., 146, 686.

George, I. M., and Fabian, A. C. 1991, M.N.R.A.S.,

George, I.M. et al. 1989, in *X-ray Astronomy: Proc. 23rd ESLAB Symposium*, eds. Hunt and Battrick, (ESA SP-296), 945.

George, I.M., Fabian, A.C., and Ross, R.R. 1990 in *Iron Line Diagnostics in X-ray Sources*. ed. A. Treves, Springer-Verlag, Berlin, (in press).

Giacconi, R., and Burg, R. 1991, (this volume).

Giacconi, R., Gursky, H., Paolini, F., and Rossi, B. 1962, Phys.Rev.Letters, 9, 439.

Giacconi, R. et al. 1979, Ap. J. (Letters), 234, L1.

Giacconi, R., and Zamorani, G. 1987, Ap. J., 313, 20.

Gioia, I. M., Maccacaro, T., Schild, R., Stocke, J., Liebert, J., Danziger, J., Kunth, D., and Lub, J. 1984, Ap. J., 283, 495.

Griffiths, R. E. et al. 1991, BAAS, 23, 957.

Guilbert, P.W. and Fabian, A.C. 1986, MNRAS, 220, 439.

Guilbert, P., and Rees, M. J. 1988, MNRAS, 233, 475.

Hamilton, T. , and Helfand, D. 1987, Ap. J., 318, 93.

Hayashida, K. 1989, Ph. D. thesis, University of Tokyo, ISAS RN 466.

Kruper, J., Urry, C.M., and Canizares, C.R. 1990, Ap. J.,(Suppl), 74, 347.

Leiter, D. and Boldt, E. 1982, Ap. J., 260, 1.

Lightman, A. P., and White, T. R. 1988, Ap. J., 335, 57.

Maccacaro, T., Avni, Y., Gioia, I.M., Giommi, P., Griffiths, R. E., Liebert, J., Stocke, J., and Danziger, J. 1983, Ap. J. (Letters), 266, L73.

Maccacaro, T., Gioia, I. M., and Stocke, J. T. 1984, Ap. J., 283, 486.

Maccacaro, T., Della Ceca, R., Gioia, I.M., Morris, S.L., Stocke, J.T., and Wolter, A. 1991, Ap. J., 374, 117.

Marshall, F. E. et al. 1980, Ap. J., 235, 4.

Marshall, H., Avni, Y., Tananbaum, H., and Zamorani, G. 1983, Ap.J., 269, 35.

Mather, J.C. et al. 1990, Ap. J. (Letters), 354, L37.

Matsuoka, M., Oda, M., Ogawara, Y., Hayakawa, S., and Kato, T. 1968, Can.J. of Phys., 46, S466.

Morisawa, K. and Takahara, F. 1989, P.A.S.J., 41, 873.

Morisawa, K. and Matsuoka, M., Takahara, F. and Piro, L. 1990, A. & Ap., 236, 299.

Mushotzky, R. F. 1984, Adv. Space Res., Vol. 3, No. 10-12, 157.

Piccinotti, G., et al. 1982, Ap. J., 253, 485.

Pounds, K. A., Nandra, K. A., Stewart, G. C., George, I. M., and Fabian, A. C. 1990, Nature, 344, 132.

Primini, F.A., Murray, SS., Huchra, J., Schild, R., Burg, R., and Giacconi, R.,

1991, Ap. J. 374, 440.

Qian, Y. 1989, unpublished Ph. D. thesis, "Quasar X-ray Spectra and Their Contribution to the Diffuse X-ray Background," SAO and Nanjing University.

Rogers, R.D. and Field, G. B. 1991a, Ap. J., 336, 22.

Rogers, R.D. and Field, G. B. 1991b, Ap. J.(Letters), 370, L57.

Schmidt, M., and Green, R.F. 1986, Ap.J., 305, 68.

Schwartz, D. A. 1969, unpublished Ph.D. thesis, "The spatial distribution of the diffuse component of Cosmic X-rays," University of California at San Diego.

Schwartz, D. A. 1970, Ap. J., 162, 439.

Schwartz, D. A. 1979, in X-ray Astronomy, L. E. Peterson and W. A. Baity, eds., (Oxford: Pergamon Press), p. 453.

Schwartz, D. A. 1990, BAAS, 22, 1220.

Schwartz, D. A. 1992, Ap. J., (in press).

Schwartz, D. A., Qian, Y, and Tucker, W. H. 1989, in *X-ray Astronomy: Proc. 23rd ESLAB Symposium*, eds., Hunt and Battrick, (ESA SP-296), 1043.

Schwartz, D. A. 1980, Physica Scripta, 21, 644.

Schwartz, D. A., and Gursky, H. 1974, in *X-ray Astronomy*, eds., H. Gursky and R. Giacconi, (Dordrecht: D. Reidel) 359.

Schwartz, D. A., Murray, S. S., and Gursky, H. 1976, Ap. J., 204, 315.

Schwartz, D. A. and Tucker, W. H. 1988, Ap. J., 332, 157.

Setti, G. and Woltjer, L. 1979, Astr.Ap., 76, L1.

Seward, F., Chodil, G., Mark, H., Swift, C., and Toor, A. 1967, Ap.J., 150, 845.

Shafer, R. A. 1983, Ph. D. thesis, University of Maryland, NASA TM-85029.

Tananbaum, H. et al. 1979, Ap. J. (Letters), 234, L9.

Titarchuk, L. G. 1987, Astrophysics, 26, 57.

Tucker, W. H. and Schwartz, D. A. 1986, Ap. J., 308, 53.

Turner, T.J., and Pounds, K.A. 1989, MNRAS, 240, 833.

Warwick, R. S., and Butcher, J. A. 1991, preprint.

Warwick, R. S., and Pye, J. P. 1978, MNRAS, 183, 169.

Warwick, R. S., and Stewart, G. C. 1989, in *X-ray Astronomy: Proc. 23rd ESLAB Symposium*, eds., Hunt and Battrick (ESA SP-296), 727.

Wilkes, B. J. and Elvis, M. 1987, Ap. J., 323, 243.

Wilkes, B. J., Masnou, J.-L., Elvis, M., McDowell, J. and Arnaud, K. 1989, in *X-ray Astronomy: Proc. 23rd ESLAB Symposium*, eds. Hunt and Battrick, (ESA SP-296), 1081.

Worrall, D. M. 1989, in *X-ray Astronomy: Proc. 23rd ESLAB Symposium*, eds., Hunt and Battrick (ESA SP-296), 719.

Zamorani, G. et al. 1981, Ap. J., 245, 357.

HEAO Study of Hard X-Ray Emission from Starburst Galaxies

D. E. GRUBER[†], Y. REPHAELI[†], AND M. PERSIC[‡]

† CASS, University of California, San Diego, ‡ SISSA and Osservatorio Astronomico, Trieste

1 ABSTRACT

A sample of 53 *IRAS*-selected candidate starburst galaxies has been searched for net x-ray emission in *HEAO* archival data. Between 0.5 keV *(HEAO-2 IPC)* and 160 keV *(HEAO-1 A-4)* net emission is detected at the 99.6% confidence level, after allowing for confusion noise in the *HEAO-1* data. Above 15 keV the confidence level is 97%. The combined spectrum is flat, with a (photon) power-law index of 1.0 ± 0.3. The contribution to the 3–50 keV cosmic X-ray background from this population is then estimated to be at least $\sim 3\%$ assuming no evolution. Moderate evolution increases this fractional contribution to $\sim 19\%$.

2 INTRODUCTION

Probably not all classes of sources contributing to the cosmic X-ray background (CXB) have been identified. In the energy range 3–50 keV the strength and hard spectrum (Marshall *et al.* 1980) of the background, different from any class of known sources, pose a special challenge. Much attention has been focused on the possibility of a major contribution from active galactic nuclei (AGN), whose mean power-law spectrum has an energy index $\alpha_{AGN} \simeq 0.7$. Two different approaches have been proposed for resolution of the spectral disagreement; with evolutionary tuning either could allow a contribution approaching 100% from AGN's. Schwartz and Tucker (1988) have proposed that the typical AGN spectrum could flatten above 30 keV and break sharply at 100–200 keV. Others (Fabian *et al.* 1990; Morisawa *et al.* 1990; Rogers and Field 1991) have pursued the effects of a reflection component in the AGN spectrum. There is growing evidence (*e. g.* Pounds et al. 1990) for such a reflection component in the spectrum of Seyfert galaxies.

Other classes of sources may still contribute substantially to the CXB. Clusters of galaxies, important at lower energies, contribute only a few percent to the CXB at 3–50 keV and probably even less at higher energies (Rephaeli, Gruber & Rothschild 1987; Rephaeli & Gruber 1988). We investigate here the possible contribution of starburst galaxies (SBG). These very likely comprise the majority of luminous *IRAS* far-infrared sources (Soifer *et al.* 1986), whose enhanced output is probably due to warm dust heated by hot, massive stars. Hard X-ray binaries and supernova remnants should also be more abundant during the starburst phase. It has been suggested earlier (Bookbinder *et al.* 1980; De Zotti 1987; Weedman 1987) that SBG emit X-rays more intensely than normal galaxies and therefore may contribute appreciably to the CXB. Weedman estimated a SBG contribution of at least 13% to the CXB flux at 2 keV. Spectral study at higher x-ray energies is motivated by this result. One such study (Persic *et al.* 1989) obtained only upper limits. Here we report the results of the largest sample yet of SBG spectra in the combined *Einstein* IPC and *HEAO-1* A-2 and A-4 energy bands, spanning the range 0.5–160 keV. We use $H_o = 50$ km s^{-1} Mpc^{-1} throughout.

3 THE SAMPLE

Evidence cited by Soifer *et al.* (1986) strongly indicates that the majority of the most luminous far-infrared *IRAS* sources are powered by star formation activity. With a rich population of young massive stars, these objects might well have X-ray emission enhanced above normal galaxies due to supernovae and their remnants. We have therefore selected our SBG sample from the *IRAS* Bright Galaxy Sample of Soifer *et al.* (1987). A luminosity-limited subset with $L_{FIR} > 10^{11} L_\odot$ was chosen to limit the sample to mostly SBG's (Soifer 1988, private communication). This choice defines a subset of 116 galaxies out of the total 324 in the *IRAS* Bright Galaxy Sample. Of this subset only 8 are in *HEAO-1* A-1 catalogue error boxes, *i. e.* these sources are faint X-ray objects. Of the 116, 19 are classified as AGN, whose CXB contribution has been accounted for separately. Therefore, the sample reduces to 97 SBG. Only 6 of these 97 galaxies are in the sample selected by Persic *et al.* (1989) for A-2 analysis, thus the two samples are essentially distinct. A cut for recession velocities lower than 6000 km s^{-1} was then applied, reducing the sample to 51. Finally, we added the two nearby SBG NGC 253 and M82, bringing the total to 53. Individual spectra were accumulated, but we report here only net spectra for the sample.

4 X-RAY DATA ANALYSIS AND RESULTS

Archival *HEAO* data now provides easy access to x-ray fluxes from much of the sky, with two full sky scans available from the A-2 and A-4 experiments of *HEAO-1* (Rothschild *et al.* 1979; Matteson 1978) and more spotty coverage from the *HEAO-2* IPC Slew Survey. The IPC images provide arc minute resolution, while A-2 and A-4 have fields of view of 9 and 28 square degrees, respectively. Thus error estimates for these higher energy results necessarily include a component for sky, or source confusion noise.

In searching *HEAO-2* data from the Imaging Proportional Counter (IPC) we have found six sources listed in IPC fields, only one of which had been identified. For another 41 positions flux estimates were obtained from the Slew Survey. There is

no IPC exposure for only six sources. The *HEAO-1* A-2 data analysis procedure
has been described by Della Ceca *et al.* (1989); objects contaminated by (i.e., < 6°
away from) bright A-2 sources were discarded, leaving 45 objects. The two Low
Energy Detectors of the A-4 experiment were sensitive in the 13–175 keV band, with
geometric area of 103 cm² each, mean efficiency of 0.7, and a field of view 1°.43 by
20° FWHM. A typical source was scanned for several thousand seconds. Background
counting rate was measured with each source transit.

Non-statistical source measurement noise arises from fluctuations of the diffuse background (or equivalently, source confusion), but also from incomplete subtraction of
detector internal background and detector gain variations. For the IPC data we have
not corrected for any such effects. Corrections for A-2 and A-4 non-statistical noise
were measured directly on blank fields. For both experiments this correction was
roughly of the magnitude of the source Poisson measurement error.

We have weighted the IPC, A-2 and A-4 measured fluxes in two different ways in
order to characterize the mean X-ray flux or luminosity. First, we weighted the
individual counting rates by z^2, to correct for the large scatter in distance. Assuming
also photoelectric absorption corresponding to $N_H = 10^{22}$ cm^{-2} (cf. Schaaf *et al.*
1989), our best-fit power law spectrum in the range 0.5–80 keV is:

$$\phi(E) = (9.48 \pm 3.61) \times 10^{-6} \left(\frac{E}{10 \ keV}\right)^{-(0.99\pm0.27)} \quad \mathrm{cm}^{-2} \ \mathrm{s}^{-1} \ \mathrm{keV}^{-1}, \qquad (1)$$

with a reduced χ^2 of 1.02. Detection confidence is P=.996. Alternatively, one might
take account of the intrinsic differences in the X properties of the sampled sources
by making the reasonable assumption that the X luminosity is proportional to L_{FIR}.
The mean sample spectrum with L_{FIR} weighting is

$$\phi(E) = (13.54 \pm 4.21) \times 10^{-6} \left(\frac{E}{10 \ keV}\right)^{-(1.05\pm0.22)} \quad \mathrm{cm}^{-2} \ \mathrm{s}^{-1} \ \mathrm{keV}^{-1}, \qquad (2)$$

with a reduced χ^2 of 0.62 and detection confidence P=.999. In both eqs. (1) and
(2), the quoted uncertainties and confidence levels include statistical and sky-noise
errors. It is possible that either spectrum could break downwards at an energy as
low as 50 keV. These mean SBG spectra are significantly flatter than the mean AGN
spectrum, whose power-law index is 1.7 ± 0.1 (Boldt 1987). We compute a mean
2–30 keV luminosity of our fiducial SBG (located at an *rms* redshift of $< z^2 >^{1/2} =$
0.0133) to be $(3.3^{+1.6}_{-1.1}) \times 10^{42}$ erg s^{-1}. Previous determinations of the X luminosity
of SBG were limited to lower energies. For example, Fabbiano (1988) calculated the
luminosities of the two nearby SBG NGC 253 and M82 to be $L_{1--10} \simeq 2.5 \times 10^{40}$ ergs
s^{-1}, and $L_{1--10} \simeq 6.5 \times 10^{40}$ ergs s^{-1}, respectively, in the *Einstein* MPC 1–10 keV
band. Although lower by roughly two orders of magnitude than the mean luminosity

determined here, the far-infrared luminosity is also much lower also for these two objects. Considering the relative x-ray bandpasses and our observed spectrum, these two sources are not significantly under-luminous if x-radiation is proportional to L_{FIR}.

5 DISCUSSION

The flat mean SBG spectra in eqs. (1) and (2) allow the possibility of a significant contribution to the CXB. The local SBG density corresponding to our sample of 53 and its cutoff at ~ 115 Mpc is $n_o = 2.4 \times 10^{-5}$ Mpc^{-3}, given the *IRAS* survey coverage of 14500 deg^2. Boldt (1987, eq. 5.53) relates the mean surface brightness to the average luminosity as

$$\frac{dI}{dE_o} = \left(\frac{c}{4\pi H_o}\right) n_o \frac{dL}{dE_o} f(\alpha, z_m, \Omega_o, Q), \qquad (3)$$

where f is a dimensionless function which depends on α, on the maximum redshift z_m, on the cosmological density parameter, Ω_o, and on a parameter Q characterizing the evolution of the sources. Integration of our measured spectrum, eq. [1], over the energy band $[E_1, E_2]$ then results in

$$I = \left(\frac{c}{H_o}\right)^3 < z^2 > n_o A f [E_2^{1.01} - E_1^{1.01}] \qquad \text{keV cm}^{-2} \text{ s}^{-1} \text{ sr}^{-1}, \qquad (4)$$

with $A = 9.6 \times 10^{-5}$, and E_1 and E_2 in keV. Assuming Q=0 (no evolution), $\Omega_o = 1$ and $z_m = 4$, which give $f = 0.61$, and using an *rms* redshift of $< z^2 >^{1/2}= 0.0133$, we calculated the integrated SBG intensity over the 3–10 and 3–50 keV energy bands to be $I_{3--10} \simeq 0.37$ and $I_{3--50} \simeq 2.5$ keV cm^{-2} s^{-1} sr^{-1}, respectively. Density and/or luminosity evolution would yield higher predicted SBG intensities: e.g., an evolution at the level suggested by Danese *et al.* (1987) for radio-selected SBG candidates, corresponding to $Q \sim 3$, would raise (see eq.[5.57] of Boldt 1987) the above intensities to $I_{3--10} \simeq 2.4$ and $I_{3--50} \simeq 16.6$ keV cm^{-2} s^{-1} sr^{-1}.

From Figure 3.6 of Boldt (1987), we then estimate the integrated intensity of the CXB to be $I_{3--10} \simeq 27$ and $I_{3--50} \simeq 88$ keV cm^{-2} s^{-1} sr^{-1}. Correspondingly, with no evolution the SBG contribution to the CXB is $\sim 1\%$ in the 3–10 keV band, and $\sim 3\%$ in the 3–50 keV band. Moderate evolution, $Q \sim 3$, would increase these values to $\sim 9\%$ and $\sim 19\%$, respectively. Since we have sampled only the brightest SBG, the total SBG contribution to the CXB may be rather higher than estimated here. Our estimates of the SBG contribution to the CXB are consistent with those of Weedman (1987) and of Griffiths & Padovani (1990) for the 0.5–2 keV band. The present results, especially above 15 keV, should nevertheless be viewed cautiously because this statistical study of hard-X emission from SBG is the first of its kind. Moreover, individual SBG have not yet been detected at these hard-X energies.

This research has been supported by NASA through grant NAG5-1385.

6 REFERENCES

Bookbinder, J., Cowie, L. L., Krolik, J. H., Ostriker, J. P., & Rees, M. 1980, *Ap.J.*, **237**, 647.

Boldt, E. 1987, *Phys. Reports*, **146**, 215.

Danese, L., De Zotti, G., Franceschini, A., & Toffolatti, L. 1987, *Ap.J. (Letters)*, **318**, L15.

Della Ceca, R., Palumbo, G. G. C., Persic, M., Boldt, E. A., De Zotti, G., & Marshall, F. E. 1990, *Ap.J. Suppl.*, **72**, 471.

De Zotti 1987, in *Proc. 7th Italian Conf. General Relativity and Gravitational Physics*, ed. U. Buzzo, R. Cianci, & E. Massa (Singapore, World Scientific), p. 331.

Fabbiano, G. 1988, *Ap.J.*, **330**, 672.

Fabian, A. C., George, I. M., Miyoshi, S., & Rees, M.J. 1990, *M.N.R.A.S.*, **242**, 14p.

Griffiths, R. E., & Padovani, P. 1990, *Ap. J.*, **360**, 483.

Lawrence, A., Walker, D., Rowan-Robinson, M., Leech, K. J., & Penston, M. V.1986, *M.N.R.A.S.*, **219**, 687.

Marshall, E. F., Boldt, E., Holt, S. S., Miller, R., Mushotzky, R. F., Rose, R. E., Rothschild, R. E., & Serlemitsos, P. 1980, *Ap.J.*, **235**, 4.

Matteson, J. L. 1978, *AIAA Conference* paper 78-35.

Morisawa, K., Matsuoka, M., Takahara, F., & Piro, L. 1990, *Astr. Ap.*, **236**, 299.

Persic, M., De Zotti, G., Danese, L., Palumbo, G. G. C., Franceschini, A., Boldt, E. A., & Marshall, F. E., 1989, *Ap.J.*, **344**, 125.

Pounds, K. A., Nandra, K., Stewart, G. C., George, I. M., and Fabian, A. C. 1990, *Nature*, **344**, 132.

Rephaeli, Y., Gruber, D. E., & Rothschild, R. E. 1987, *Ap.J.*, **320**, 139.

Rephaeli, Y., & Gruber, D. E. 1988, *Ap.J.*, **333**, 133.

Rogers, R. D. and Field, G. B. 1991, *Ap. J. (Letters)*, **370**, L57.

Rothschild, R. E. *et al.* 1979, *Space Sci. Instr.*, **4, 269.**

Schwartz, D. A., & Tucker, W. H., 1988, *Ap.J.*, **322**, 157.

Schaaf, R., Pietsch, W., Biermann, P. L., Kronberg, P. P., & Schmutzler, T. 1989, *Ap.J.*, **336**, 722.

Soifer, B. T., Sanders, D. B., Neugebauer, G., Danielson, G. E., Lonsdale, C. J., Madore, B. F., & Peterson, S. E. 1986, *Ap.J. (Letters)*, **303**, L41.

Soifer, B.T. *et al.* 1987, *Ap.J.*, **320**, 238.

Weedman, D. W. 1987, in *Star Formation in Galaxies*, ed. C.J. Lonsdale Persson (NASA Conf. Publ. 2466), p. 351.

Implications of the Optical and Soft X-Ray Source Counts on the X-Ray Spectra of AGNs

JOSÉ MARíA MARTíN-MIRONES

Dpto. de Física Moderna, Univ. de Cantabria, Santander, Cantabria, España
and
Osservatorio Astronomico di Padova, Padova, Veneto, Italia

ABSTRACT

The discrepancy between the observed soft X-ray (0.3 - 3.5 keV) Log $N(> S_X)$ - Log S_X relation for AGNs (in the EMSS) and the predicted one from optical data [optical luminosity functions obtained by Schmidt and Green (1983) (quasars), Danese et al. (1985) (quasars) and Cheng et al. (1985) (Seyfert nuclei)] is studied. We show that the two spectral distributions derived from the different recent spectral surveys built from the data obtained with the *Einstein Observatory* (IPC and SSS), *EXOSAT* and *GINGA* (one based on the fits to uniform absorption models for the soft X-ray absorption and the another based on the fits to partial covering models for that absorption) are compatible with the soft X-ray source counts and the optical data.

1 PROCEDURE OF ANALYSIS AND DATA. AGN SPECTRAL DISTRIBUTIONS

The main objective of this analysis is the study of the discrepancy between the observed soft X-ray (0.3 - 3.5 keV) Log $N(> S_X)$ - Log S_X distribution for AGNs and the predicted one from the data in the optical band, determining if the distributions of AGN spectra derived from the different recent spectral surveys can explain such a discrepancy. The analysis procedure (Martín-Mirones et al. 1991) has 4 steps:

a) Building of the bivariate function $f(M_B, L_X)$, which gives the distribution of AGNs with the soft X-ray (0.3 - 3.5 keV) luminosity L_X for each bin of magnitude M_B.

b) $f(M_B, L_X)$ is applied to the optical luminosity function under consideration in order to obtain the soft X-ray luminosity function.

c) The soft X-ray luminosities are converted into fluxes S_X taking into account the K-correction and that the individual X-ray spectra of AGNs are composed by 4 broad-band components: power-law non-thermal continuum, soft X-ray absorption, soft X-ray excess and high energy bump. We have considered spectral distributions in which Log $f_{2\ \text{kev}}$ ($f_{2\ \text{keV}}$ being the ratio of the soft X-ray component to the hard one absorbed by cold gas), Log N_H^1 and Log N_H^2 (N_H^1 and N_H^2 being the column densities of the soft X-ray absorption and of the absorption generating the high energy bump respectively) are uniformly distributed around a mean value with a determined width

Δ and the soft and hard X-ray spectral indices (α_S and α_H) and the covered fractions (f^1_{cov} and f^2_{cov}) are fixed parameters (α_H is fixed to the canonical value 0.7 in all cases). *d)* Distribution of sources (per square degree) according to the X-ray flux S_X [Log $N(> S_X)$ - Log S_X relation] and comparison with the observed distribution.

In the quasar magnitude range ($-30 \leq M_B \leq -23$), we have used the Schmidt and Green (1983) optical luminosity functions obtained from the *Bright Quasar Survey* (BQS) and other complete samples [where they consider Luminosity-Dependent Density Evolution (LDDE) and two types of universe: $\Omega = 1.0$ and $\Omega = 0.2$] and the Danese et al. (1985) luminosity functions obtained from the data used by the previous authors complemented with the Braccesi faint sample and the Koo and Kron (1982) very deep sample [where they have considered LDDE evolution and Pure Luminosity Evolution (PLE) and two types of universe: $\Omega = 1.0$ and $\Omega = 0.2$]. In the Seyfert nuclei range ($-23 \leq M_B \leq -18$), we have used the Cheng et al. (1985) optical luminosity functions obtained from all confirmed Seyfert 1 and 1.5 galaxies lying in the sky area defined by the first 9 Markarian lists. The main reference sample used in the calculation of $f(M_B, L_X)$ is composed by all the sources of the BQS that have been observed with the IPC out of the declination range $30° < |\delta| < 60°$ and sources of the *Cheng et Al. (1985) Survey* that have been observed with IPC, HRI or *UHURU*.

The observational soft X-ray (0.3 - 3.5 keV) Log $N(> S_X)$ - Log S_X relation for AGNs is built from the 427 identified AGNs (Maccacaro et al. 1991) of the *Extended Medium Sensitivity Survey* (EMSS) obtained with the IPC.

2 RESULTS AND CONCLUSIONS

1) As starting point, we have used the spectral distribution compatible with several results (hardness ratios for different surveys, high flux soft and hard X-ray source counts, X-ray background spectral shape, ...) directly obtained from observational X-ray data (XRD distribution) (Franceschini et al. 1991): $\alpha_H = 0.7$, $\langle$Log $f_{2\,keV}\rangle = -1.5$ ($\Delta = 1.0$), $\alpha_S = 3.5$, $\langle$Log $N^1_H\rangle = 22.1$ ($\Delta = 2.0$), $f^1_{cov} = 1.0$, $\langle$Log $N^2_H\rangle = 24.1$ ($\Delta = 1.5$) and $f^2_{cov} = 0.8$. The result appears in Figure 1 for the Schmidt and Green (1983) models and in Figure 2 for the Danese et al. (1985) models (maximum redshift for the presence of quasars $z^{qua}_{max} = 3.0$). We see that the XRD spectral distribution cannot reproduce the observational soft X-ray Log $N(> S_X)$ - Log S_X relation independently of the optical local luminosity function and evolution.

2) Secondly, we have changed the maximum redshift for the presence of quasars z^{qua}_{max} from 1.0 up to 3.0 (see Fig. 3). If we assume that the quasars are not very far ($z^{qua}_{max} \sim 1.0 - 1.5$), the XRD distribution is compatible with the observational source counts only if an important very low-flux population of sources different to the standard quasars and Seyferts 1 and 1.5 exists [may be Seyferts 1.8, 1.9 and 2, LINERS, low-luminosity quasars ($M_B > -23$), high-redshift Seyferts 1 and 1.5 ($z > 0.1$), ...]. On the contrary, if $z_{max}^{qua} > 1.5$, the predicted source counts remain above the observational ones.

3) Finally, we have focussed on the possibility that the true spectral distribution of the optically selected AGNs be different from the XRD distribution, since the results used to obtain the latter distribution have been predominantly derived from high X-ray flux AGNs, whereas the optically selected AGNs used by us also contain low X-ray flux sources.

a) Firstly, we have investigated the influence of the change of each physical component of AGNs on the source counts determining the bounds imposed on the spectral parameters by the observational relation (fixing the parameters that we do not vary at the values of the XRD distribution):

a1) The soft excesses must be strong: $\langle$Log $f_{2\ keV}\rangle > -0.5$ (with $\Delta = 1.0$ and $\alpha_S = 3.5$) or $\alpha_S > 5.0$ ($\langle$Log $f_{2\ keV}\rangle = -1.5$ with $\Delta = 1.0$). On the other hand, the change of the width Δ of the distribution of Log $f_{2\ keV}$ does not yield significant variations in the predicted source counts (at least, if $\langle$Log $f_{2\ keV}\rangle$ does not take very great values). Obviously, in those cases in which the predicted source counts are less than the observational ones ($\langle$Log $f_{2\ keV}\rangle \gg -0.5$, $\alpha_S \gg 5.0$), we can account for the difference by means of the presence of new populations of X-ray sources (similar to those needed in the case of $z_{max}^{qua} < 1.5$).

a2) The soft X-ray absorption must be small: $\langle$Log $N_H^1\rangle < 21.5$ - 21.7 ($\Delta = 2.0$ and $f_{cov}^1 = 1.0$) or $f_{cov}^1 < 0.7$ - 0.8 ($\langle$Log $N_H^1\rangle = 22.1$, $\Delta = 2.0$). Now, the change of Δ (width of the distribution of Log N_H^1) produces important variations in the source counts. In the concrete, if we consider the values of the XRD distribution for all the other parameters, we can reproduce the observational curve (at the 99% confidence level) only when there is not dispersion of N_H^1, i. e., $\Delta = 0$.

a3) By changing only the high energy absorption (N_H^2 and f_{cov}^2) within physical values, it is impossible to reproduce (at the 99% confidence level) the observational soft X-ray source counts. The change of Δ (width of the distribution of Log N_H^2) does not solve the problem.

b) These results have led us to use two spectral distributions whose parameters are derived from the different recent spectral surveys (Turner and Pounds 1989; Kruper et al. 1990; Reichert et al. 1985; Morisawa et al. 1990) built from the data obtained with the *Einstein Observatory* (IPC and SSS), *EXOSAT* and *GINGA* (Martín-Mirones et al. 1991):

b1) Spectral distribution based on fits to uniform absorption models: $\langle$Log $N_H^1\rangle = 21.25$ ($\Delta = 2.95$), $f_{cov}^1 = 1.0$ and

b2) Spectral distribution based on fits to partial covering absorption models:

$\langle$Log $N_H^1\rangle = 22.81$ ($\Delta = 1.35$), $f_{cov}^1 = 0.84$,

where the remaining values of the parameters are the same in both cases: $\alpha_H = 0.7$, $\alpha_S = 4.22$, $\langle$Log $f_{2\ keV}\rangle = -1.5$ ($\Delta = 1.0$), $\langle$Log $N_H^2\rangle = 24.1$ ($\Delta = 2.26$) and $f_{cov}^2 = 0.53$. In Figure 4, we see that, in both cases, the predicted counts are compatible with the observed ones.

So, the problem of the apparent inconsistencies between the source counts of X-ray selected AGNs and those predicted on the basis of X-ray observations of optical samples can be satisfactorily understood in terms of the K-correction. Setti (1987) tried to solve the discrepancies using this approach, but the ignorance of the true spectral shapes of AGNs [following the results by Elvis et al. (1986), he used an effective power-law of spectral index $\simeq 1.0$] led him only to diminish the X-ray fluxes without reaching the observed values. Now, the spectral surveys recently built from the data obtained with the different X-ray satellites [*Einstein Observatory* (IPC and SSS), *EXOSAT*, *GINGA*] have allowed us to clarify the origin of the inconsistencies.

REFERENCES

Cheng, F.-Z., Danese, L., De Zotti, G., and Franceschini, A. 1985, *M. N. R. A. S.*, **212**, 857.

Danese, L., De Zotti, G., and Franceschini, A., 1985, *Astr. Ap.*, **143**, 277.

Elvis, M., Green, R. F., Bechtold, J., Schmidt, M., Neugebauer, G., Soifer, B. T., Matthews, K., and Fabbiano, G. 1986, *Ap. J.*, **310**, 291.

Franceschini, A., Martín-Mirones, J. M., De Zotti, G. and Danese, L. 1991, in preparation.

Koo, D. C., and Kron, R. G. 1982, *Astr. Ap.*, **105**, 107.

Kruper, J. S., Urry, C. M., and Canizares, C. R. 1990, *Ap. J. Suppl.*, **74**, 347.

Maccacaro, T., Della Ceca, R., Gioia, I. M., Morris, S. L., Stocke, J. T., and Wolter, A. 1991, *Ap. J.*, **374**, 117.

Martín-Mirones, J. M., De Zotti, G., Franceschini, A., and Danese, L. 1991, in preparation.

Morisawa, K., Matsuoka, M., Takahara, F., and Piro, L. 1990, *Astr. Ap.*, **236**, 299.

Reichert, G. A., Mushotzky, R. F., Petre, R., and Holt, S. S. 1985, *Ap. J.*, **296**, 69.

Schmidt, M., and Green, R. F. 1983, *Ap. J.*, **269**, 352.

Setti, G. 1987, in *IAU Symposium 124, Observational Cosmology*, eds. A. Hewitt, G. Burbidge, and L. Z. Fang (Dordrecht: Reidel), p. 579.

Turner, T. J., and Pounds, K. A. 1989, *M. N. R. A. S.*, **240**, 833.

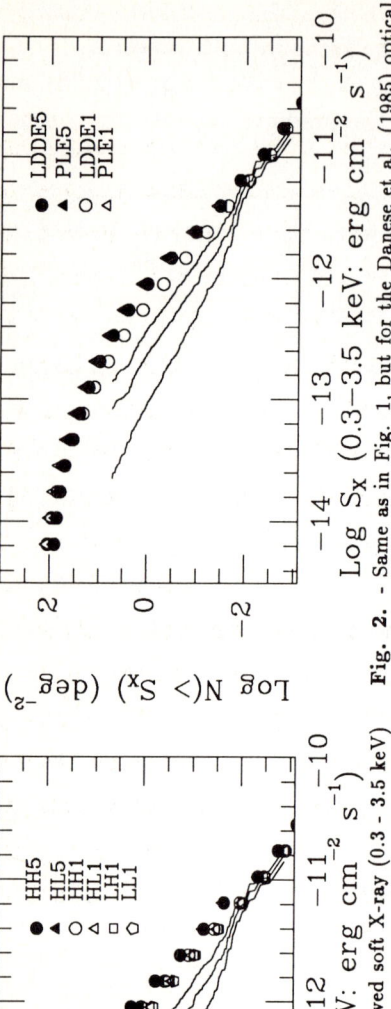

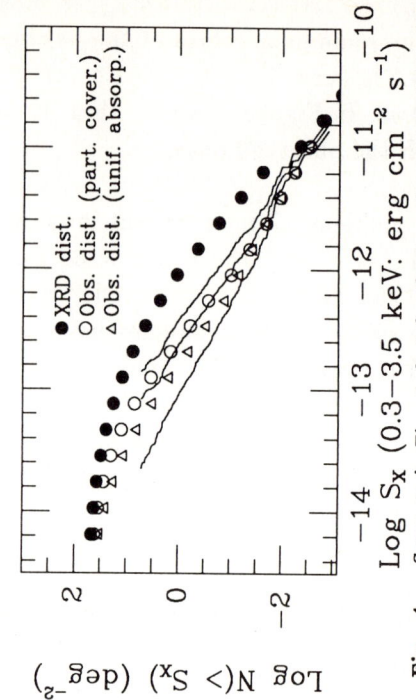

Fig. 2. - Same as in Fig. 1, but for the Danese et al. (1985) optical luminosity functions for quasars.

Fig. 4. - Same as in Fig. 1, but for the two observational distributions described in §2 (*open symbols*) and the XRD distribution (*filled dots*) [HL5 optical luminosity function for quasars of (Schmidt and Green 1983) and $z_{max}^{qua} = 3.0$].

Fig. 1. - Comparison between the observed soft X-ray (0.3 - 3.5 keV) source counts (EMSS) and the predicted ones for the XRD spectral distribution (see §2) and the Schmidt and Green (1983) optical luminosity functions for quasars ($z_{max}^{qua} = 3.0$). The upper and lower solid lines show the 99% bounds of the observed relation.

Fig. 3. - Same as in Fig. 1, but changing the maximum redshift for the presence of quasars z_{max}^{qua} in the case of the HL5 quasar optical luminosity function (Schmidt and Green 1983).

Optical and Radio Properties of Faint Sources Contributing to the Cosmic X-ray Background

D.M. WORRALL, M. BIRKINSHAW, S.S. MURRAY

Harvard-Smithsonian Center for Astrophysics, Cambridge, MA, U.S.A.

ABSTRACT

We report on our X-ray, radio, and optical studies of a survey field in Eridanus. Radio mapping can locate faint X-ray sources well, and assist in the spectroscopic identification of a component of the soft cosmic X-ray background (CXRB). Joint X-ray/radio measurements can be compared with results for brighter sources in the context of AGN unification models.

1 INTRODUCTION

The *Einstein* Observatory and ROSAT have imaged small regions of sky to faint X-ray flux densities ($\sim$ a few nJy at 1 keV). By resolving the individual sources, such deep exposures provide the most direct way of studying the discrete-source component of the soft CXRB. Burg, Hamilton, Griffiths, and Hasinger have also discussed X-ray deep surveys at this workshop.

To understand which classes of source dominate at faint X-ray fluxes, it is necessary to perform optical spectroscopic observations of candidates falling in the error circle of each X-ray source. The identification is based on plausibility arguments which use our knowledge of the X-ray properties of the various types of optical source, and the probability of chance coincidence. The distribution of mJy-level radio sources is sufficiently sparse that if one falls in an *Einstein* or ROSAT error circle it is almost certain to correspond to the X-ray source. Thus mapping the same regions of sky to faint flux levels in both the radio and X-ray locates X-ray sources associated with radio emission to the $\sim$ arcsec accuracy of the radio map, often reducing the number of optical candidates from several to one. Such dual-band surveys also allow X-ray to radio correlations to be studied with few of the biases implicit in the use of incomplete samples selected from different parts of the sky.

What objects make the largest contribution to the CXRB at faint flux densities? This important and controversial issue relates not only to the origin of the CXRB but also to the evolution of many types of radio and X-ray source (*e.g.*, Windhorst

1986; Kellermann and Wall 1987; Griffiths and Padovani 1990). For example, Griffiths and Padovani suggested that a large fraction of the CXRB at faint X-ray fluxes should be due to starburst galaxies. This is supported by claims of $\sim 2\sigma$ X-ray excesses near faint radio sources (Helfand *et al.* 1989; Hamilton, this workshop). Moreover, since X-ray/radio correlations are important evidence behind "unified schemes" that relate radio core-dominated and lobe-dominated quasars and radio galaxies through orientation effects, deep surveys can be used to make the important test of whether such correlations extend to faint sources.

2 THE *EINSTEIN* DEEP-SURVEY (EDS) FIELDS

The EDS comprises six long exposures in widely spaced regions of the sky. Giacconi *et al.* (1979), Murray (1981) and Griffiths *et al.* (1983) present preliminary analyses of the fields and a definitive re-analysis of the entire survey is reported by Primini *et al.* (1991); see also Hamilton *et al.* (1991) for an alternative analysis of four fields. X-ray source lists from the Primini *et al.* analysis, including sources of lower significance than in the published work, have been generally available since January 1990, and provide the *Einstein* X-ray fluxes used in our continuing radio, optical, and ROSAT studies of these fields. Note that overlapping *Einstein* and ROSAT deep surveys are crucial for combining source-count information from the two missions since the telescopes have significantly different energy responses.

For the remainder of this paper we will concentrate on the central 1520 sq arcmin region of the Eridanus EDS field, where some of the best *Einstein* data were taken. This region contains 18 *Einstein* Imaging Proportional Counter (IPC) X-ray sources detected to a positional accuracy of $\sim 1'$ and at a confidence such that 16 of them are real. Seven of these sources are located to $\sim 5''$ through detection with the *Einstein* High Resolution Imager (HRI).

3 RADIO MAPPING OF THE ERIDANUS EDS FIELD

In February 1990 we mapped the central part of the Eridanus EDS field with the VLA in D-array at 5 GHz, using a mosaic of 37 pointings, and detected 9 sources brighter than the 800μJy completeness limit. This is about half the number of sources expected on the basis of previous deep radio surveys (Kellermann and Wall 1987): there is a Poisson probability of only $\sim 2\%$ that our small number of detections is a chance coincidence. There is one particularly interesting 'missing' radio source. 'Source 6' of Primini *et al.* (1991), identified with a z=0.52 quasar, was reported by Giacconi *et al.* (1979) to be coincident with a 5 GHz Parkes source of 10 ± 2 mJy, and its absence in our survey indicates an unusual level of radio variability.

4 X-RAY/RADIO FLUX CORRELATION

Four of the nine 5 GHz radio sources fall in X-ray error circles (three IPC and one HRI). Since only 0.3 coincidences would have been expected by chance, these associations are likely to be real. For the three IPC sources, the radio position reduces the number of optical candidates from many to one or two. The HRI source had already yielded a single optical candidate.

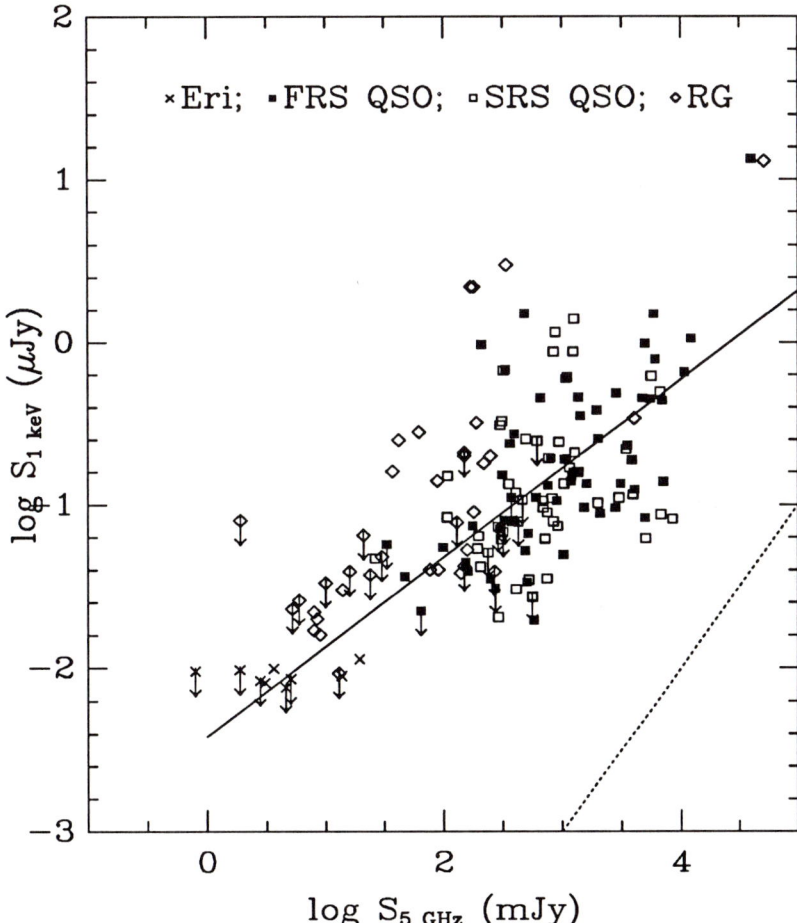

Figure 1 The X-ray detections and upper limits for the nine Eridanus 5 GHz radio sources (crosses) extend to faint flux densities the correlation observed for brighter flat radio spectrum (FRS) and steep radio spectrum (SRS) quasars (squares; Worrall *et al.* 1987) and radio galaxies (diamonds; Fabbiano *et al.* 1984). Comparing the data with the dotted line, which is of slope unity, we see that faint radio sources appear to be higher in X-ray to 5 GHz emission than brighter radio sources.

The X-ray flux densities (or upper limits) for the nine Eridanus 5 GHz radio sources are compared with those for brighter radio-loud quasars and radio galaxies in Figure 1. The dashed line is the maximum-likelihood power-law fit to all the data assuming the upper limits and detections are from the same population, and assuming Gaussian residuals (Avni and Tananbaum 1986). The formal slope, 0.55, is flatter than unity (although the preponderance of upper limits at low flux densities suggests caution) which implies that radio sources several times fainter than those plotted, but still readily accessible to the VLA, would be detectable in the X-ray only slightly below the present X-ray threshold. A second argument also suggests that radio observations are a good way to locate faint X-ray sources. At a radio threshold of 1 mJy, the radio detection rate of the Eridanus sources (4/16) is comparable to the $\sim 20\%$ rate for the *Einstein* Extended Medium Sensitivity Survey (EMSS; Gioia *et al.* 1990), even though a typical Eridanus X-ray source is about 20 times weaker than a typical EMSS source. If some radio sources < 1 mJy emit X-ray flux densities above the Eridanus EDS threshold (but see Figure 1), then faint radio sources may become a relatively larger fraction of the total X-ray source population at faint X-ray flux levels.

5 WORK IN PROGRESS

5.1 Radio
We have recently completed our 5 GHz map of Eridanus down to a uniform flux density limit of 500μJy, and have increased the number of radio sources by a factor of ~ 2. The 1.5 GHz data we also acquired indicates that the 5 GHz radio sources divide fairly evenly between steep and flat spectrum.

5.2 X-ray
We have just received the data from our long ROSAT observations of Eridanus with the Positional Sensitive Proportional Counter (PSPC) and HRI. Preliminary inspection shows that the PSPC has roughly tripled the number of detected X-ray sources, some of which are located to a few arcsec accuracy with the HRI. We will be comparing the *Einstein* and ROSAT samples in detail to derive information on the spectra of the faint sources across the joint ROSAT/*Einstein* energy band.

5.3 Optical
We have a continuing program to identify spectroscopically the X-ray and radio sources in Eridanus. The astrometry is performed by superimposing the radio and X-ray error regions on a digitized and position-calibrated UK Schmidt J plate with limiting magnitude ~ 22 (from the Guide Star program of the Space Telescope Science Institute, and provided to us by Daniel Golombek). Earlier work identified

four of the IPC sources (with quasars at z=0.95 and z=0.52, a G star, and a K star) and presented photometric and spectroscopic data for candidate identifications of some of the other sources (Giacconi *et al.* 1979; Primini *et al.* 1991).

6 CONCLUSIONS

At 5 GHz the Eridanus EDS field is underpopulated with $\sim$ 1 mJy radio sources. A particularly interesting case is a z=0.5 quasar which varied from 10 ± 2 mJy (Giacconi *et al.* 1979) to < 0.8 mJy in our observations. There is an indication that the X-ray to 5 GHz ratio of the Eridanus radio sources is higher than for radio-brighter quasars and radio galaxies, and radio sources may become a relatively larger fraction of the total X-ray source population at faint X-ray flux levels. Radio mapping of deep X-ray fields is a productive way to locate X-ray sources more precisely and reduce the number of candidates for the optical identification. Our continuing work will address the importance of starburst galaxies to the CXRB, and will study the X-ray/radio correlations as a function of optical classification within the context of AGN unification models.

This work was supported by NASA contract NAS8-30751. DMW also thanks the Smithsonian Institution Research Opportunities Fund and the workshop organizers for travel support.

Avni, Y., and Tananbaum, H. 1986, *Ap. J.*, **305**, 83

Fabbiano, G., Miller, L., Trinchieri, G., Longair, M. and Elvis, M. 1984, *Ap. J*, **277**, 115.

Giacconi, R. *et al.* 1979, *Ap. J. (Letters)*, **234**, L1.

Gioia, I.M., Maccacaro, T., Schild, R.E., Wolter, A., Stocke, J.T., Morris, S.L., and Henry, J.P. 1990, *Ap. J. Suppl*, **72**, 567.

Griffiths, R.E. *et al.* 1983, *Ap. J.*, **269**, 375.

Griffiths, R.E., and Padovani, P. 1990, *Ap. J.*, **360**, 483.

Hamilton, T.T., Helfand, D.J. and Wu, X. 1991, *Ap. J.*, **379**, 576.

Helfand, D., Hamilton, T.T., Rich, R.M., Becker, R., and White R. 1989, *BAAS*, **21**, 1220.

Kellermann, K.I., and Wall, J.V. 1987, in *Observational Astronomy*, eds. A. Hewitt *et al.*, (Reidel), p. 545.

Murray, S.S. 1981, in *X-ray Astronomy with the Einstein Satellite*, ed. R. Giacconi, p. 281.

Primini, F., Murray, S.S., Huchra, J., Schild, R., Burg, R. and Giacconi, R. 1991, *Ap. J.*, **374**, 440.

Windhorst, R.A. 1986, in *Highlights of Astronomy*, ed. J.P. Swings (Reidel), p. 355

Worrall, D.M., Giommi, P., Tananbaum, H. and Zamorani, G. 1987, *Ap. J.*, **313**, 596.

Chapter V

Models for sources of

the X–ray background

Source Models of the X-Ray Background

G. SETTI

University of Bologna
Istituto di Radioastronomia CNR, Italy

and

European Southern Observatory
W-8046 Garching bei München, Germany

ABSTRACT

The contribution of AGNs (quasars and Seyfert type 1 nuclei) to the extragalactic X-ray background (XRB) is discussed in the light of new observational results and theoretical models. It is concluded that low luminosity AGNs (and related populations of objects) are still the most likely candidates to supply the dominant contribution to the XRB intensity over the complete X-ray band, but the actual fraction that is contributed may be different in different parts of the spectrum. In particular, AGNs are the only known class of extragalactic sources which may provide an adequate description of the detailed spectral intensity of the hard XRB where most of the energy is concentrated.

1 INTRODUCTION

The compatibility between the typical source spectral shape and that of the XRB has been frequently used as an argument to discriminate amongst potential sources of the XRB intensity. Although this argument must be basically correct, it should be treated with care. For instance, the discovery that the observed XRB spectrum in the 3–50 keV energy interval can be so well approximated by an optically thin thermal bremsstrahlung at a temperature $kT \simeq 40$ keV (Marshall *et al.*, 1980) has been considered by many as strong evidence in favour of an earlier conjecture (Cowsik and Kobetich, 1972) that the XRB is due to a hot diffuse intergalactic gas (IGG). Although there are well known difficulties associated with this interpretation, such as the rather extreme energy requirements and a baryon gas density in excess of present upper limits from primordial nucleosynthesis models, it has only very recently been definitely ruled out following the very accurate measurements made by the FIRAS instrument on board the COBE (Mather *et al.*, 1990). The absence of a detectable deviation from a pure black-body of the 2.7 cosmic background sets an upper limit to the Comptonization parameter $y < 10^{-3}$, while $y \simeq 10^{-2}$ is required by the hot IGG models (Guilbert and Fabian, 1986), and as a consequence a hot diffuse IGG cannot contribute more than a few percentage points to the hard XRB.

Strong clumping can in principle help in circumventing the difficulties associated with a diffuse hot IGG model. However, the temperature fluctuations induced in the cosmic (microwave) background (CMB) are bound by stringent upper limits on (sub)arcminute angular scales leading to a somewhat unrealistic picture whereby a large number of small size ($<$ few tens kpc) hot clumps would be kept in pressure equilibrium by an even hotter but much less dense gas component (Barcons and Fabian, 1988). Likewise, proposed models, in which the gas has been heated to the required temperature to explain the XRB by infall into deep potential wells of condensates composed of baryonic (or dark) matter at large redshifts, are severely constrained by the upper limits on the fluctuations of the XRB and of the CMB leading to a minimum number of condensates such that too many gravitationally lensed quasars are predicted (Rogers and Field, 1991a). It is also found that if the gas is not gravitationally bound, but heated only once, the expansion time is much shorter than the radiative cooling time and, therefore, these types of models cannot provide an adequate explanation of the XRB.

As a matter of fact, several authors had pointed out that contrary to expectation the very precise fit of the XRB data above 3 keV with the 40 keV optically thin thermal bremsstrahlung spectrum provides in itself the strongest evidence against any interpretation in terms of thermal emission from a hot IGG. This is because any reasonable subtraction of the integrated contributions from known classes of extragalactic X-ray sources would destroy the almost perfect spectral shape (e.g., Giacconi and Zamorani, 1987), which, as a consequence, should be considered as accidental, a conspiracy of nature.

Among the classes of extragalactic objects known to emit substantially in the X-ray domain the AGNs, and related objects, have been for a long time (Setti and Woltjer, 1973) the most likely candidates to provide the bulk of the XRB both because they represent the strongest sources of hard X-rays and because they are known to have been more numerous and/or more powerful emitters at early epochs, a prerequisite essential to explain the global energetics associated with the XRB. The plausibility of this interpretation, however, was shaken when the power law fits to the spectra of a bright hard X-ray selected sample of AGNs, mostly nearby Sy 1 galaxies, indicated an average spectral index ($< \alpha > \simeq -0.65$) much steeper than that of the XRB ($\alpha \simeq -0.4$) in the 2–10 keV energy interval (Mushotzky, 1984). No more than $\sim 30\%$ of the XRB in the 2–10 keV interval could be contributed by any population of sources having hard X-ray spectra typical of the nearby sample of AGNs otherwise the ensuing spectral wiggles would be incompatible with the general smoothness of the XRB spectrum (Danese *et al.*, 1986). Subsequent measurements both in the hard and soft X-rays confirmed the generally steeper slopes of the Sy 1 and quasars spectra further strengthening the notion of a discrepancy frequently referred to as the "spectral paradox" (Boldt, 1987). However, the recent results from the X-ray mission GINGA (to be discussed later) have shown that the hard X-ray spectra of AGNs are much more complex and not inconsistent with the production of the XRB spectrum once the cosmological evolution and redshift effects are properly taken into account. Thus, the "paradox" has vanished and once again one has been reminded that the inferences made on the basis of the spectral shape alone in a restricted passband may not be meaningful.

The spectrum of the XRB at lower energies (< 3 keV) is much less known largely because the contributions due to galactic components become substantial and difficult to separate out. The results of an extensive analysis of *Einstein Observatory* IPC fields in a region of the sky where the background intensity is the lowest indicate an XRB intensity well above (a factor ~ 2.5 on average) the downward extrapolation of the hard XRB following a power law of spectral index $\alpha = -0.4$ (Wu *et al.*, 1991; Micela *et al.*, 1991). This excess, but not necessarily its detailed shape, has been confirmed by the much cleaner measurements made with ROSAT, while at the same time it has been found that a large fraction ($> 35\%$) of the XRB in the 1–2 keV passband is resolved into sources, mostly AGNs (Hasinger *et al.*, 1991a; Shanks *et al.*, 1991).

In what follows we shall discuss AGN models of the XRB. Because of the differences in the observational material which has been gathered in the soft and hard X-rays it is accordingly convenient to split the discussion of these two energy domains.

2 THE SOFT XRB AND THE AGN X-RAY/OPTICAL COUNTS

In the last decade much of the discussion concerning the contribution of the AGNs to the XRB has been largely driven by the analysis of the data obtained with the *Einstein Observatory*, able to access a much larger volume of space compared to other X-ray space missions. For this reason the estimates have been frequently referred to a monochromatic energy of 2 keV, well within the *Einstein Observatory* band but high enough to try to match the information on the XRB and the sources observed in the hard X-ray band. The estimates of the AGNs contribution to the nominal 2 keV XRB (obtained by smoothly extrapolating downward the 40 keV optically thin thermal bremsstrahlung fit) have spanned a wide range from a minimum of 37% to the full complement of the 2 keV XRB depending on the X-ray source counts, the average spectral index and the assumptions made on the cosmological evolution of the sources (Setti, 1990, and references therein; Schmidt, 1989). Recently Maccacaro *et al.* (1991) have made an estimate based on a sample of more than 400 AGNs identified in the *Einstein Observatory* Extended Medium Sensitivity Survey (EMSS). The sample is large and deep enough to allow a determination of the evolution of the

X-ray luminosity function as a function of cosmic time independent from the optical data. By parametrizing the cosmological evolution as a pure luminosity evolution out to a cut-off redshift $z \simeq 3$ and by adopting an average energy spectral index $\alpha \simeq -1.0$ applicable to these sources in the EMSS IPC band (Maccacaro *et al.*, 1988), these authors find that the AGNs contribute $\sim 36\%$ of the nominal 2 keV XRB, and proportionally less if the true 2 keV XRB is in excess of the nominal one.

This result should be contrasted with the latest findings of the ROSAT Medium Sensitivity Survey (MSS). As reported by Hasinger *et al.* (1991a,b) the analysis of eight high galactic latitude ROSAT PSPC fields has revealed that, depending on exposure time, up to 45% of the XRB in the 1–2 keV energy interval is resolved into sources down to a sensitivity limit about a factor 3–5 fainter than the *Einstein Observatory* Deep Sensitivity Survey, EDS (Giacconi *et al.*, 1979).

No optical identifications of the 184 sources found in the 8 ROSAT fields are available yet. Hasinger *et al.* argue that by comparing the optical identification content of the EMSS and EDS surveys the contamination by galactic sources should be $< 10\%$. This is corroborated by the optical identifications of the sources resulting from a similarly exposed ROSAT PSPC field in a region already searched for UVX objects (Shanks *et al.*, 1991). Only 4 out of 39 X-ray sources were identified with stars down to an optical limiting magnitude $B = 22.3$, while $\sim 62\%$ were identified with quasars whose redshift distribution is essentially comprised between 0.5 and 2.5 with a mean $z \simeq 1.5$. This should be compared with the redshift distribution found in the EMSS where $\sim 50\%$ of the AGNs have redshifts < 0.27 and the mean redshift is ~ 0.4 (Maccacaro *et al.*, 1991). In fact, only 3 of the identified quasars have low luminosities typical of Sy 1 nuclei ($M_B > -23$; $q_0 = 0$, $H_0 = 50$). Nine sources were not identified either because of lack of optical counterparts down to the limiting magnitude of the plate or because they were too faint for spectroscopic identification. Thus in this field the AGNs content could be as high as 85%.

The low instrumental background of the PSPC has also permitted a direct comparison of the spectra. According to the two references mentioned above the average energy spectral index of the sources (mostly AGNs) in the 1–2 keV interval is $\alpha = -1.3 \pm 0.2$, while that of the residual XRB is $\alpha \simeq -0.9$, definitely much steeper than that of the downward extrapolated hard XRB in the same energy interval. The source spectrum appears to be steeper than that of the XRB, but given the small leverage of the energy interval the possibility that the spectral indices are the same cannot be rejected (Shanks *et al.*). It is interesting to note that a steepening of the XRB below 3 keV is predicted in models, to be discussed later, which have been proposed to account for the hard XRB (Setti and Woltjer, 1989).

That most sources in the ROSAT PSPC fields should be AGNs, of which a large fraction are quasars, can also be substantiated by the following argument:

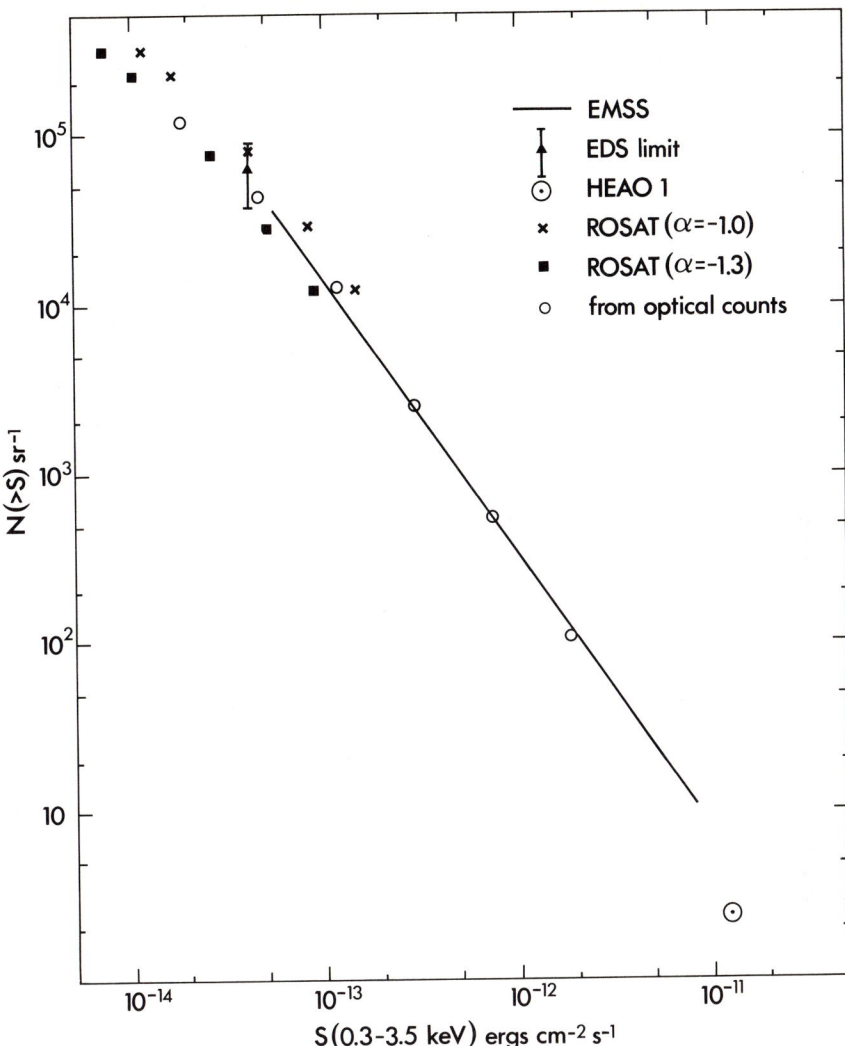

Fig. 1 – The integral source counts for various samples versus the fluxes in the *Einstein Observatory* IPC band 0.3–3.5 keV. The straight line represents the best fit to the AGN counts in the EMSS (Gioia and Maccacaro, 1991). The filled triangle represents the *Einstein* Deep Survey (EDS) limit of Giacconi *et al.* (1979) assuming a mean source spectral index $\alpha = -1.0$. The crosses represent the ROSAT MSS source counts adapted from Hasinger *et al.* (1991) with a mean spectral index $\alpha = -1.0$, while the filled squares are the same counts but with a mean $\alpha = -1.3$. The open circles represent the counts predicted on the basis of the AGN optical counts from Setti (1987). The ($\odot$) is the AGN surface density in the Piccinotti *et al.* (1982) sample making use of the conversion factor from the HEAO–1 A2 band to the IPC band given by Martin-Mirones *et al.* (1990).

The X-ray integral counts for the AGN sample from the EMSS are plotted in Fig. 1 as a function of the flux in the 0.3–3.5 keV interval together with the EDS limit, the ROSAT MSS source counts and, at the bright end, the total number of AGNs observed in the HEAO–1 A2 survey (Piccinotti *et al.*, 1982). The best fit to the EMSS $\log N - \log S$ has a slope of -1.61 and it agrees within the errors with the $\log N - \log S$ of a much smaller number of AGNs previously found in the MSS (Gioia and Maccacaro, 1991). The EDS point has been set by adopting an average energy spectral index $\alpha = -1.0$, in conformity with the quoted mean spectral index of the EMSS AGNs, which entails at least a 10% reduction of the flux limit derived by Giacconi *et al.* (1979) on the basis of an assumed average spectral index $\alpha = -0.5$ for the EDS sources (Tananbaum *et al.*, 1986).

The HEAO–1 point has been set by converting the Piccinotti *et al.* survey flux limit with the conversion factor 1.8 (corrected to the 0.3–3.5 keV interval) derived by Martin-Mirones *et al.* (1990) for this AGN sample. It is seen that it conveniently falls by a factor larger than 2 below a straight extrapolation of the EMSS counts to brighter fluxes with the slope given above.

The ROSAT source counts (crosses) have been converted to the *Einstein* IPC energy interval with a power law spectral index $\alpha = -1.0$ already used by Hasinger *et al.* (1991a) in conformity with the EMSS to define the flux scale of the counts in the 0.1–2.4 keV energy interval. In the region of overlap the ROSAT counts are a factor ~ 1.5–1.6 above those of the AGNs in the EMSS. However, if the observed mean spectral index $\alpha = -1.3$ is applied then the flux scale of the ROSAT source counts in the IPC band should be reduced by $\sim 36\%$ and now the best fit to the ROSAT counts falls below that of the AGN sample of the EMSS by $\sim 23\%$ (filled squares). This in itself provides a strong indication that most sources at the bright end of the ROSAT survey should be identified with AGNs.

It is of interest to compare the X-ray counts with those obtained from the optical counts. In Fig. 1 one has plotted the X-ray counts (open circles) directly derived from the optical counts of AGNs (quasars and Sy type 1 nuclei) in a way which is as far as possible model independent (Setti, 1987). While one is referred to the original paper for details concerning this derivation, several points relevant to the present discussion must be recollected. The adopted optical counts have been represented by two power laws down to a limiting magnitude $B = 23$ with a break at $B \simeq 20$. While at the bright end the adopted integral counts (slope $= -2.12$) are consistent with all available data, at the faint end ($B \simeq 22$) they may underestimate the true surface density for AGNs by as much as 30% according to the most recent deep optical surveys (Zitelli *et al.*, 1991). The calibration of the X-ray to optical ratios has been done making use of the *Einstein Observatory* observations of a sub-sample of the Bright Quasar Survey (Tananbaum *et al.*, 1986), and therefore the derived X-ray counts can be directly compared with the *Einstein* surveys. Finally, the flux scale has been derived from that at 2 keV in Setti (1987) first by integrating over the 0.3–3.5 interval with a spectral index $\alpha = -0.5$ used in the original derivation and then by applying a 10% reduction in the flux scale because of the steeper spectral index ($\alpha = -1.0$) required for a comparison with the EMSS counts. That a 10% reduction in the flux scale should be applied has also been verified internally to the EMSS AGN sample (R. Della Ceca, private communication).

It is seen that the X-ray counts derived from the optical ones are in very good agreement with the EMSS counts. [It should be noted that the flattening of the slope from the much steeper optical counts has been obtained by applying two effects: the observed inverse correlation between the X-ray to optical luminosity ratios and the optical luminosities (Avni and Tananbaum, 1986) and a relative K-correction term such that the relevant part of the optical continua are on the average flatter, $|\Delta\alpha| = 0.5$, than the *Einstein* IPC source spectrum.] Further, it has been anticipated that the counts should start to flatten well before the EDS limit is reached and this is what is being confirmed by the ROSAT counts. At the faint end the derived X-ray counts have a slope $\simeq -1.0$, essentially dictated by the optical counts, in good agreement with the ROSAT findings, and fall just above the ROSAT source counts with the steeper source spectrum. These results provide independent strong evidence that the sources found down to the limit of the ROSAT MSS are largely composed of AGNs.

By (arbitrarily) assuming that all AGNs cut off at a redshift $z = 3$ it was found (Setti, 1987) that at a flux corresponding to the EDS limit approximately 75% of the AGNs are quasars ($M_B < -23$), while the percentage of low luminosity nuclei ($M_B > -23$) should increase with decreasing X-ray flux. This suggests that a large proportion of the unidentified sources in the Shanks *et al.* field could be Sy 1 nuclei. By adapting Setti's estimate of the quasar contribution to the nominal 2 keV XRB, it is found that quasars down to $B = 23$ contribute $\gtrsim 27\%$ of the 1–2 keV XRB (or $\gtrsim 10\%$ in the 0.3–3.5 keV passband) assuming that the XRB observed by ROSAT turns up to the ~ -1 slope at 2 keV. The contribution can be increased by an additional $\sim 5\%$ before the quasars are cut off.

It is immediately clear that if the X-ray counts could be extrapolated following a power law of slope ~ -1.0 by a factor ~ 100 below the ROSAT flux limit of Fig. 1 then the full soft XRB could be accounted for (e.g., Giacconi and Zamorani, 1987). This would involve low luminosity AGNs to be observed down to very faint magnitudes ($B \sim 28$-29) and with intrinsic luminosities as low as $M_B \sim -19$. However, a recent spectroscopic analysis of a deep quasar field (Zitelli *et al.*, 1991) has essentially confirmed an earlier estimate (Setti, 1984) indicating the presence of a considerable cosmological evolution for Sy type 1 nuclei. These authors find that low luminosity nuclei ($M_B > -23$) show a steep $\log N - B$ relationship such that at $B \sim 22$ their surface density is about the same as that of quasars with $z < 2.2$. It is claimed that if the steep counts continue at fainter magnitudes the low luminosity nuclei would outnumber the quasars at $B \gtrsim 23$. In this case from the discussion previously made one would expect a turn up of the X-ray source counts at a flux level about a factor 2, or so, below the lowest ROSAT flux quoted in Fig. 1. It is possible that the very deep ROSAT fields may substantiate the existence of such an important effect.

3 MODELS OF THE HARD (> 3 keV) XRB

As has been mentioned in the Introduction, although the AGNs have been generally considered to be the prime source candidates able to satisfy the huge energy requirements of the XRB, the fact that the mean spectral index of the AGNs is much steeper (Williams *et al.*, 1991; Comastri *et al.*, 1992, for recent discussions) than that of the XRB in the 2–10 keV energy interval has been a major obstacle to any detailed modeling of the hard XRB. This has led a number of authors either to hypothesize the existence of a completely new class of extragalactic hard X-ray sources or to appeal to cosmic evolution effects such that the AGN spectra were much harder at earlier epochs (Leiter and Boldt, 1982). Recently, Schwartz and Tucker (1988) have shown that if the mean AGN spectrum would flatten beyond ~ 10 keV then it is possible to model the flatter portion of the hard XRB without invoking any cosmic evolution of the spectra of the AGNs and, therefore, a basic change in the physical parameters of these sources. Basically the profile of the XRB is obtained by redshifting the mean spectrum together with an appropriate time evolution in the volume emissivity of the sources which in turn defines the shape of the mean source spectrum beyond ~ 10 keV to be constrained within the observational limits. By parametrizing the model these authors obtained a good representation of the XRB spectrum in the 3–30 keV interval with a cut-off in the evolution of AGNs volume emissivity at a (reasonable) redshift $z = 3\text{–}4$, while the steepening of the XRB spectrum beyond ~ 30 keV could be achieved in principle by properly choosing a high energy cut-off in the mean AGN spectrum.

The conjecture of a flattening of the hard X-ray spectrum of AGNs has been dramatically confirmed by GINGA and has been variously interpreted either as partial coverage of an underlying X-ray power law continuum by cold matter with typical column densities (solar composition) in the range $2 - 6 \times 10^{24}$ cm^{-2} (Piro *et al.*, 1989; Matsuoka *et al.*, 1990) or as reprocessed X-rays (usually referred to as *reflection*) from a thick slab of relatively cold ($< 10^6$ K) matter (Pounds *et al.*, 1990) or, sometimes, as a combination of these two mechanisms.

By assuming a mean AGN spectrum based on a handful of Seyfert 1 galaxies with observed humps at $\gtrsim 10$ keV and a partial coverage of an underlying continuum of slope $\alpha = -0.7 \div -0.75$, a luminosity function of Seyfert 1 nuclei consistent with that of Piccinotti *et al.* (1982), a pure luminosity evolution cut-off at a redshift $z = 3$ and a cut-off in the source spectra at 110–125 keV, then, consistently with the findings of Schwartz and Tucker (1988), it is possible to work out a reasonably good fit to the spectrum (after subtracting a 5% contribution at 2 keV from clusters of galaxies) of the XRB from 3 to ~ 100 keV (Morisawa *et al.*, 1990). However, there may be some problems for this type of models with the predicted source number counts in the soft X-ray band. For instance, from the paper of Morisawa *et al.* (1990; Fig. 4) I find that the surface density of sources down to a flux of 10^{-14} erg cm^{-2} s^{-1} (0.3–3.5 keV) exceeds that found by ROSAT by a factor ~ 1.5 and, moreover, the source spectra are on the average much flatter than those of the ROSAT sources in the 1–2 keV interval.

Fabian *et al.* (1990) argued that the rather sharp break of the XRB spectrum at ~ 30 keV must correspond to some basic physical process occurring in the sources (rather than *ad hoc* assumed cut-offs) and identified such a process in the Compton down-scattering of high energy photons reflected from relatively cold gas ($< 10^6$ K, located in the vicinity of the central source) which is known to steepen by ~ 1 the spectral index of an incident power law spectrum above ~ 120 keV. These authors proposed a new class of sources, possibly identifiable as early phases in the time evolution of AGNs, for which more than 90% of the energy flux goes into the Compton reflected component, while at most 10% of the direct flux (assumed to be a power law continuum of spectral index $\alpha = -0.7$) is allowed to escape. With the further assumption that the comoving volume emissivity of the sources increases as $(1 + z)^4$ out to a redshift $z \simeq 5$ it has been possible to extend the fit of the background data up to $\gtrsim 10^3$ keV meeting the bump at a few MeV, while the predicted intensity below ~ 8 keV is significantly lower than that of the XRB. Recently, Rogers and Field (1991b,c) have reanalyzed the Fabian *et al.* (1990) model and concluded that the fit to the intensity of the XRB at $\gtrsim 3$ keV can be improved by assuming a much steeper primary source spectrum. They give parameter ranges within which it is possible to model the XRB up to several hundred keV, typical values being: a primary source spectral index $\alpha = -1.1$, a comoving volume emissivity of the sources increasing as $(1 + z)^{2.8}$ out to a redshift $z = 4.6$ and 89% of the energy flux in the reflection component.

The total surface density of the sources in this model is constrained to be $> 10^2$ deg^{-2} by the surface brightness fluctuations of the XRB above 3 keV and, therefore, their luminosity should be $\lesssim 10^{45}$ erg s^{-1} keV^{-1} at 10 keV (Fabian *et al.*, 1990). Aside from the fact that the energy flux in the reflected component is a much larger fraction of the total flux than observed in the Seyfert 1 galaxies and that this fraction appears to become vanishingly smaller in high luminosity objects (Williams *et al.*, 1991), considerations which do not necessarily apply to a new class of objects, it should be carefully checked that the total number of sources together with their cosmic evolution out to large redshifts is consistent with the number counts in the soft X-ray domain. For instance, it can be easily seen that a model with the minimum number of sources given above would largely outnumber the EMSS counts. Obviously, this can be cured by increasing the total number of sources and correspondingly decreasing their luminosity. For example, by assuming a simple model such that the density of sources (assumed to have the same luminosity) increases as $(1 + z)^4$ out to a redshift $z = 5$ it is found (Setti *et al.*, in preparation) that, in order not to violate the ROSAT counts at the lowest flux level of 10^{-14} erg cm^{-2} s^{-1}, there should be a minimum of 5000 sources deg^{-2}, but their spectra are flatter than the average source spectrum of the ROSAT PSPC fields and the counts continue to rise steeply at fainter fluxes. Clearly, much more complicated models should be considered together with a reasonable interpretation of the nature of these hypothetical objects.

On the other hand Terasawa (1991) has investigated a number of models based on a mean AGN spectrum resulting from a primary power law spectrum, a Compton reflection component from a cold thick gas and a cold uniform absorber. He assumes the same local luminosity function of Seyfert type 1 nuclei as that of Morisawa *et al.* (1990), but unlike these authors, and following the basic suggestion of Fabian *et al.* (1990), the sharp steepening of the source spectra at high energies is due to the Compton down-scattering of the high energy photons (note the approximate coincidence of the best fit value of the high energy cut-off derived in the Morisawa *et al.* model with the steepening of the Compton reflected spectrum at $\sim$ 120 keV). As in the Morisawa *et al.* paper a pure luminosity evolution of the local luminosity function is assumed out to a maximum redshift $z = 3$. The best fit values of the parameters resulting from fitting the XRB are compared for consistency with those obtained from the spectral fittings of the Seyfert 1 sample observed by GINGA. Terasawa's preferred model is that of a primary power law spectrum of spectral index $\alpha \simeq -1$, a Compton reflected spectrum (claimed to be consistent with GINGA data but with $\sim$ 80% of the flux in the reflection component), a uniform absorber of cold gas with a mean column density $N_H \simeq 2 \times 10^{21}$ cm^{-2} and a partial coverage with column densities $N_H \simeq 2 \times 10^{23}$ cm^{-2}. It provides a good fit of the data up to 1 MeV. The expected $\log N - \log S$ in the 0.3–3.5 keV interval appears to be consistent with that of the AGNs from the EMSS. Also, from Fig. 4 of Terasawa's paper it is seen that the expected surface density of sources down to a flux of 10^{-14} erg cm^{-2} s^{-1} is consistent with that found by ROSAT surveys. However, as can be judged from Fig. 2 of Terasawa's paper the mean spectral index of his sources in the derived $\log N - \log S$ is much flatter than those of the AGNs in the EMSS and of ROSAT MSS sources.

Setti and Woltjer (1989) have discussed a very different model based on the X-ray properties of unified schemes of AGNs. Specifically, reference has been made to the unified scheme proposed by Barthel (1989) for which the radio-loud quasars and the strong radio galaxies are members of the same population, the radio-loud quasar phenomenon showing up whenever the source axis is aimed toward us within a specified angle from the line of sight. Similarly, it has been proposed that the radio-quiet quasars and luminous IR galaxies can be unified, the apparent morphological differences being attributable only to the orientation of the sources with respect to the observer. Notably, this type of models has been introduced for the first time by Antonucci and Miller (1985) to explain the polarized Sy 1 type emission seen in Sy 2 galaxies. According to this scheme Setti and Woltjer assumed that the AGNs remain hidden by optically thick tori of surrounding absorbing material and showed that for very reasonable values of the tori masses and sizes also the X-rays emitted by the central sources can be effectively absorbed up to 20 keV, or more, whenever the line of sight is not favourably placed. The X-ray properties of the (unfavourably oriented) parent populations above the absorption cut-offs should be the same as those of the AGNs, unless the X-ray emission itself is largely beamed. A recent analysis of the observations of a sample of Sy 2 galaxies obtained with the LAC instrument on board GINGA provides strong supportive evidence for this hypothesis.

By adopting straight power law source spectra with a mean slope $\alpha = -0.7$, very simple models for the fractional distribution of the absorption cut-offs and assuming further a cosmic evolution (out to $z = 3$) of the local AGN luminosity function such to produce the full intensity of the nominal 2 keV XRB, it has been shown that the spectral intensity of the XRB in the 3–20 keV interval can be easily attained within 5% accuracy with a number of "absorbed" sources approximately equal to that of the "unabsorbed" ones. No attempts have been made to fit the XRB at energies > 20 keV. Incorporating the new results of the AGN spectra obtained by GINGA should lead to a much improved model.

It should be noted that in Setti and Woltjer's model the AGNs parent populations by definition would not contribute to the source counts of ROSAT and only negligibly to those of the *Einstein* IPC band. On the other hand they would certainly contribute to the source counts in the hard X-ray passbands and, therefore, it is tempting to speculate that they might account for the spatial fluctuations of the XRB measured by GINGA in the 2–10 keV interval (Warwick and Stewart, 1989; Hayashida, 1991). This, however, appears to be difficult to reconcile with the spectral properties of the fluctuations indicating a rather steep spectral index ($\alpha \simeq -0.8$) and relatively low column densities of absorbing material.

4 CONCLUSIONS

The AGNs are still the most likely candidates to provide the dominant source of the XRB intensity over the complete X-ray band. However, the still unknown total contribution from other classes of extragalactic sources (Setti, 1990; De Zotti *et al.*, 1990; for recent reviews) may not be negligible and may affect diversely the different parts of the XRB spectrum. It also follows that any attempt to discriminate amongst models on the basis of a very fine tuning to the observed XRB spectral intensity may not be very significant at present.

Both new observational results and recent theoretical models have contributed to remove the apparent spectral incompatibility between the mean AGN spectra and that of the XRB in the 2–10 keV interval which has been the main obstacle to the interpretation of the hard XRB in terms of the AGN population. This is basically achieved as a result of (enhanced) broad-band emission features at energies > 10 keV in the source frame being redshifted into the observer rest frame in combination with an appropriate cosmic evolution of the AGN volume emissivity. Current models involve either the observed flattening of the spectra of low luminosity AGNs at $\gtrsim 10$ keV, interpreted as a Compton reflected component and/or absorption of the continuum by cold gas, or "absorbed" AGNs in the framework of unified schemes of AGNs, or a new class of powerful AGNs whose X-ray emission is largely dominated by a Compton reflection component. Compton down-scattering of high energy photons by relatively cold gas ($< 10^6$ K) may provide a natural and elegant explanation of the steepening of the XRB spectrum beyond a ~ 40 keV and of its shape up to the MeV bump.

It is important to note that these models can be checked in relation to the predicted source counts in the soft X-ray band. The EMSS and the ROSAT MSS source counts, together with the observed mean spectral indices of the sources in these two surveys, already provide severe constraints for most models proposed to explain the hard XRB on the basis of a typical source spectrum. On the other hand, models which resort to highly "absorbed" AGN populations, for which there is fresh observational evidence, possess the required flexibility to effectively uncouple the hard from the soft X-ray region of the XRB spectrum.

The X-ray source counts derived from the optical AGN counts are in good agreement with the AGN counts in the EMSS and with the ROSAT MSS source counts, including the predicted flattening of the counts at about the EDS limit, further strengthening the indication based on the optical identification content of one ROSAT field that the large majority of the ROSAT MSS sources are AGNs.

A more detailed comparison of the source counts requires a rather precise knowledge of the source spectra in the X-ray domain and also in the optical domain when comparison is made with the optical counts. This type of knowledge is partially available and it should be possible, particularly after the completion of the source identification programmes of the ROSAT fields, to work out more detailed models.

Extrapolation of the source counts below the ROSAT MSS limit by a factor ~ 100 with an integral slope ~ -1.0 could provide the full complement of the soft XRB. This would involve low luminosity AGNs being observed down to very faint optical magnitudes ($B \simeq 28 - 29$), while quasars ($M_B < -23$) can account for only 10 to 15% of the soft XRB. Clearly, a somewhat difficult perspective from the observational standpoint. However, if the low luminosity AGNs ($M_B > -23$) start to outnumber the quasar counts at $B \gtrsim 23$, one can predict a steepening of the X-ray source counts at a flux level a few times fainter than the ROSAT MSS limit and this may be evidenced by the very deep ROSAT surveys.

ACKNOWLEDGEMENTS

It is a pleasure to thank I. Gioia and T. Maccacaro for permission to quote the best fit to the AGN integral counts from the EMSS, and L. Danese and G. Zamorani for discussions.

REFERENCES

Antonucci, R.R.J., and Miller, J.S. 1985, *Astrophys. J.*, **297**, 621

Avni, Y., and Tananbaum, H. 1986, *Astrophys. J.*, **305**, 83

Barcons, X., and Fabian, A.C. 1988, *Mon. Not. R. astr. Soc.*, **230**, 189

Barthel, P.D. 1989, *Astrophys. J.*, **336**, 606

Boldt, E. 1987, *Phys. Reports*, **146**, No. 4, 215

Comastri, A., *et al.* 1992, *Astrophys. J.*, **384**, 1st January issue

Cowsik, R., and Kobetich, E.J. 1972, *Astrophys. J.*, **177**, 585

Danese, L., De Zotti, G., Fasano, G., and Franceschini, A. 1986, *Astrophys. J.*, **161**, 1

De Zotti, G., Martin-Mirones, J.M., Franceschini, A., and Danese, L. 1990, Proc. XIth Moriond Astrophysics Meeting, to be published

Fabian, A.C., George, I.M., Miyoshi, S., and Rees, M.J. 1990, *Mon. Not. R. astr. Soc.*, **242**, 14p

Giacconi, P., and Zamorani, G. 1987, *Astrophys. J.*, **313**, 20

Giacconi, R., *et al.* 1979, *Astrophys. J.*, **234**, L1

Gioia, I.M., and Maccacaro, T. 1991, private communication

Guilbert, P.W., and Fabian, A.C. 1986, *Mon. Not. R. astr. Soc.*, **220**, 439

Hasinger, G., Schmidt, M., and Trümper, J. 1991a, *Astron. Astrophys. Letters*, **246**, L2

Hasinger, G., Schmidt, M., and Trümper, J. 1991b, these proceedings

Hayashida, K. 1991, Ph.D. Thesis, ISAS RN 466, and these proceedings

Leiter, D., and Boldt, E. 1982, *Astrophys. J.*, **260**, 1

Maccacaro, T., Gioia, I.M., Wolter, A., Zamorani, G., and Stocke, J.T. 1988, *Astrophys. J.*, **326**, 680

Maccacaro, T., Della Ceca, R., Gioia, I.M., Morris, S.L., Stocke, J.T., and Wolter, A. 1991, *Astrophys. J.*, **374**, 117

Marshall, F.E., *et al.* 1980, *Astrophys. J.*, **235**, 4

Martin-Mirones, J.M., De Zotti, G., Franceschini, A., and Danese, L., 1990, Proc. of the Joint Texas/ESO-CERN Symposium, Brighton, Dec. 1990, New York Academy of Sciences, to be published

Mather, J., *et al.* 1990, *Astrophys. J.*, **354**, L37

Matsuoka, M., Piro, L., Yamauchi, M., and Murakami, T. 1990, *Astrophys. J.*, **361**, 440

Micela, G., Harnden Jr., F.R., Rosner, R., Sciortino, S., and Vaiana, G.S. 1991, *Astrophys. J.*, **380**, 495

Morisawa, K., Matsuoka, M., Takahara, F., and Piro, L., 1990, *Astron. Astrophys.*, **236**, 299

Mushotzky, R.F. 1984, in COSPAR/IAU Symp. on "High-Energy Astrophysics and Cosmology", G.F. Bignami and R.A. Sunyaev, *Adv. Space Res.*, **3**, 157

Piccinotti, G., Mushotzky, R.F., Boldt, E.A., Holt, S.S., Marshall, F.E., Serlemitsos, P.J., and Shafer, R.A. 1982, *Astrophys. J.*, **253**, 485

Piro, L., Matsuoka, M., and Yamauchi, M. 1989, Proc. 23rd ESLAB Symp. on "Two-Topics in X-Ray Astronomy", ESA SP-296, p. 819

Pounds, K.A., Nandra, K.A., Stewart, G.C., George, I.M., and Fabian, A.C. 1990, *Nature*, **344**, 132

Rogers, R.D., and Field, G.B. 1991a, *Astrophys. J.*, **366**, 22

Rogers, R.D., and Field, G.B. 1991b, *Astrophys. J.*, **370**, L57

Rogers, R.D., and Field, G.B. 1991c, *Astrophys. J.*, **378**, L17

Schmidt, M. 1989, Proc. 23rd ESLAB Symp. on "Two-Topics in X-Ray Astronomy", ESA SP-296, p. 853

Schwartz, D.A., and Tucker, W.H. 1988, *Astrophys. J.*, **332**, 157

Setti, G. 1984, in "X-Ray and UV Emission from Active Galactic Nuclei", eds. W. Brinkmann and J. Trümper, MPE Report 184, p. 243

Setti, G. 1987, Proc. IAU Symp. No. 124 on "Observational Cosmology", eds. A. Hewitt, G. Burbidge, and L.Z. Fang, Reidel: Dordrecht, p. 579

Setti, G. 1990, Proc. IAU Symp. No. 139 on "Galactic and Extragalactic Background Radiation", eds. S. Bowyer and C. Leinert, Reidel: Dordrecht, p. 345

Setti, G., and Woltjer, L. 1973, in Proc. IAU Symp. No. 55 on "X- and Gamma-Ray Astronomy", eds. H. Bradt and R. Giacconi, Reidel: Dordrecht, p. 208

Setti, G., and Woltjer, L. 1989, *Astron. Astrophys.*, **224**, L21

Shanks, T., Georgantopoulos, I., Stewart, G.C., Pounds, K.A., Boyle, B.J., and Griffiths, R.E. 1991, *Nature*, **353**, 315

Tananbaum, H., Avni, Y., Green, R.F., Schmidt, M., and Zamorani, G. 1986, *Astrophys. J.*, **305**, 57

Terasawa, N. 1991, *Astrophys. J.*, **378**, L11

Warwick, R.S., and Stewart, G.C. 1989, Proc. 23rd ESLAB Symp. on "Two-Topics in X-Ray Astronomy", ESA SP-296, p. 727

Williams, O.R., *et al.* 1991, *Astrophys. J.*, in press

Wu, X., Hamilton, T., Helfand, D., and Wang, Q. 1991, *Astrophys. J.*, **379**, 564

Zitelli, V., Mignoli, M., Zamorani, G., Marano, B., and Boyle, B.J. 1991, *Mon. Not. R. astr. Soc.*, in press

Contribution of Clusters to the X-ray Background.

A. BLANCHARD (1), K. WACHTER (2), A.E. EVRARD (3), J.SILK (4)

(1) Observatoire de Meudon, DAEC, Place Janssen, F-92190 Meudon, France.
(2) Dpt of Statistics, University of California, Berkeley, CA 94720, U.S.A.
(3) Dpt of Physics, University of Michigan, Ann Arbor, MI 48109, U.S.A.
(4) Dpt of Astronomy, University of California, Berkeley, CA 94720, U.S.A.

1 INTRODUCTION

The contribution of clusters to the X-ray background has been analysed several times (Silk, Tarter, 1973; Stottlemeyer, Boldt, 1984; Schaeffer, Silk, 1988). Hot gas in clusters emits bremsstrahlung radiation, but because of their moderate temperature (of the order of 7 keV for the rich clusters) their contribution to the observed hard component is negligible. At softer energies, as in the *ROSAT* bands, the contribution of the potentials of small clusters and groups (with typical temperatures of the order of 1keV) might be important. We investigate what this contribution could be in different models of galaxy formation (CDM and power law spectra). As thermal bremsstrahlung emission is dominated by the highest density region, our discussion applies directly to the possibility of producing the soft x-ray background from thermal emission of hot gas as recently suggested (Cen et al, 1991).

2 CONSTRAINTS ON GALAXY FORMATION THEORIES

2.1 Clusters properties

By clusters, we actually mean any collapsed region into which some baryonic gas has fallen. The gas is expected to be at the virial temperature of the dark matter. This temperature could be easily determined in the top-hat model provided that the final radius is taken to be 60% of the turn-around radius :

$$T_X = 3.94(1 + z)M_{15}^{2/3} \text{keV}$$

in order to agree with numerical simulations by Evrard (1990) (the final gas distribution is assumed to remain isothermal). This allows a useful comparison of the theoretical temperature distribution function with the observations.

It is tempting to try to derive a similar relation for the luminosity. However the luminosity of clusters is totally dominated by the core radius, and the validity of this scaling relation for the luminosity therefore critically depends on whether the core radius of the X-ray emitting gas will follow such a scaling law. In the following, we assume a relation between mass and bolometric luminosity:

$$L_X \propto M^p (1 + z)^s$$

where p, s and the normalization are free parameters (in the standard scaling solution, $p = 4/3$ and $s = 3.5$).

2.2 Mass function

In order to compute the overall background arising from clusters, the mass function is needed. In the following we use the Press and Schechter recipe (Press and Schechter, 1974):

$$N(M) = \sqrt{\frac{2}{\pi}} \frac{\bar{\rho}}{M^2} \frac{f_c}{\sigma} \frac{dln\sigma}{dlnM} e^{-\nu^2/2}$$

Numerical simulations actually show that it provides an accurate description of the mass function in the self-similar regime (Efstathiou, *et al.* 1988). This is also supported by theoretical considerations (Bond et al., 1991; Blanchard, A., Valls-Gabaud D., G.Mamon, 1991). This mass function, for the CDM spectrum, is in good agreement with the optical luminosity function of clusters (Schaeffer and Silk, 1988)

2.3 Comparison with observed properties of clusters

The $M–T_X$ and $M–L_X$ relations introduce some free parameters into the theory. However the $M–T_X$ relation can be determined accurately from numerical simulations. The temperature distribution can then be compared confidently to the observations (Blanchard and Silk, 1991). The distribution function expected in CDM is in rather poor agreement with the observations: the high temperature clusters are too rare in CDM, especially when the recent discovery of a 14 keV cluster is taken into account (Arnaud et al, 1991), while low temperature clusters are overabundant by a factor 4. A power-law spectrum $\sigma \propto M^{-\alpha}$ with $\alpha = (3 + n)/6$. The value $n = -2$ then provides a much better fit to the data.

The parameter p in the relation between mass and luminosity can be specified by requiring the luminosity function of clusters to agree with the observed one. For the CDM spectrum (or for the $n = -1$ powerlaw spectrum) the value $p = 3$ is preferred, while for $n = -2$ powerlaw spectrum, a smaller value is required $p = 2$ (Blanchard and Silk, 1991). Another way to determine the value of p is obtained by noticing that the above relations imply a relation between the luminosity and the temperature:

$$L_X \propto T_X^{3p/2}$$

The observed relation is $L_X \propto T_X^{2.98}$. This leads to $p \approx 2$.

Clusters contribution

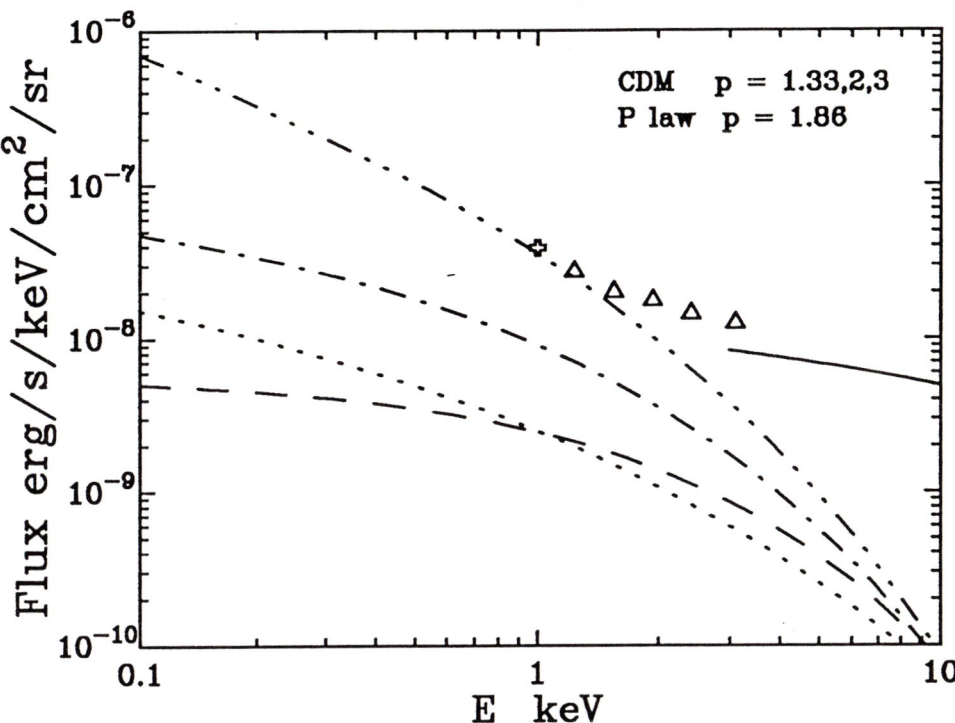

Figure : Resulting background in different models compared to the observations (full line corresponds to the 40 keV bremsstrahlung, triangles are the value inferred by Wu, Hamilton, Helfand and Wang (1991), and the cross has been derived from *ROSAT* spectrum) : the dashed line, dashed-dotted line, dashed-dotted-dotted-dotted line represent CDM model with $p = 1.33, 2., 3$, the dotted line represents the power law $n = -2$ and $p = 1.86$.

2.4 The resulting background

The contribution to the overall background can then be easily estimated

$$\phi_E = \int_0^{+\infty} dV(z) \int_0^{+\infty} dM \ \Phi(M, z) \frac{L_{E(1+z)}(M, z)}{4\pi D_{\text{lum}}^2(z)}$$

This background depends sensitively on p and is less sensitive to the evolution properties of the models. In the figure, we have plotted the spectrum resulting from various models (CDM with $p = 4/3, 2, 3$, power law model with $n = -2$ and $p = 3$). Models consistent with the observations lead to a quite sensitive fraction of the background arising from clusters.

2.5 Cluster number counts

If clusters contribute significantly to the background, as the above calculations seem to indicate, they should be detected in deep X-ray surveys.

Up to now, point sources identified in both *Einstein* and *ROSAT* deep counts were not identified as clusters. This is surprising at first as the contribution from clusters should come from relatively bright sources. However one should keep in mind that these sources are extended, therefore they are likely to have been missed by algorithms suited for point source detection. The recent discovery of the so-called "blotch"(Hasinger, Schmidt and Trümper, 1991) and its identification with a nearby loose group of galaxies provide evidence that the contribution of clusters (i.e. potential wells containing hot gas) with small temperature might be underestimated. The expected number counts predict that clusters as faint as the blotch should be quite common (Blanchard, Wachter, Evrard and Silk, 1991). Furthermore one should worry whether cluster samples available up to now do not suffer from systematic biases: low temperature clusters ($T < 3$keV) could have been missed by both by *HEAO-1*, because of the low energy cut-off of this experiment, while they could have been missed by *ROSAT* and *Einstein* because of their spatial extension. If the local density of low temperature clusters has been underestimated, their contribution to the X-ray background in our calculation has been underestimated as well.

3. CONCLUSION

The contribution of clusters to the soft X-ray background could easily be as large as 50% of the extragalactic flux. This contribution does actually depend on the luminosity mass relation. From the available observations, the temperature distribution function indicates that the power law index of the initial fluctuations should be close to $n = -2$ and that the parameter p entering in the relation between mass and luminosity should be of the order of 2 in order to explain both the luminosity function and the observed T_X–L_X relation. The scaling model ($p = 1.33$) seems to be confidently ruled out by the observations, as is the $n = -1$ powerlaw spectrum. With these constraints, the contribution of clusters to to the background could be of the order of 30 % of the unresolved component of the extragalactic background at 2 keV.

REFERENCES

Arnaud, M., et al. 1991, preprint.

Blanchard, A., Valls-Gabaud, D., Mamon, G. 1991, to be published.

Blanchard, A., Silk, J. 1991, to appear in the proceedings of the *XXVIth Rencontres de Moriond in Astrophysics*, Editions Frontières, Gif-sur-Yvette, France.

Blanchard, A., Wachter, K., Evrard, A.E., Silk, J. 1991, *Astrophys.J.*, to be published.

Boldt, E. 1987, *Physics Reports*, **146**, 215.

Bond, J.R., Cole, S., Efstathiou, G., Kaiser, N. 1991, *Astrophys.J.*, **370**, 440.

Cen, R.Y., Jameson, A., Liu, F., Ostriker, P.J. 1991, *Astrophys.J.*, **362**, L41.

Edge, A.C., Steward, G.C., Fabian, A.C., Arnaud, K.A. 1990, *Mon. Notices Roy. Astron. Soc.*, **245**, 559.

Efstathiou, G., Frenk, C.S., White, S.D.M., Davis, M. 1988, *Mon. Notices Roy. Astron. Soc.*, **235**, 715 .

Evrard, A., E. 1990, in *Clusters of Galaxies*, eds. W.R. Oegerle, M.J. Fitchett and L. Danly, (Cambridge: Cambridge Univ. Press), p. 287.

Hasinger, G., Schmidt, M., Trümper,J 1991, *Astrophys. Astron.*, **246**, L2.

Kaiser, N. 1986, *Mon. Notices Roy. Astron. Soc.*, **222**, 323.

Press, W., Schechter, P. 1974, *Astrophys.J.*, **187**, 425.

Silk, J., Tarter, J. 1973, *Astrophys.J.*, **183**, 387.

Schaeffer, R., Silk, J. 1988, *Astrophys.J.*, **333**, 509.

Stottlemeyer, A.R., Boldt, E.A. 1984, *Astrophys.J.*, **279**, 511.

Wu, X., Hamilton, T.T., Helfand J.J. and Wang, 1991, to be published.

Remnant Black Holes Associated with AGN Models for the High- and Low-Energy X-Ray Backgrounds

R. A. DALY

Department of Physics: Joseph Henry Laboratories, Princeton University

ABSTRACT

If AGN contribute substantially to either of the X-ray backgrounds, a significant mass density of remnant black holes must be present in the nuclei of galaxies. The energy density of each background is used to estimate the mean mass density of remnant black holes, which implies a black hole mass per galaxy that had an X-ray AGN phase. A lower bound on the fraction of L^* galaxies which must have had an X-ray AGN phase can be deduced from the isotropy of the background under consideration. These bounds are discussed and applied to estimate the masses of the remnant black holes and the fraction of present day L^* galaxies which are likely to contain massive black hole remnants.

1 INTRODUCTION

The X-ray background is observed to extend from a fraction of a keV to about 200 keV. The dominant contribution to the energy density of the X-ray background arises from photons with energies of about 30 to 40 keV. The observational bounds on the 1 to 3 keV X-ray background differ significantly from those on the 3 to 100 keV X-ray background, and it is therefore expedient to refer to these as two separate backgrounds. At the present time it is not known whether there are two populations of sources, one contributing primarily to the low energy background (the 1 to 3 keV background) and the other contributing to the high-energy background (the 3 to 100 or 200 keV background), or whether one population of sources simultaneously produces both backgrounds, that is, each source contributes significantly to both the high- and low-energy backgrounds. Hence, AGN models for the X-ray backgrounds divide into three categories, which are illustrated in Figure 1. Of course, there may be different types of AGN. If some AGN contribute to the high-energy background and others contribute to the low-energy background, they should be considered to constitute two separate classes of AGN, and are discussed as such here.

2 THE CONSTRAINTS

Figure 1.   One class of AGN may produce both the high- and low-energy X-ray
backgrounds, in which case the spectrum
of an individual AGN should resemble that
of the dotted line, or that of the dotted
line at energies greater than about 5 keV
plus an extrapolation of the dotted line to
lower energies. Alternatively, one class of
AGN would produce the high-energy background but not the low-energy background
if the spectrum is absorbed at low-energies,
as indicated by the solid curve. Another
class of sources, or of AGN, could contribute to the low-energy background but
not to the high-energy background if the
sources have steep spectra, as indicated by
the solid line.

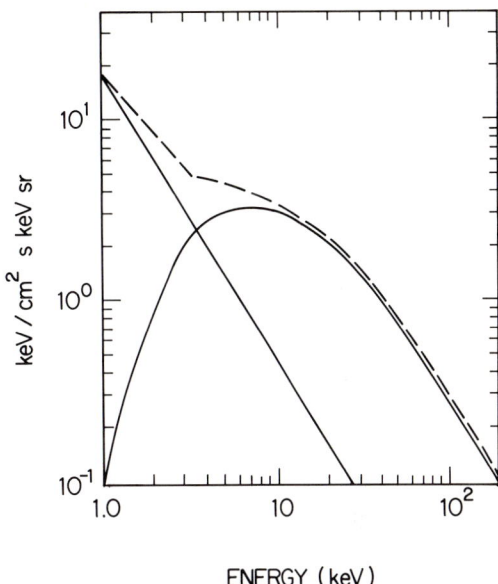

2.1 The Energy Constraints

In AGN models for the origin of the backgrounds the X-ray emission arises from
the active nucleus and is associated with processes occuring in the vicinity of the
black hole. As matter falls onto the black hole, the mass of the black hole grows
and energy is released in the form of radiation. Given that the AGN emits a total
energy E_x in X-rays with an efficiency ϵ, a lower bound on the black hole mass
is $M_{bh} \simeq E_x \, \epsilon^{-1} \, c^{-2}$. The energy density ρ produced by AGN at a characteristic
redshift z_x, with a comoving number density n_o, that would be detected at a redshift
of zero is $\rho = n_o \, E_x (1 + z_x)^{-1}$. In order to compare the comoving number density
of this class of AGN with that of present day L^* galaxies, their comoving number
density is parameterized by δ: $n_o = 3 \times 10^{-3} \, \delta \, h^3 \, \mathrm{Mpc}^{-3}$, where Hubble's constant
is $H_o = 100 \, h \, \mathrm{km \, s^{-1} \, Mpc^{-1}}$. The comoving number density of L^* galaxies at the
current epoch is obtained with $\delta \simeq 1$.

Combining the expressions for the black hole mass, the comoving number density
of sources, and an energy density of about 10^{-16} erg cm^{-3} for the high-energy background (e.g. Daly and Turner 1988), the minimum black hole mass in some fraction
δ of L^* galaxies at the current epoch is $M_{bh,heb} \sim 10^7 \, h^{-3} \, \delta^{-1} (w_x/3) \, (\epsilon/0.1)^{-1} M_\odot$,
where $w_x \equiv 1 + z_x$. The energy density of the low energy background in about one
tenth of that of the high-energy background, hence, if AGN produce this background
the requisite black hole mass is $M_{bh,leb} \sim 10^6 \, h^{-3} \, \delta^{-1} (w_x/3) \, (\epsilon/0.1)^{-1} M_\odot$. Note that
in Compton reflection models the masses of the remnant black holes may be quite
large, since $\epsilon \sim 0.01$ (e.g. White *et al.* 1988; Fabian *et al.* 1990; Rogers and Field

1991).

The black hole mass is strongly dependent on the value of Hubble's constant, being proportional to h^3. A lower bound on the age of the universe at the present epoch is about 12.5×10^9 yr, which implies that in a flat matter-dominated universe $h \lesssim 0.5$, in an open universe $h \lesssim 0.75$, and in a flat universe dominated by a cosmological constant $h \lesssim 0.95$ (for a detailed discussion of this and other points see Daly 1991 and Cowie 1989). The black hole remnants may be distributed such that most L^* galaxies contain black holes of "modest" mass, or a small fraction of L^* may contain very massive black holes. A lower bound on the fraction of L^* galaxies that contain black holes arises from the isotropy constraint of the background. This is to be compared with the current constraints on and detections of massive black holes in galactic nuclei as discussed, for example, by Kormendy (1992) and Lauer (1992).

2.2 The Isotropy Constraints

The isotropy constraint of the 1 to 3 keV background is substantially tighter than that of the high-energy background. Sources producing 100% of the flux of the 1 to 3 keV background must number at least 5000 per square degree, as discussed by Hamilton and Helfand (1987), and this number increases if the sources are clustered. However, if the typical low-energy spectrum of the sources is an extrapolation of the high-energy spectrum through the low-energy band, then these sources will produce about half of the flux of the low-energy background. In this case the isotropy limit is relaxed, and the sources must number at least 1000 per square degree, as discussed by Daly (1991). The isotropy of the high-energy background indicates that there must be at least 50 sources per square degree contributing to this background (Shafer 1983).

For a given comoving number density of AGN, the surface density of sources is largest when each source produces X-rays for a relatively long period of time, longer than the Hubble time when the sources begin to produce X-rays. In this case the surface density of sources depends on the fraction δ of L^* galaxies that go through an X-ray AGN phase, the redshift interval of the X-ray AGN phase, and the cosmological model. Allowing the X-ray emission to occur over the generous redshift interval from about 1 to 5, the surface density of sources depends on δ: about $10^4 \delta$ per square degree in a flat matter-dominated universe, and about $10^5 \delta$ per square degree in an open universe (see Daly 1991).

Hence, if the sources produce about one half of the flux of the 1 to 3 keV background, they must number at least 1000 per square degree, indicating that $\delta > 0.1$ in a flat matter-dominated universe, and $\delta > 0.01$ in an open universe. If the sources comprise all of the flux of the 1 to 3 keV background, then they must number at least

5000 per square degree indicating that $\delta > 0.5$ in a flat matter-dominated universe, and $\delta > 0.05$ in an open universe. If the sources are to produce the high-energy background but not the low-energy background, the surface density of sources is at least 50 per square degree, indicating that $\delta > 5 \times 10^{-3}$ in a flat matter-dominated universe, and $\delta > 5 \times 10^{-4}$ in an open universe.

3 ONE CLASS OF AGN PRODUCES BOTH BACKGROUNDS

The energy density of the backgrounds is dominated by the high-energy background. Therefore, the black hole mass in a fraction δ of present day L^* galaxies if one class of AGN produces both backgrounds is $M_{bh,heb} \sim 10^7 \, h^{-3} \, \delta^{-1} (w_x/3) \, (\epsilon/0.1)^{-1} M_\odot$. Let us assume that the high-energy spectrum extrapolates through the low energy band, in which case the isotropy constraint requires that there be 1000 sources per square degree. In a flat matter-dominated universe the present age of the universe implies that $h = 0.5$, indicating a black hole mass of about $M_{bh} \sim 10^8 \, \delta^{-1} \, (\epsilon/0.1)^{-1} M_\odot$; the isotropy constraint implies that $\delta > 0.1$, and, for $\epsilon \lesssim 0.1$, observations of local galaxies suggest that $\delta < 0.5$. Therefore, if one class of AGN produces both the high- and low-energy backgrounds in a flat matter-dominated universe the remnant black hole mass should lie within the limits given by $0.5 \gtrsim \delta \gtrsim 0.1$: between about one half of L^* galaxies contain black holes of mass $M_{bh} \sim 2 \times (10^8 \text{ to } 10^9) M_\odot$ (for $\delta \simeq 0.5$ and $\epsilon \sim 0.1$ to 0.01), and about one out of ten L^* galaxies contain black holes of mass $M_{bh} \sim (10^9 \text{ to } 10^{10}) M_\odot$ (for $\delta \simeq 0.1$ and $\epsilon \sim 0.1$ to 0.01). Note that Compton reflection models have the positive aspect that the spectrum resembles that of the high-energy background, however, in these models $\epsilon \sim 0.01$, and the mass of the remnant black hole is very large. In an open universe $h = 0.75$, and the mass of the remnant black hole is $M_{bh} \sim 2 \times 10^7 \, \delta^{-1}(\epsilon/0.1)^{-1} M_\odot$; the isotropy constraint implies that $\delta > 0.01$. Observations of L^* galaxies imply that $\delta \lesssim 1$ for this mass, hence the remnant black holes should lie within the limits given by $1 \gtrsim \delta \gtrsim 0.01$: between about each L^* galaxy has a black hole of mass $M_{bh} \sim 2 \times (10^7 \text{ to } 10^8) M_\odot$ (for $\delta \simeq 1$ and $\epsilon \sim 0.1$ to 0.01) and about one out of every 100 L^* galaxies has a black hole with mass $M_{bh} \sim 2 \times (10^9 \text{ to } 10^{10}) M_\odot$ (for $\delta \simeq 0.01$ and $\epsilon \sim 0.1$ to 0.01).

4 AGN PRODUCE THE HIGH-ENERGY BACKGROUND

If AGN produce the high-energy background then the mass of the remnant black hole in a fraction δ of L^* galaxies is $M_{bh,heb} \sim 10^7 \, h^{-3} \, \delta^{-1}(w_x/3) \, (\epsilon/0.1)^{-1} M_\odot$. In a flat matter-dominated universe the isotropy bounds imply that $\delta > 5 \times 10^{-3}$ and the present age of the universe requires $h \lesssim 0.5$. For $h = 0.5$ observations of local galaxies suggest that $\delta < 0.5$ (see Daly 1991), hence the black holes masses and number densities should lie between the limits given by $0.5 \gtrsim \delta \gtrsim 5 \times 10^{-3}$: between about one half of L^* galaxies contain black holes with mass $M_{bh} \gtrsim 2 \times (10^8 \text{ to } 10^9) M_\odot$ (for $\delta \simeq 0.5$ and $\epsilon \sim 0.1$ to 0.01) and about one out of 200 L^* galaxies contain black holes of mass $M_{bh} \sim 2 \times (10^{10}$ to $10^{11})M_\odot$ (for $\delta \simeq 1/200$ and $\epsilon \sim 0.1$ to 0.01). In an open universe the isotropy constraint implies that $\delta \gtrsim 5 \times 10^{-4}$, and in such a universe $h \lesssim 0.75$. Therefore, in an open universe the energy and isotropy constraints imply that the black hole remnants should lie between the limits given by $1 \gtrsim \delta \gtrsim 5 \times 10^{-4}$: every L^* galaxy has a black hole with mass $M_{bh} \gtrsim 2 \times (10^7$ to $10^8)M_\odot$ (for $\delta \simeq 1$ and $\epsilon \sim 0.1$ to 0.01) and one out of every 2000 L^* galaxies has a black hole with mass $M_{bh} \sim 4 \times (10^{10}$ to $10^{11})M_\odot$ (for $\delta \simeq 1/2000$ and $\epsilon \sim 0.1$ to 0.01).

5 AGN PRODUCE THE LOW-ENERGY BACKGROUND

If AGN produce the total energy density of the 1 to 3 keV background then they must have a surface density of at least 5000 sources per square degree, and the remnant black holes must satisfy $M_{bh,leb} \sim 10^6 \; h^{-3} \; \delta^{-1} \; (w_x/3) \; (\epsilon/0.1)^{-1} \; M_\odot$. In a flat matter-dominated universe the present age of the universe implies that $h = 0.5$, and the isotropy constraint implies that $\delta > 0.5$, so $1 \gtrsim \delta \gtrsim 0.5$ and about half of all L^* galaxies must have a black hole with a mass $M_{bh} \sim 2 \times (10^7$ to $10^8)M_\odot$ (for $\delta \simeq 0.5$ and $\epsilon \sim 0.1$ to 0.01), or every L^* galaxy has a black hole with a mass of about $M_{bh} \sim (10^7$ to $10^8)M_\odot$ (for $\delta \simeq 1$ and $\epsilon \sim 0.1$ to 0.01). In an open universe $h = 0.75$, and the isotropy constraint implies that $\delta > 0.05$. Hence, in an open universe the energy and isotropy constraints imply that the density and mass of the remnant black holes lie between the limits given by $1 \gtrsim \delta \gtrsim 0.05$: between about every L^* galaxy has a black hole with mass $M_{bh} \sim 2 \times (10^6$ to $10^7)M_\odot$ (for $\delta \simeq 1$ and $\epsilon \sim 0.1$ to 0.01), and one out of 20 L^* galaxies has a black hole with mass $M_{bh} \sim 4 \times (10^7$ to $10^8)M_\odot$ (for $\delta \simeq 0.05$ and $\epsilon \sim 0.1$ to 0.01).

It is a pleasure to thank Xavier Barcons and Andy Fabian for organizing the meeting, and the British and Spanish Governments for their support. This work was supported in part by the U.S. National Science Foundation.

Cowie, L.L. 1989, in Two-Topics in X-ray Astronomy, ed. J.J. Hunt & B. Battrick
    (Noordwijk: ESA), 797
Daly, R. A. 1991, ApJ, 379, 37
Daly, R. A. & Turner, E. L. 1988, Comments Ap., 12, 219
Fabian, A. C., George, I. M., Miyoshi, S., & Rees, M. J. 1990, MNRAS, 242, 14p
Hamilton, T. T., & Helfand, D. J. 1987, ApJ, 318, 97
Kormendy, J. 1992, in Testing the AGN Paradigm, (New York: AIP), in press
Lauer, T. R. 1992, in Testing the AGN Paradigm, (New York: AIP), in press
Rogers, R. D., & Field, G. B. 1991, ApJ(Letters), 370, L57
Shafer, R. A. 1983, Ph.D. Thesis, University of Maryland
White, T. R., Lightman, A. P., & Zdziarski, A. A. 1988, ApJ, 331, 939

A Compton Reflection Model for the Cosmic X-Ray and Gamma-Ray Backgrounds

R.D. ROGERS and G.B. FIELD

Harvard-Smithsonian Center for Astrophysics

1 ABSTRACT

In this paper we show that Active Galactic Nuclei (AGN) whose X-ray spectra are dominated by a Compton reflected component can explain both the X-ray and gamma-ray backgrounds. Such a model requires that the X-rays be produced by inverse-Compton scattering of soft photons from an accretion disk by relativistic electrons located in a thin layer above the disk. We explore some implications of this model.

2 INTRODUCTION

The discovery that *Ginga* spectra of AGN are better fit by a Compton reflected power law model than by a pure power law (Pounds *et al.* 1990) led to the interesting suggestion that Compton-reflection-dominated AGN may contribute significantly to the X-ray background (XRB: Fabian *et al.* 1990). This model has been explored in detail, and it has been found that such a model can explain the spectrum of the 3 - 500 keV XRB for a broad range of model parameters (Fabian *et al.* 1990; Rogers and Field 1991a). In addition, we have examined the possibility that the $0.5 - 10$ MeV gamma-ray background (GRB) may also be explained by such a model (Rogers and Field 1991b). In this paper we describe this research.

3 COMPTON REFLECTION MODEL

In the Compton reflection model, an initial spectrum of X-rays impinges on a mass of cold gas ($T < 10^6$ K) which reprocesses the spectrum (Lightman and White 1988; White, Lightman and Zdziarski 1988). Here we assume that this gas is located in an accretion disk. The reflected spectrum is cut off at energies below about 15 keV due to photoelectric absorption. If the initial spectrum is a power law, then the reflected spectrum is also steepened by ≈ 1 in the power law index at energies above about 120 keV due to energy losses from repeated Compton scatterings. The observed spectrum from a Compton reflected source is then a superposition of the initial spectrum and the reprocessed spectrum. The 3 - 500 keV range in Figure

1 shows such a spectrum for an initial power law energy index, $\alpha = 1.1$. In this spectrum a fraction, $f = 0.89$, of the initial power law energy impinges upon cold gas.

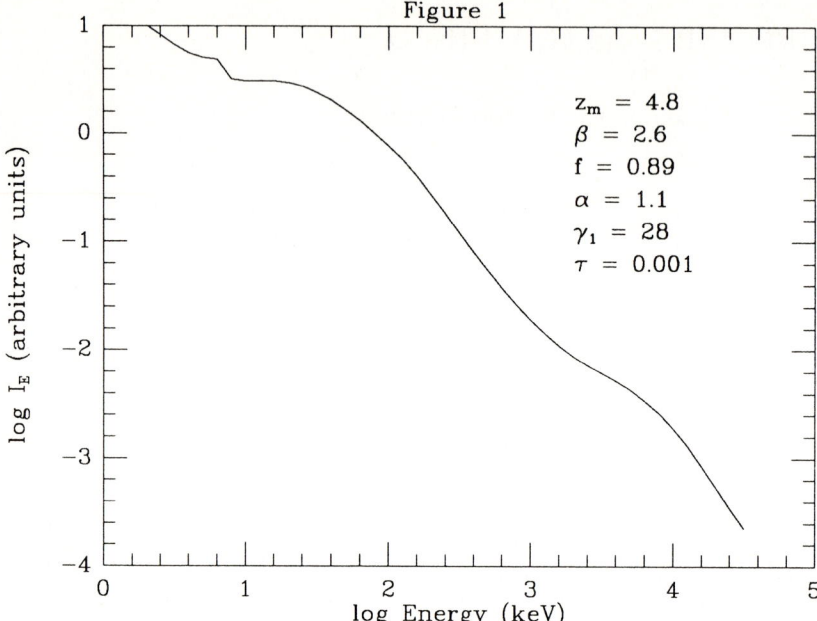

Figure 1   X-ray spectrum of an AGN in the Compton reflection model. Note the flattening at energies above 1 MeV due to second-order inverse-Compton scattering.

4 X-RAY BACKGROUND

In order to compare a model spectrum, such as that shown in Figure 1, to the observed XRB spectrum we must integrate the emission from AGN over a range of redshifts using an appropriate evolution function. In this work we assume that the volume spectral emissivity, S_E, of these sources can be written

$$S_E = L_E n_o (1+z)^{3+\beta} \tag{1}$$

where L_E is the spectral luminosity of a source, n_o is the current number density of sources and β is the evolution parameter. Integrating this emissivity over redshift up to a maximum redshift, z_m, yields the intensity, I_E, of the XRB,

$$I_E = \frac{3ct_o}{8\pi} \int n_o L_{E(1+z)} (1+z)^{-2.5+\beta} dz. \tag{2}$$

We then fit this model to the XRB, allowing the parameters α, f, z_m and β to vary. The resulting best-fit model is shown in Figure 2, where the model parameters are

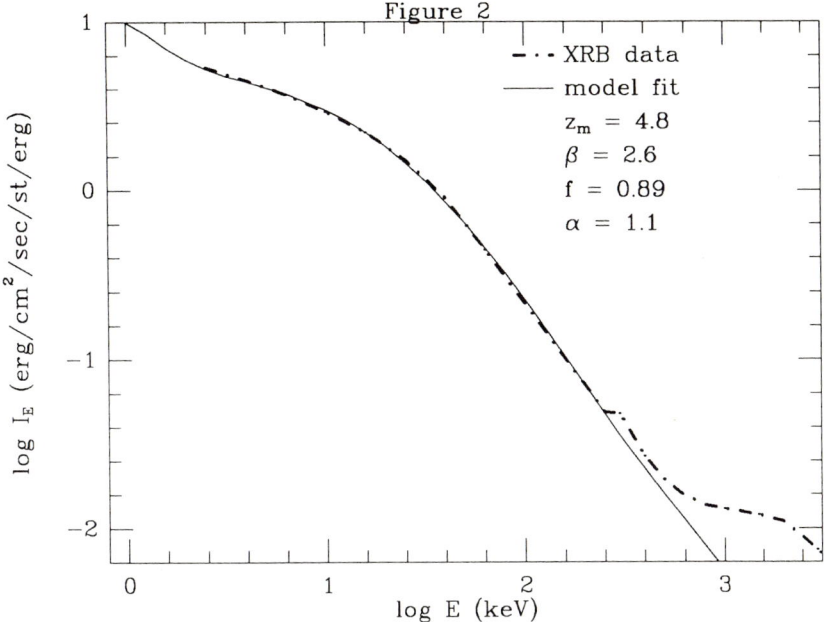

Figure 2 Comparison of observed high energy background spectrum and Compton reflection model without second-order inverse-Compton scattering component.

$\alpha = 1.1$, $f = 0.89$, $z_m = 4.85$, and $\beta = 2.6$. Note that the fit is excellent over the entire 3 - 500 keV energy range of the XRB.

5 GAMMA-RAY BACKGROUND

If the initial power-law X-rays are due to inverse-Compton scattering of soft photons by relativistic electrons, then some X-rays may themselves be scattered up to gamma-ray energies. In order to calculate the detailed spectrum in this case, we need to know the energy distribution and Thomson depth of the relativistic electrons. Here we assume that the electron energy distribution is

$$\frac{dN}{d\gamma} = K_e \gamma^{-p} \qquad \gamma > \gamma_1 \tag{3}$$

where γ is the relativistic Lorentz factor of the electrons, p is the electron power law index, K_e gives the density normalization and γ_1 is a low energy cutoff.

Defining τ to be the optical depth of the electon layer, we can then calculate the AGN spectrum from 3 keV to tens of MeV. The resulting spectrum, with $p = 3.2$, $\gamma_1 = 28$ and $\tau = 0.001$ is shown in Figure 1, where the flattening at energies greater than about 1 MeV is due to second-order inverse-Compton scatterings. Performing the same integration over redshift as for the XRB, and keeping the XRB best-fit parameters, we find the model high-energy background spectrum shown in Figure 3. Note that the model now fits the cosmic background spectrum over the entire 3 keV to 10 MeV energy range.

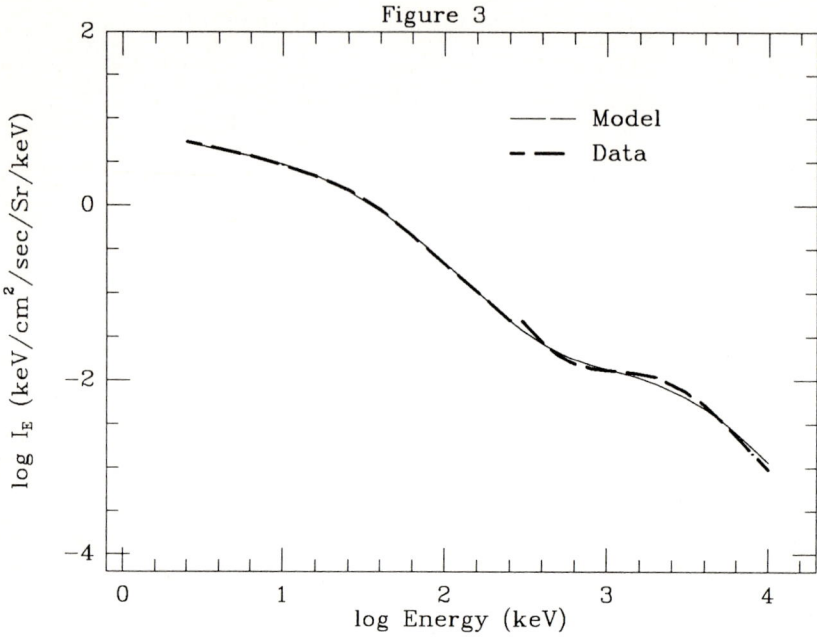

Figure 3 Comparison of Compton reflection model, including second-order inverse-Compton scattering, with the observed high energy background spectrum.

6 DISCUSSION

It can be seen in Figure 3 that the model is capable of explaining the spectrum of the high energy background. By calculating a least squares fitting parameter for various models, we have put constraints on the allowed ranges of these parameters. These allowed ranges are given in the Table below.

Table
Allowed Ranges of Model Parameters

Parameter	Value	Parameter	Value
z_m	≥ 2.8	f	$0.85 - 0.95$
β	$1.3 - 5.3$	τ	$(3 - 10) \times 10^{-4}$
α	$0.9 - 1.2$	γ_1	$15 - 30$

An important result is that the covering factor, f, must be greater than 0.85, significantly larger than the geometrical covering of an accretion disk (≈ 0.5) and the covering observed in nearby AGN by *Ginga* (0.4 - 0.5: Pounds *et al.* 1990). There are two factors which may lead to large values of f. First, the inverse-Compton radiation from an anisotropic radiation field is also anisotropic, peaked toward the initial photon source (Ghisellini *et al.* 1990; Rogers 1991). Second, if the relativistic electrons are accelerated by reconnection at the tops of magnetic field loops, then there will be more electrons moving toward the disk than away, leading to enhanced emission in the direction of the disk. If an accretion disk is responsible for the Compton reflection, then both effects will lead to a large covering factor.

We have not yet found an explanation for the low energy cutoff, but it is not implausible that such a cutoff would exist in the electron distribution. Finally, we make a comment on the magnetic structure alluded to above. We expect that ordered magnetic fields will play a large role in these objects for two reasons. First, a magnetic field will confine relativistic electrons to the region just above the accretion disk, and second, magnetic reconnection is a plausible means of transferring energy from the gravitational potential energy of infalling material to the acceleration of relativistic electrons. If magnetic fields dominate, then it is possible to self-consistently calculate the entire AGN spectrum from IR to gamma-ray energies (Field and Rogers 1991).

We conclude that Compton reflection in AGN can account for the X-ray and gamma-ray backgrounds, and leads to interesting predictions for the spectra and structures of high redshift AGN.

7 REFERENCES

Fabian, A. C., George, I. M., Miyoshi, S., and Rees, M. J. 1990, M.N.R.A.S., **242**, 14p.

Field, G. B., and Rogers, R. D. 1991, in preparation.

Ghisellini, G., George, I. M., Fabian, A. C., and Done, C. 1990, M.N.R.A.S., **248**, 14.

Lightman, A. P., and White, T. R. 1988, Ap.J., **335**, 57.

Pounds, K. A., Nandra, K., Stewart, G. C., George, I. M., and Fabian, A. C. 1990, preprint.

Rogers, R. D. 1991, Ap.J., **383**, 550.

Rogers, R. D., and Field, G. B. 1991a, Ap.J. (Letters), **370**, L57.

Rogers, R. D., and Field, G. B. 1991b, Ap.J., **366**, 22.

White, T. R., Lightman, A. P., and Zdziarski, A. A. 1988, Ap.J., **331**, 939.

On a Possible Contribution to XRB and the Heating of the IGM

I. E. DE SALAMANCA

Obs. de Meudon, DAEC, Paris 7

Abstract A possible explanation of the XRB, in connection with the Gamma-Ray Background - GRB -, proposed by Collin-Souffrin,(1991a and b) is discussed. We see that with this model we are able to reproduce the XRB and GRB.

1 RECALLS ON XRB AND GRB

Let us recall the basic points on the XRB and GRB:

.Between 1 - 10 kev it can be explained by the sum of discrete sources, like AGN, clusters of galaxies, starburst galaxies, etc.

.In the hard range (10 to 100 kev), however, a satisfactory explanation is not found. We know that a diffuse intergalactic medium cannot be responsible for this emission, so we must look for very numerous discrete sources. One type of candidate is local AGN, with a spectrum given by the Reflection Model, discussed in this meeting. Another possibility, proposed by Collin-Souffrin (1991a and b) is that the hard X-ray Background is due to large redshift objects emitting in the gamma ray range. We will see later what it does mean exactly.

.And finally, between 100 kev and 10 Mev, we have the Gamma- ray Background (GRB), whose most remarkable features are the similarity with the spectrum of the XRB, and the bump at a few Mev. Its origin is also unknown, but it seems likely that it is due to local AGN. (Bassani and Dean, 1982, Mereghetti, 1990)

2 ACTIVE AND INACTIVE GALACTIC NUCLEI: MASSIVE BLACK HOLES

It is thought that AGN are fueled by accretion onto a massive black hole (Rees, 1984). So the mass accumulated in AGNs is equal to:

$$M_{BH} = \frac{1}{\eta c^2} \frac{4\pi}{c} \int dz (1+z) \int dS S n(S, z)$$

(Soltan, 1983)

where S is the received flux, n is the density of sources, z is the redshift, η is the efficiency of the mass-energy conversion, and c is the speed of light.

Soltan calculated this mass for quasars, e.g. the mass accumulated in black holes at z

1-3, and obtained $510^{13} M_o Gpc^{-3}$, a value revised by Padovani, Burg, Edelson (1990). If we compare this mass with the total mass in local AGN (at $z = 0$), $10^{11} Gpc^{-3}$, we find that there is at least a factor 100 of difference between the two masses !

The immediate question is: **where are the "lost" black holes?** And the logical answer is that they must be in inactive galactic nuclei (IGN) (Cavaliere and Padovani, 1990) Some observational evidences support this idea: for example the existence of LINERS, which can be explained by a small black hole accreting at a very low rate, in the core of Radiogalaxies, and the large mass concentration (with an M/L ratio 100) in the cores of some inactive galaxies, as M31, M32, M104 ...

3 HYPOTHESIS
The following hypothesis about these black holes are made by Collin-Souffrin:
1 They are formed at hight redshift ($z > 5$)
2 They are able to convert an important fraction of their mass into energy, i.e. $\eta M c^2$, with $\eta \sim 0.1$ and $M \sim 210^5 M_o$ during their growing phase ($t_{Edd} \sim 10^8$ years)
3 This energy is radiated mainly in the gamma range.
(Remember the model of Boldt & Leiter, 1982, in which AGN progenitors radiate a gamma-ray thermal spectrum)
4 They radiate the same spectrum as local AGN:

$$I(E) = E^{-\alpha_x}[1 + \frac{E}{E_{break}}^{\alpha_\gamma - \alpha_x}]^{-1}$$

With this simple model we are able to reproduce the spectrum of both backgrounds, the XRB being accounted by the growing black holes at $z > 5$, and the GRB by the local AGN. We can add the contribution of others sources, as local AGNs, clusters of galaxies, etc., to the XRB, and find a rather good general fit. (cf figures 1 and 2). An impressive success of the model is that the background intensity is obtained assuming that a proportion of $\sim 75\%$ of the Eddington luminosity of the growing black holes is converted into energy, with an efficiency $\eta = 0.1$.

4 RESULTS
With this model, I have reproduced the spectrum of the XRB, adding the contribution of AGN and clusters of galaxies, as they are the most important known sources in the soft X ray range.
The results are summarized in the following figures, which are obtained with this model,for $H_o = 75 Km s^{-1} Mpc^{-1}$ and $\Omega = 1$. For the rest of the parameters, see the notes on the figures. The first figure is from Collin-Souffrin, which shows both backgrounds, the XRB and GRB, but takes into account only black holes and local AGN. All the rest are only for the XRB, as I had concentrated in this part of the spectrum.

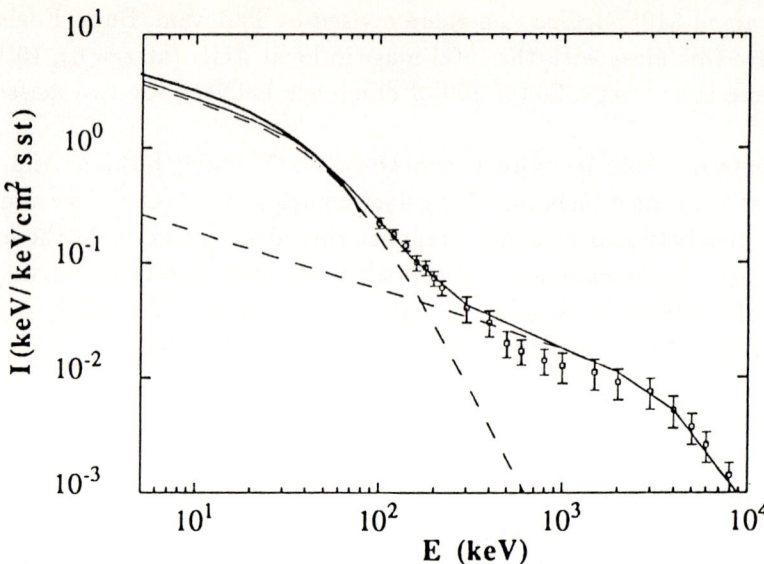

Figure 1: Computed and observed XRB and GRB, from Collin-Souffrin (1991b). Dashed lines: separated contribution of black holes and AGN. Full bold line: fit of the observations by Marshall at al. Full thin line: computed spectrum. The parameters are: $\alpha_x = 0.5$, $\alpha_\gamma = 3$, $E_{break} = 500$keV, fraction of L_{EDD} radiated: 75 %, $z_{init} = 24$, $z_{fin} = 5$

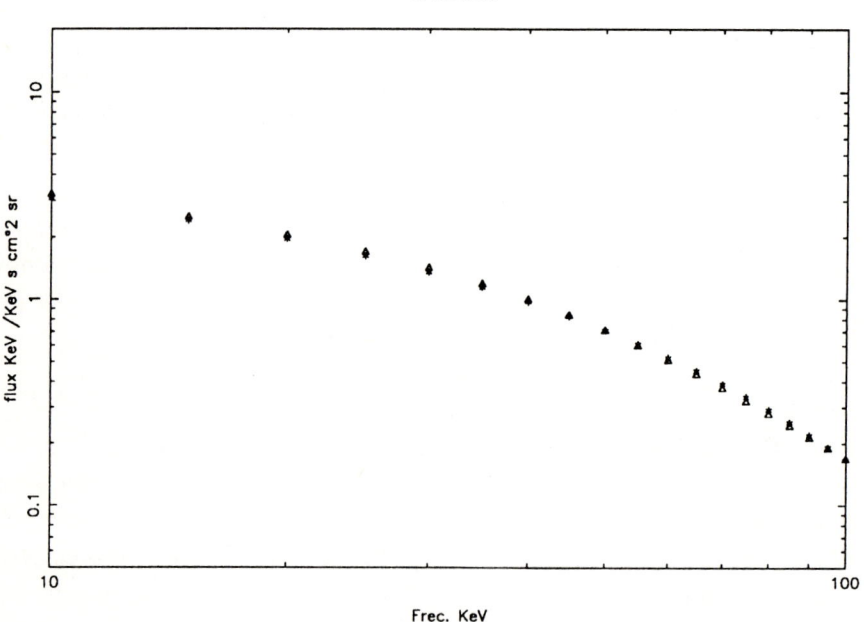

Fig. 2a

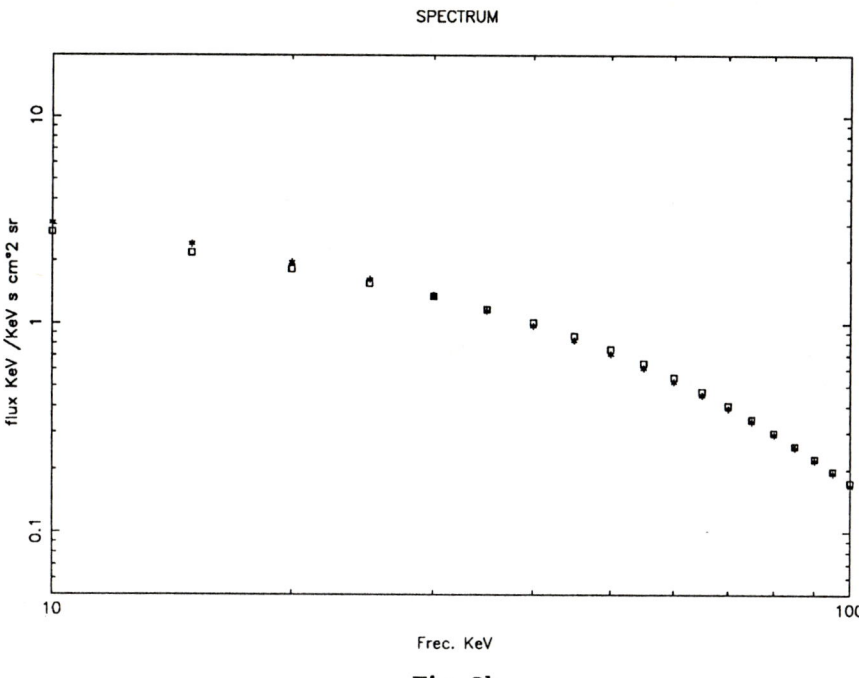

Fig. 2b

Figures 2: Calculated and observed XRB spectrum for two set of parameters. Crosses: observed XRB spectrum. Others: computed spectrum. 2a) $\alpha_x = 0.5$, $\alpha_\gamma = 3$, $E_{break} = 500\text{keV}$, fraction of L_{EDD} radiated: 50 %, $z_{init} = 15$, $z_{fin} = 8$ 2b) $\alpha_x = 0.6$, $\alpha_\gamma = 3.5$, $E_{break} = 700\text{keV}$, fraction of L_{EDD} radiated: 50 %, $z_{init} = 24$, $z_{fin} = 9$

References

Bassani L., Dean A.J., 1983, Space Science Review 35, 367

Bassani L., Dean A.J., 1984, Astrophysics and Space Science 100, 457

Cavaliere A., Padovani P., 1989, Ap.J. 340, L5

Collin-Souffrin S., a, 1991, A & A 243,5

Collin-Souffrin S., b, 1991, to be published in the proceedings of the Moriond Conference on Diffuse Backgrounds, Mars 1991, Eds. Deharveng and Rocca-Volmerange.

Marshall et al., 1980, Ap.J. 235, 4

Maccacaro et al.,1991,Ap.J. 374, 117

Mereghetti S., 1990, Ap.J.354, 58

Padovani P., Burg R., Edelson R.A., 1990, Ap.J. 353, 438

Piccinoti et al. 1982, Ap.J. 253, 485

Rees M.J., 1984, Ann. Rev. Astr. Ap. 22, 471

Soltan A., 1982, Mon. Not. R. Astr. Soc. 200, 115

I would like to thanks to Suzy Collin and Marc Lachièze-Rey, who helped me to do all this work.

Massive X-Ray Binaries in Star-Forming Galaxies : Contribution to the X-Ray Background

M.TREYER(1,2), M.MOUCHET(2), A.BLANCHARD(2) & J.SILK(1)

(1) U.C. Berkeley, Astronomy Dept., CA 94720, U.S.A.
(2) Observatoire de Meudon, DAEC, F-92190 Meudon, France.

INTRODUCTION :

Massive X-Ray Binaries (MXRB) are composed of a compact source (neutron star or black hole) accreting matter from a very closeby O-B (Be) star. Hard X-ray emission results from this mass transfer. These objects are expected to be produced in large number during bursts of star formation, and might then be the main source of the X-ray luminosity of starburst galaxies. It is therefore plausible that MXRB in starburst galaxies be responsible for a substantial fraction of the X-ray background (Griffiths 1989, Griffiths & Padovani 1990).

We present a model in which starburst galaxies are powered by the X-ray emission of MXRB, and numerous and/or bright enough in the past to generate flux intensities comparable to those of the observed XRB. The initial assumptions are :
(i) the typical X-ray spectrum of starburst galaxies, that we will assume to be a MXRB composite spectrum.
(ii) the X-ray luminosity function of starburst galaxies, as a function of time.
We focus on 3 observational constraints : the spectrum of the XRB, the small-scale anisotropies, and the energetics constraints.

(i) The MXRB spectral energy distribution : sed(E)dE

We have constructed a MXRB composite spectrum from a homogeneous sample of eight galactic massive X-ray binaries (White, Swank and Holt 1982). As there is no clear correlation between spectral characteristics and absolute luminosity, all spectra were arbitrarily normalised at 3 keV and attributed a similar weight. The composite spectrum is best fitted by a double power-law with parameters :

$$E_c = +19.28 \ keV$$
$$E \lesssim E_c, \quad \alpha_1 = -0.217$$
$$E > E_c, \quad \alpha_2 = -2.666$$

Uncertainties in the mean spectrum were estimated by bootstrap resampling method. We have computed two extreme representative spectra corresponding to the mean spectrum modified by $\pm 2\sigma$. Despite the small number of objects in the sample, the dispersion is very weak.

(ii) The X-ray luminosity function : $\phi(L_X, z) dlog L_X$

Since most of the blue light from young massive stars is reprocessed into the far-infrared by dust heated by these stars (e.g. Young, Kleinmann, and Allen 1988), far-infrared emission is probably the best indicator of active star formation. Griffiths & Padovani (1990) derived a relationship between X-ray (in the 0.5-3.0 keV band) and 60μm luminosities from a sample of IRAS galaxies, and a group of starburst/interacting galaxies. Following their approach, we use this X-ray/IR correlation, along with the IRAS luminosity function of star-forming galaxies (Saunders *et al.* 1990), to derive the X-ray luminosity function of these sources. The contribution of local starburst galaxies to the XRB is negligeable (less than 2% of the 1-10 keV XRB). Their X-ray light density must have been a lot larger in the past to be able to provide a substantial fraction of the XRB. Deep IRAS counts suggest that starburst galaxies were either brighter or more numerous in the past (Lonsdale & Hacking 1989, Saunders *et al.* 1990). We model these two types of evolution in the standard way :

$$L_X(z) = L_X(0) \times \exp\left(t_{lb}(z)/\tau\right) \tag{A}$$

$$\phi_\star(z) = \phi_\star(0) \times (1 + z)^{\mathrm{D}} \tag{B}$$

t_{lb} refers to the lookback time. τ and D are fitting parameters, the required values of which we discuss in the following section.

1 THE X-RAY BACKGROUND SPECTRUM

At a given energy E, the X-ray flux due to MXRB in star-forming galaxies is the integrated emission of all these sources over luminosity and space. The general equation can be written :

$$\frac{dI(E)}{dE} = \int_z \int_{L_X} \frac{\mathrm{sed}(E(1+z))(1+z)}{4\pi D_L^2(z)} L_X \phi(L_X, z) d\log L_X(z) dV(z) \tag{1}$$

$D_L(z)$ and $dV(z)$ are the luminosity distance and the volume element per unit solid angle respectively. We assume a standard cosmological model with $\Omega_0 = 1$, $\Lambda = 0$ and $\mathrm{H}_0 = 50$ km/s/Mpc.

The evolution parameters are fixed so that the MXRB contribute 100 % of the observed flux at 3 keV. The spectrum of the background is accurately measured above this value but poorly known below.

The epoch z_{max} up to which galactic evolution is traced back in time strongly influences the shape of the integrated spectrum. If most of the X-ray emission comes from very high redshifts, the energetic part of the MXRB spectrum is shifted to low energies. If on the contrary we assume that the source redshift distribution peaks at a relatively late epoch, which we can crudely model by a low redshift cutoff z_{max}

above which $\phi(L_X, z > z_{max}) \equiv 0$, then the shape of the redshifted MXRB spectrum stays in better agreement with the observed spectrum, at least up to ~ 20 keV. We have selected two 'acceptable' models, in terms of fit to the data. The resulting spectra are shown below. The shaded areas account for the $\pm 2\sigma$ error range in the MXRB composite spectrum. The dot-dashed line is the observed XRB spectrum. MXRB seem able to produce if not the whole, at least a dominant fraction of the 3-10 keV XRB, but clearly fail to explain the hard tail of the observed spectrum.

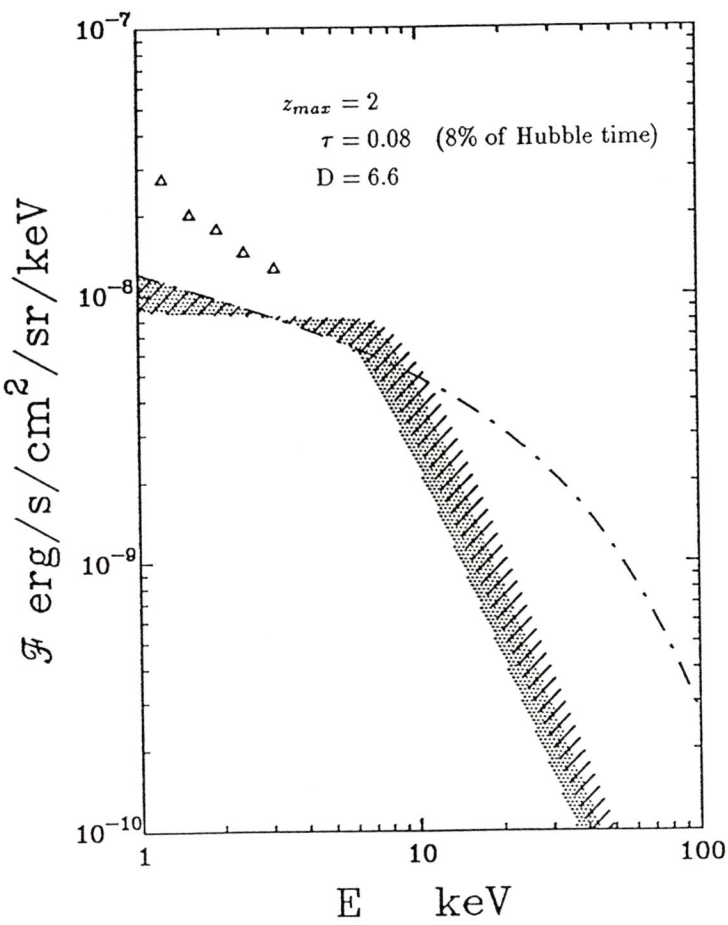

2 SMALL SCALE ANISOTROPIES

The auto-correlation function of the soft XRB was recently derived by Hasinger *et al.* (1991-a) from the ROSAT Medium Sensitivity Survey (0.5 - 1.6 keV). The small amplitude of the observed fluctuations is a strong constraint on the discrete source contribution and degree of clustering. We investigate whether the correlated component of the XRB due to star-forming galaxies is consistent with the observed amplitude of the fluctuations. The starburst galaxy spatial correlation function can be written :

$$\xi_*(r, z = 0) \; = \; b_*^2 \; \left(\frac{r}{r_0}\right)^{-\gamma} \tag{2}$$

where $\gamma \approx 1.8$, $r_0 \approx 5.5\,h^{-1}\mathrm{Mpc}$, and b_* is a bias parameter accounting for the uncertainty on the starburst galaxy correlation length. The way in which this function evolves with time can be parameterized as followed :

$$\xi_*(r, z) = (1 + z)^{-(3+\epsilon)} \; \xi_*(r, 0) \tag{3}$$

with $\epsilon = 0$ to -1.2. Most scenarios of galaxy formation and clustering sit within these two limiting cases (Efstathiou *et al.* 1991).

From all previous assumptions, we can derive the angular correlation function of the XRB surface brightness fluctuations contributed by the correlated star-forming galaxies : $c(\theta) = A \, \theta^{1-\gamma}$. A is a non trivial function of γ, r_0, and $\phi(L_X, z)$.

The auto-correlation function ($\mathrm{ACF}_{\mathrm{XRB}}(\theta)$) determined by Hasinger *et al.* (1991-a) quantifies the fluctuations of the still unresolved fraction of the soft XRB (f_{diffuse}). If f_{MXRB} is the fraction contributed by MXRB in undetected star-forming galaxies, then we have :

$$\mathrm{ACF}_{\mathrm{XRB}}(\theta) = c(\theta) \left(\frac{f_{\mathrm{MXRB}}}{f_{\mathrm{diffuse}}}\right)^2 \tag{4}$$

We have computed the amplitude of $c(\theta)$ for increasing values of the evolution parameters τ and D. Amplitude A decreases very steeply from no evolution to moderate evolution ($\tau \sim 20\%$ of H_0^{-1}), and then varies very little up to the drastic values necessary to produce 100% of the 3 keV flux. Consequently, we can use relation (4) to constrain f_{MXRB} (in the ROSAT band, namely ~ 1 keV). With $f_{\mathrm{diffuse}} \lesssim 55\%$ (Hasinger *et al.* 1991-b), the fluctuations in the model fall within ROSAT limits when :

$$b_* \; f_{\mathrm{MXRB}}(1\,keV) \lesssim 20\% \tag{5}$$

Starburst galaxies are therefore either less clustered than 'normal' optical galaxies, or else their contribution to the soft XRB is negligeable.

3 ENERGETICS

The MXRB hypothesis raises an energetic problem (Daly 1991). The X-ray production is related to that of massive stars, and thus implies a large production of heavy elements in the Universe.

The X-ray energy density generated by the MXRB in star-forming galaxies is the integral over time of their luminosity density : $E_X = \int_z \rho(L_X, z)dt(z)$. This energy represents a fraction ε of the MXRB initial mass energy : $E_X = \varepsilon \, \rho_{\mathrm{MXRB}}c^2$. We

may evaluate of rough order of magnitude of ε from known quantities in the Galaxy. A 'typical' MXRB has luminosity $L_X \approx 5 \times 10^{38}\,\mathrm{erg\,s^{-1}}$, mass $M \approx 20\,\mathrm{M_\odot}$ and lifetime $t_X \lesssim 10^6$ years. So $\varepsilon = E_X/Mc^2 \sim 10^{-5} - 10^{-4}$ at the present time. To be consistent with the model initial assumptions, we must expect this efficiency to be higher in the past, but the mean value over time is unlikely to exceed 10^{-3}. With respect to the critical density ρ_c, both evolution models (A) and (B) yield :
$\Omega_{\mathrm{MXRB}} = \rho_{\mathrm{MXRB}}/\rho_c \approx 4 \times 10^{-8}\varepsilon^{-1}$.

An OB star with mass $M \approx 20\,\mathrm{M_\odot}$ releases about $4\,\mathrm{M_\odot}$ of metals (mostly oxygen) into the interstellar medium after explosion (Woosley & Weaver 1986). Our model produces $\rho_Z \approx 4 \times 10^{-38}\,\varepsilon^{-1}\,\mathrm{g/\,cm^3}$. The observed metal density is $\sim 1 - 6 \times 10^{-34}\,\mathrm{g/\,cm^3}$ (Songaila *et al.* 1990). $\varepsilon \approx 10^{-4}$ implies that all the metals in the Universe be produced by MXRB.

Given the uncertainties the model can not really be ruled out, but it seems nevertheless seriously endangered by all three constraints.

REFERENCES

Daly, R.A. (1991) *Astrophys. J.*, **379**, 37.

Efstathiou, G., Bernstein, G., Katz, N., Tyson, J.A. and Guhathakurta, P., (1991) Preprint

Griffiths, R.E. (1989) *The Epoch of Galaxy Formation*, eds. C.S. Frenk *et al.* ., Kluwer Academic Publishers vol. 264, p.235

Griffiths, R.E. and Padovani, P. (1990) *Astrophys. J.*, **360**, 483.

Hasinger, G, Schmidt, M., Trümper, J., (1991-a) *Astron. Astrophys.*, **246**, L2..

Hasinger, G, Schmidt, M., Trümper, J., (1991-b), this volume

Lonsdale, C.J. and Hacking, P.B., (1989) *Astrophys. J.*, **339**, 712.

Saunders, W., Rowan-Robinson, M., Lawrence, A., Efstathiou, G., Kaiser, N., Ellis, R.S. and Frenk, C.S., (1990) *Mon. Not. R. astr. Soc.*, **242**, 318.

Woosley, S.E. and Weaver, T.A. (1986) *Ann. Rev. Astron. Astrophys.*, **24**, 205.

Young, J.S., Kleinmann, S.G. and Allen, L.E., (1988) *Astrophys. J. Lett.*, **334**, L63.

Chapter VI

Recent observational results

ROSAT Deep Surveys

G. HASINGER

Max-Planck-Institut für extraterrestrische Physik

8046 Garching, Germany

ABSTRACT

A series of 20 deep pointed observations with the ROSAT PSPC is discussed. 530 X-ray sources with 0.5-2 keV fluxes down to $3 \cdot 10^{-15}$ $erg\ cm^{-2}\ s^{-1}$ have been discovered in 5.9 deg^2. The N($>$S) relation of the sources selected in the 0.5-2 keV band shows a density in excess of 200 deg^{-2} at the faintest fluxes and a flattening below $2 \cdot 10^{-14}$ $erg\ cm^{-2}\ s^{-1}$. The average spectrum of those sources is a power law with energy index 1.2±0.1. The absorption column densities are consistent with the galactic HI columns. More than 50% of the 1-2 keV background has been resolved into discrete sources in the deepest field. The total background spectrum shows an emission line feature around 0.65 keV, most probably due to OVII-OVIII from a $2 \cdot 10^6 K$ plasma. Above $\sim 1\ keV$ the background is dominated by a power law spectrum with a normalization of 13.4±0.2 $keV\ cm^{-2}\ s^{-1}\ sr^{-1}\ keV^{-1}$ and a slope 1.2±0.1, i.e. considerably steeper than the extrapolation from higher energies. An angular correlation function in the 0.9-2 kev band has been constructed, averaged over 14 fields. 1σ upper limits for structure in the background on scales of 2-10 arcmin are 10-5%.

1 INTRODUCTION

Deep imaging studies of the X-ray background have first been performed with the *Einstein* IPC and HRI (Giacconi et al., 1979, Griffiths et al., 1983, Primini et al., 1991). These observations were able to resolve about 20% of the background at 2 keV into sources, the majority of which turned out to be extragalactic. The PSPC (Pfeffermann et al., 1986) aboard ROSAT (Trümper, 1983) comprises a few features which make it particularly useful for the study of the soft X-ray background: its extremely low intrinsic background (Snowden et al., 1991), its good energy resolution, temporal gain stability and spatial homogeneity. In addition, the large collecting area and large field of view of the ROSAT telescope are particularly useful. Consequently, every pointed PSPC observation contains a number of serendipitous sources. About half of all PSPC pointings reach sensitivity levels comparable to or fainter than the *Einstein* Deep Surveys.

First results on medium-deep pointed observations have already been published elsewhere (Hasinger et al., 1991, hereafter *paper I*; Shanks et al., 1991). Here I report on further results from early long PSPC observations at high galactic latitudes (RMSS) and on the recent ROSAT Deep Survey, obtained in the *Lockman Hole*, a region with the absolutely lowest neutral hydrogen interstellar column density (Lockman et al., 1986). This project is performed in collaboration with R.Burg, R.Giacconi (both STScI), G.Hartner, J.Trümper (both MPE), M.Schmidt (Caltech) and G.Zamorani (Bologna). Section 2 gives an overview of the selection of survey fields. In section 3 the X-ray logN-logS relation is derived. Section 4 and 5 describe analytic fits to the spectra of X-ray sources and the background, repectively. Section 6 discusses the angular correlation function of the X-ray background.

Table 1: Field Selection

Field Name	Gal. Long. [deg]	Gal. Lat. [deg]	Column Density [cm^{-2}]	Exp. Time [s]	"Clean" Part [%]	Nr. Src.	S_{min} (0.5-2 keV) [erg/cm²s]	Fraction Resolved [%]
α Boo*,+	15	69	$2.5 \cdot 10^{20}$	17100	-	31	$0.93 \cdot 10^{-14}$	34.8
β Leo+	251	71	$3.1 \cdot 10^{20}$	20160	31	25	$0.76 \cdot 10^{-14}$	34.5
δ Leo	224	67	$1.1 \cdot 10^{20}$	21630	-	37	$0.83 \cdot 10^{-14}$	42.6
EF Eri*,+	214	-58	$1.8 \cdot 10^{20}$	15840	-	23	$0.87 \cdot 10^{-14}$	28.7
EX Hya*	303	34	$6.1 \cdot 10^{20}$	15620	32	16	$1.44 \cdot 10^{-14}$	22.9
HR 857*,+	192	-58	$3.2 \cdot 10^{20}$	10680	-	7	$1.16 \cdot 10^{-14}$	31.2
HZ 43+	54	84	$1.0 \cdot 10^{20}$	13100	3	32	$0.70 \cdot 10^{-14}$	44.6
HZ 43P	54	84	$1.0 \cdot 10^{20}$	22530	41			
LHS 2924*	55	68	$1.1 \cdot 10^{20}$	10680	21	27	$0.51 \cdot 10^{-14}$	32.3
Lockman	149	53	$0.6 \cdot 10^{20}$	78320	23	92	$0.28 \cdot 10^{-14}$	50.7
Lynx*	176	53	$2.7 \cdot 10^{20}$	66900	34			
Meaty	111	53	$4.0 \cdot 10^{20}$	23400	-	21	$0.60 \cdot 10^{-14}$	47.0
NEP+	96	30	$3.9 \cdot 10^{20}$	49200	11	41	$0.44 \cdot 10^{-14}$	37.8
NEPN	96	30	$3.9 \cdot 10^{20}$	41760	-	44	$0.62 \cdot 10^{-14}$	43.0
Nower1	202	55	$1.8 \cdot 10^{20}$	41760	18	40	$0.44 \cdot 10^{-14}$	34.3
Nower2	136	68	$1.1 \cdot 10^{20}$	30590	10	38	$0.57 \cdot 10^{-14}$	29.2
Pavo*	319	-77	$5.3 \cdot 10^{20}$	25080	-	29	$0.99 \cdot 10^{-14}$	32.0
SEP+	276	-30	$5.3 \cdot 10^{20}$	20220	8	16	$0.79 \cdot 10^{-14}$	34.8
VW Hyi*,+	285	-38	$8.5 \cdot 10^{20}$	8760	27	11	$1.47 \cdot 10^{-14}$	32.6
Warlock*	161	53	$0.7 \cdot 10^{20}$	34860	53			

* Courtesy AO-1 PIs   + *paper I*

2 FIELD SELECTION

Table 1 summarizes the selected survey fields, all at high galactic latitudes and with exposure times greater than 8 ksec. Limiting fluxes in the 0.5-2 *keV* band have been calculated assuming a power law spectrum with energy index of -1. Fields marked with an asterisk have been included in the analysis by courtesy of the principal investigators owning the data rights. Analysis of eight of the current fields has been published in paper I. A more thorough report on the *Lockman Hole* data is in preparation, as is a detailed description of the fits to the X-ray background spectrum.

Figure 1 shows the image of the *Lockman Field* in the 0.1-2.4 *keV* band. The data have been background-subtracted and smoothed with a Gaussian with a FWHM of 28". About 90 discrete X-ray sources have been detected in the inner, most sensitive part of the field of view, marked by the circular shadow of the PSPC window support structure at a radius of 20 arcmin.

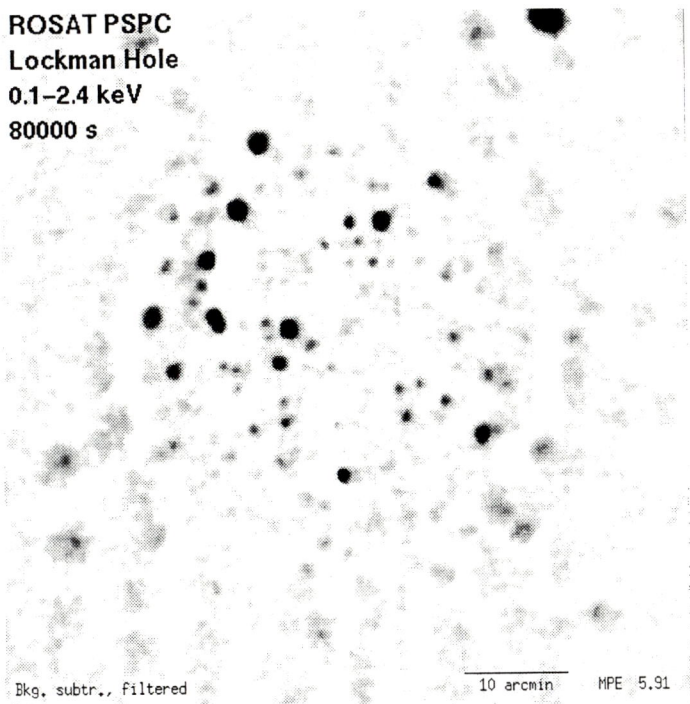

Figure 1: X-ray Image of the Lockman Hole in the 0.1-2.4 keV band. The data is background-subtracted and filtered with a Gaussian of 28" FWHM.

3 LOGN-LOGS FUNCTION

The source detection process has been described in some detail in *paper I*. The same sliding box algorithm has been applied here, however separately for 4 energy bands: S (0.1-0.4 *keV*), H (0.4-2.4 *keV*), T (0.1-2.4 *keV*) and J (0.9-2.4 *keV*). In the previous analysis source counts were accumulated in a ring of radius 40" around each source. In the meantime more calibration data became available, and it turned out that the above radius was too small, in particular at larger off-axis angles and that the source fluxes have been systematically underestimated by 30% on the average. For the current flux and significance determination a 90% radius has been determined separately for each source as a function of off-axis angle and average energy band (eg. between 29" and 69" in the H-band). The source detection threshold was set at a likelihood of $L = -ln(P) = 10$, corresponding to not more than a couple of spurious sources in the whole survey. A total of 530 discrete sources was detected in the inner 20' of 17 fields ($5.9deg^2$) in at least one of the above energy bands. Source count rates were converted to incident fluxes in the 0.5-2 keV band assuming a power law spectrum with energy index -1 and galactic H_I absorption.

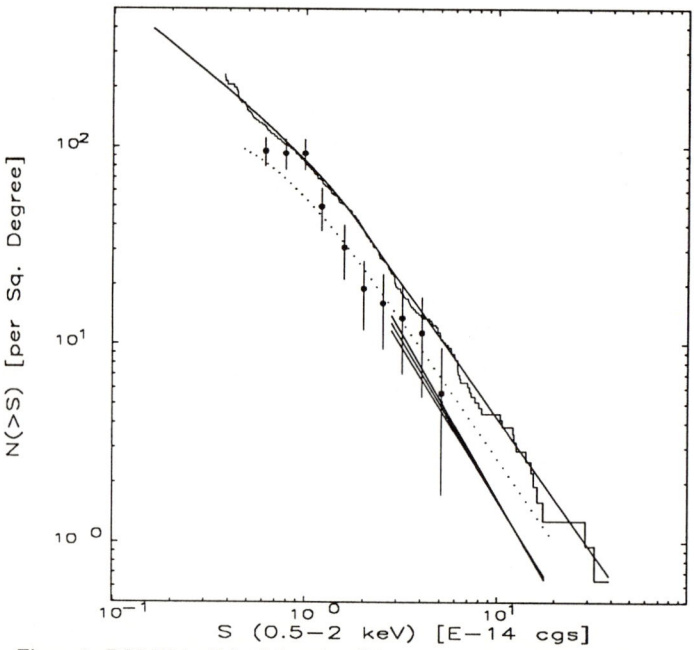

Figure 2: ROSAT logN-logS function (histogram and solid line) compared to the EMSS(two crossing solid lines) and the data of Shanks et al. (filled circles). Note that the contribution of brighter fluxes has not been added to the latter data.

For the construction of a preliminary logN-logS function the sample has been restricted to detections and fluxes in the H-band and at off-axis angles less than 15'. A sensitivity histogram (surveyed area as a function of limiting flux) was constructed taking into account the variing sensitivity as a function of off-axis angle. 317 sources in the flux interval $3 \cdot 10^{-15} - 4 \cdot 10^{-13}$ erg cm^{-2} s^{-1} remained in the sample. In figure 2 the integral $N(> S)$ function derived in this work (histogram) is displayed together with the estimate from *paper I* (dotted line). Due to the above 30% systematic flux error, the two curves are shifted with respect to each other. A broken power law with the two slopes α_1, α_2 and a break flux S_B was fit to the differential source counts. The solid line shows the integrated best fit function with the parameters $\alpha_1 = 2.4 \pm 0.2$, $\alpha_2 = 1.7 \pm 0.3$, $S_B = 2.2 \pm 0.9 S_{14}$ (90% errors) and normalization of 89.6 at $S_{14} = 10^{-14}$ $erg\ cm^{-2}\ s^{-1}$. The flattening of the source counts, as already indicated in *paper I* could be confirmed here, however, it appears less dramatic. An extrapolation of the logN-logS function to fainter fluxes can account for $80 \pm 20\%$ of the total 1-2 keV background.

Because spectroscopic identifications are not yet available the dataset contains galactic as well as extragalactic objects. However, preliminary X-ray/optical correlations and the first spectroscopic identifications (Shanks et al., 1991) indicate that the stellar content is small ($\sim 10\%$), compared to about 25% in the EMSS (Gioia et al., 1990). Figure 2 also contains the source counts of the first spectroscopically identified ROSAT field (Shanks et al., 1991; filled dots) and the fit to the EMSS AGN (Gioia et al., 1990, straight solid lines). In order to convert the EMSS fluxes (0.3-3.5 keV) to our 0.5-2 keV passband a power law source spectrum with energy index -1 has been assumed. The mismatch between these two datasets and the new ROSAT logN-logS, in particular at bright fluxes may be partially due to the stellar content and due to the fact that Shanks et al. (1991) did not add the contribution of fluxes brighter than $6 \cdot 10^{-14}$ $erg\ cm^{-2}\ s^{-1}$ to their counts. Also a systematic flux difference between the EMSS and the RMSS (20-30%) may be present.

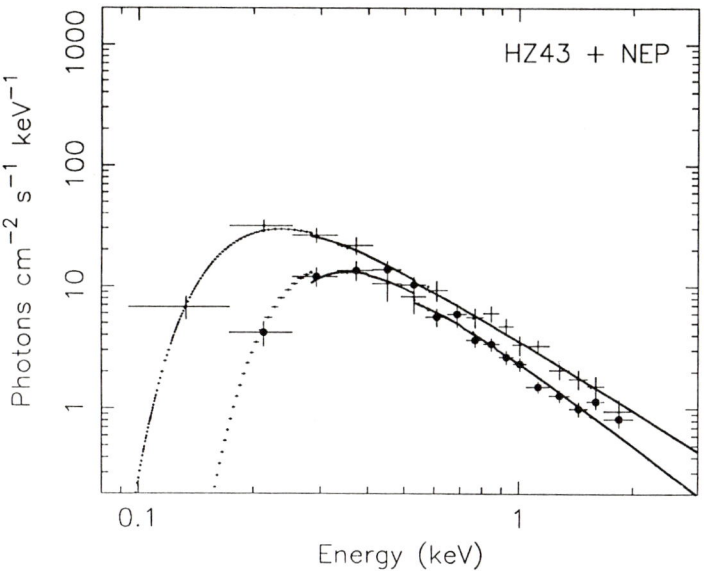

Figure 3: Average spectra of the sources in the HZ43-field (crosses) and in the NEP-field (filled circles). Single power law spectra with galactic absorption have been fit.

4 AVERAGE SPECTRA OF RESOLVED SOURCES

Sources brighter than 10^{-13} *erg cm^{-2} s^{-1}* have been omitted from the analysis to minimize statistical fluctuations. Average source spectra were determined by accumulating the counts in rings containing more than 95% of the photons of all sources detected in the hard band in the inner region (15' radius) of each field. The average background (sky plus particle), determined from the remaining adjacent pixels, was subtracted. A simple power law model with cold gas absorption could be fit to the average source spectrum of every individual field in the energy range 0.1-2 *keV*. Figure 3 compares the fits to the spectra of the HZ43 and the NEP field. The power law slopes derived in the individual fields agree very well with each other, yielding an average value of $\alpha_E = 1.16 \pm 0.08$. The derived values of the absorption column densities N_H are consistent with the values derived from 21-cm maps of the galactic neutral hydrogen. Spectral fits to individual sources in general agree very well with this model, in a small number of cases, however, significantly different spectra are present. At least two heavily absorbed sources and one object with a comparatively soft spectrum have been found in the *Lockman hole*.

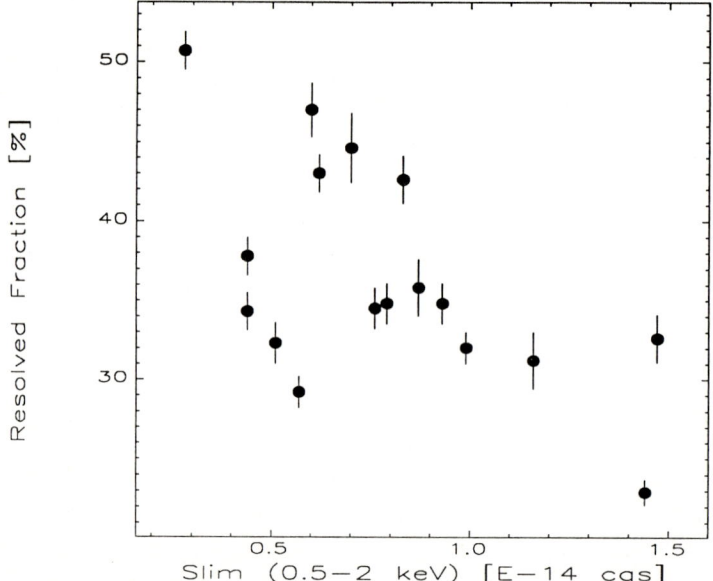

Figure 4: Fraction of counts in sources in the 1-2 keV band vs. measured total background in the same band.

The integral flux of the average source spectra directly yields the contribution of sources with fluxes between the sensitivity limit of the corrresponding field and the upper flux limit (10^{-13} *erg cm^{-2} s^{-1}*) to the total background in the 1-2 *keV* band. In order to quantify the total amount of background already resolved into sources a correction for the contribution of sources brighter than the upper flux threshold has to be applied. This was derived from an extrapolation of the above N(>S) relation to the bright end. In table 1 and in figure 4 the fraction of background resolved in the 1-2 *keV* band is given. In the *Lockman field* the total resolved flux corresponds to a power law normalization of 6.8 ± 0.3 *keV* cm^{-2} s^{-1} sr^{-1} keV^{-1} at 1 *keV*.

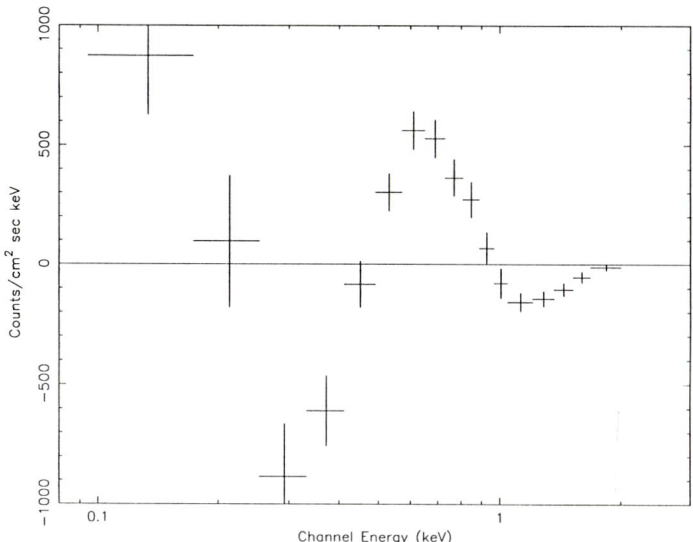

Figure 5: Residuals to a fit of the total background spectrum in the HZ43P field with a power law plus a 1 million degree thermal spectrum. A line feature is clearly seen the range 0.5-0.7 keV.

5 THE COSMIC X-RAY BACKGROUND SPECTRUM

To minimize the contamination by non-cosmic background components, mainly the particle background, which is strongest at high geomagnetic latitudes (Snowden et al., 1991), and the solar scattered light plus a geocoronal oxygen line at 0.53 *keV*, which both can be very strong on the dayside of the satellite orbit (see eg. Fink et al., 1988), restrictive selection criteria have been applied. Only observation intervals at geographic latitudes between -30 and 30 degrees (*tropic*) and with Sun-Earth-satellite angles larger than 120 degrees (*night*) have been accepted. On average the tropic night data comprises about 25% of the net observing time, but there are quite a few observations without useful clean observation periods (see table 1). To improve on the statistical quality of the data background spectra have generally been accumulated over the whole PSPC field- of-view (0-57' radius), however, if a bright target (with broad scattering wings) was observed, or if the boron filter (covering a central circle with radius $\sim 25'$) was used, the analysis was restricted to the outer field-of-view (27'-57').

The particle background with the spectral shape given by Snowden et al. (1991) was subtracted, normalized to the highest energy channels (2.4-2.7 *keV*), where the sky contribution can be neglected. Some fields indicate a further, slowly time variable background component with a relatively soft spectrum which is as yet not understood. Affected times were excluded as far as possible from the analysis. Fields on or close to the north-polar spur have also been discarded.

First a relatively simple spectral fit was attempted with a $10^6 K$ thermal plasma for the carbon-band (0.1-0.3 keV) emission (see McCammon and Sanders, 1990) and a single power law with fixed galactic absorption for the higher energies. Figure 5 shows the residuals to this fit for the HZ43P field, which display a significant excess in the 0.5-0.7 keV range, resembling an emission line with a centroid energy of ~ 0.65 keV, i.e. higher than that of a monochromatic oxygen $K\alpha$ line, and therefore most probably due to ionized oxygen (OVII-OVIII). This line, which is present in practically all fields, confirms the existence of hot plasma at temperatures of $2 - 3 \cdot 10^6 K$ (Inoue et al., 1980; Rocchia et al., 1984), distributed over the whole sky.

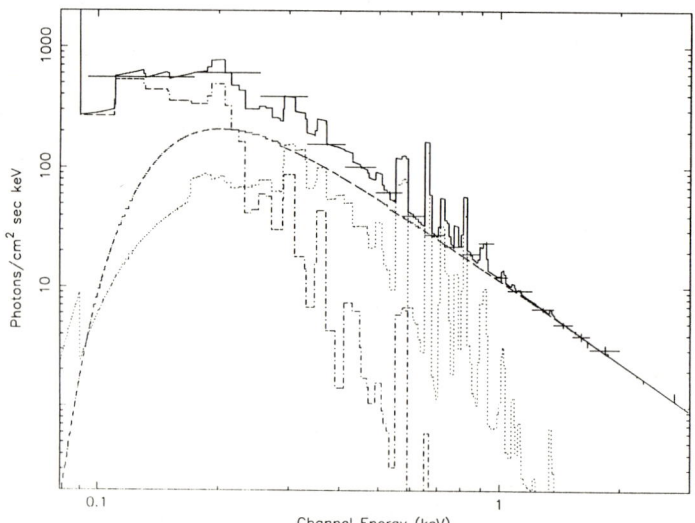

Figure 6: Three-component fit to the total background in the HZ43P field: a 1 million K thermal plasma, a 2-3 million degree thermal plasma and a power law for the higher energies.

Therefore, a more complicated spectrum had to be fit, with two thermal line spectra at one million and 2-3 million degrees, respectively, plus a power law absorbed through the galaxy for the higher energy extragalactic component. The best-fit absorption for the hot thermal component is strongly correlated with the flux of the soft thermal component and therefore was fixed at a rather arbitrary value of $5 \cdot 10^{19} cm^{-2}$. This fit yielded acceptable χ^2 values for all analysed fields. The best-fit values of the power law index and flux, as well as the temperature and flux of the hot thermal component are consistent with being constant in all fields, while the flux of the soft thermal component varies strongly over the sky. The following average values have been derived for the different parameters: power law energy index $\alpha_E = 1.12 \pm 0.07$, power law norm 13.4 ± 0.3 $keV\ cm^{-2}\ s^{-1}\ sr^{-1}\ keV^{-1}$ (at $1\ keV$), temperature of hot component $0.20 \pm 0.01\ keV$.

Figure 7 (adapted from McCammon and Sanders 1990) shows a comparison of the ROSAT background spectrum (power law component only; two crossing solid lines) with the Wisconsin data and the the extrapolation of the HEAO-1 spectrum (lower dashed line; Marshall et al., 1980). The power law, dominating above $1\ keV$, meets the HEAO-1 extrapolation at roughly $2\ keV$, but is considerably steeper. The ROSAT flux density at $1\ keV$ is about 10% lower than that of the Wisconsin survey, consistent with the uncertainties in the relative calibration of the two instruments. The figure also contains the average spectrum of the resolved sources with the normalization taken from the *Lockman Hole* data (long dashes); the two arrows indicate that this is a lower limit of the total source contribution to the X-ray background.

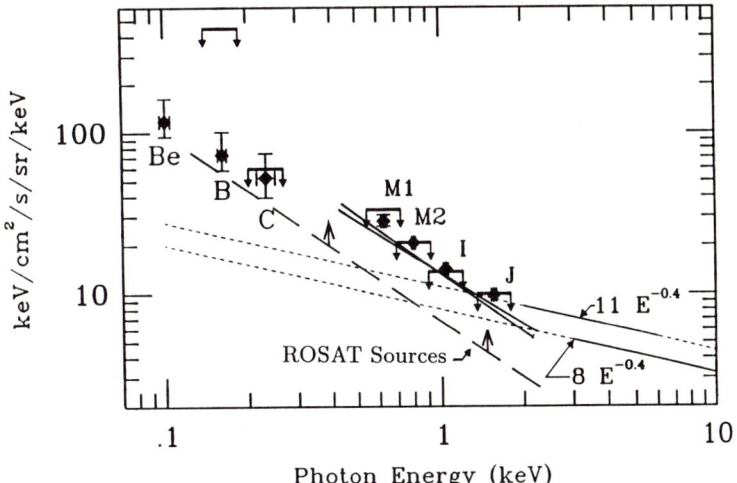

Figure 7: ROSAT background and source spectrum compared to the Wisconsin data (adapted from McCammon and Sanders (1990).

6 ANGULAR CORRELATION FUNCTION

The discovery of a diffuse, low surface brightness structure on a scale of 10-20 arcmin (called 'the NEP blotch') in the residual background of the North-ecliptic pole field has been discussed in *paper I*. Its large angular scale led to the interpretation of this structure in terms of clustering of unresolved background sources in a supercluster at a reshift larger than 0.3. In the meantime Burg et al. (1992) have detected a normal cluster of galaxies at $z = 0.09$ coinciding with the peak of the X-ray surface brightness distribution. In addition there is evidence, both from the X-rays and from optical spectroscopy, for the existence of a supercluster close to the NEP, also at a redhift of 0.09. The presence of the supercluster might be responsible for the apparent large size and low surface brightness of the NEP blotch.

Nevertheless, as long as the sources making up the X-ray background are connected to galaxies (AGN, starburst galaxies, clusters etc.) clustering and therefore structure in the background is expected. Upper limits on or the detection of a significant signal in the angular autocorrelation function of the background can provide stringent constraints on the contributors to the X-ray background (see eg. Danese et al., 1992).

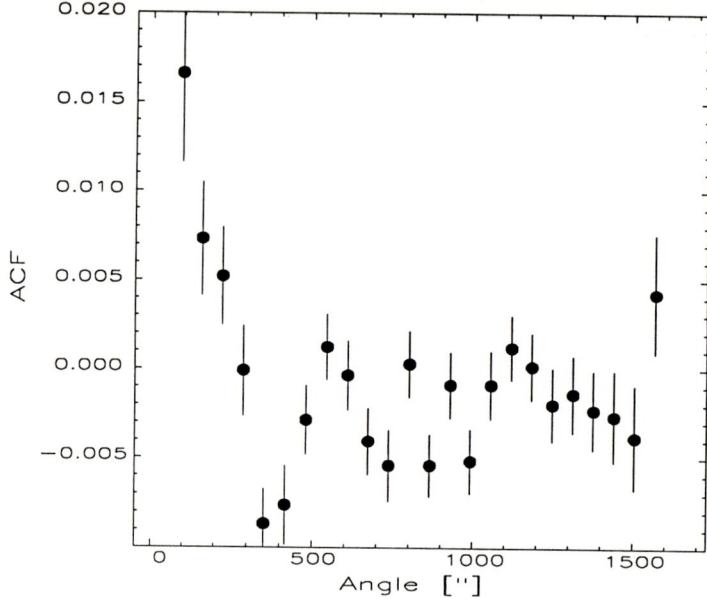

Figure 8: Angular Correlation Function averaged over 14 fields

The derivation of an average angular correlation function was restricted to the J-band (0.9-2.4 keV) because of the uncertain contribution of galactic and geocoronal backgrounds at lower energies. This has the advantage that the denominator of the angular correlation function (see Barcons and Fabian 1989) can be replaced by the actually measured total background in the field so that no uncertain extrapolations need to be made. Fields containing significant contamination by extended sources or the mirror scattering halo of a bright central target were rejected (including the NEP and α Boo data). Each of the 14 remaining fields underwent the following procedure: The J-image was binned into coarse (64 × 64") pixels. Pixels close to one of the detected sources (judged by the above 95% criterion) and those at larger off-axis angles than 15' were removed. For the remaining pixel an angular autocorrelation function (ACF) was calculated, following the procedure outlined by Barcons and Fabian (1989) and in *paper I*. Finally the ACFs from the 14 individual fields were averaged and the ACF of the PSPC flat field taking care of the vignetting function (see *paper I* was subtracted. Figure 8 shows the resulting average ACF. There is some power with a correlation length of ~ 5 arcmin, but its significance is yet unclear. An upper limit for the correlation, taken from the average of the datapoints plus their 1σ error at 2-3 arcmin and at 9-10 arcmin is 9 and 4 %, respectively.

REFERENCES

Barcons X. & Fabian, A.C. *Mon. Not. R. astr. Soc.,* **247**, 119 (1989).

Burg R. et al. *Nature,* (1992) submitted.

Danese L. (1992) this issue.

Fink H.H., Schmitt J.H.M.M. & Harnden F.R.,Jr. *Astr. Astroph.,* **193**, 345 (1988).

Giacconi R. et al. *Astrophys. J.,* **234**, L1 (1979).

Gioia I.M. et al. *Astrophys. J. Suppl.,* **72**, 567 (1990).

Griffiths R.E. et al. *Astrophys. J.,* **269**, 375 (1983).

Hasinger G., Schmidt M. & Trümper J. *Astr. Astroph.,* **246**, L2 (1991) (*paper I*).

Inoue H. et al. *Astrophys. J.,* **238**, 886 (1980).

Lockman F.J., Jahoda K. & McCammon D. *Astrophys. J.,* **302**, 432 (1986).

Marshall F.E. et al. *Astrophys. J.,* **235**, 4 (1980).

McCammon D. & Sanders W.T. *Ann. Rev. Astr. Astrophys.,* **28**, 657 (1990).

Primini F.A. et al. *Astrophys. J.,* **374**, 440 (1991).

Pfeffermann E. et al. *Proc. SPIE* **733**, 519 (1986).

Rocchia R. et al. *Astr. Astroph.,* **130**, 53 (1984).

Snowden S. et al. *Astrophys. J.,* (1991) submitted.

Shanks T. et al. *Nature,* **253**, 315 (1991).

Trümper J. *Adv. Space Res.* **2**, no.4, 241 (1983).

ACKNOWLEDGEMENTS

I am grateful to the ROSAT team, my collaborators and all AO-1 PIs allowing me to use their data in advance of publication.

BBXRT Observations of the DXRB: First Results

K. Jahoda, P. J. Serlemitsos, K. A. Arnaud[1], E. Boldt, S. Holt, F. E. Marshall, R. F. Mushotzky, R. Petre, W. T. Sanders[2], A. Smale[3], J. Swank, A. Szymkowiak, K. Weaver[1], S. Yamauchi[4]

Goddard Space Flight Center, Laboratory for High Energy Astrophysics; [1]also Univ. of Maryland; [2]Univ. of Wisconsin; [3]also University Space Research Association; [4]Nagoya University

1 INTRODUCTION

The Broad Band X-Ray Telescope (BBXRT) was part of the 9 day mission of the ASTRO-1 observatory, carried above the atmosphere on the Space Shuttle Columbia mission STS-35. The BBXRT is a high throughput, moderate energy resolution, spectrometer with significant response from ~0.5 to 10 keV. The original mission plan was to observe about 150 targets brighter than about 10^{-12} ergs cm^{-2} sec^{-1} in the 2-10 keV band. About 300 individual pointings were scheduled for a 10 day mission. Actual operations were less efficient than originally hoped due to unanticipated complications with the BBXRT pointing system and complications experienced by the three ultraviolet telescopes which made up the rest of the ASTRO-1 payload (Blair and Gull 1990). However, BBXRT successfully observed about 60 targets, and the instrument functioned well during the numerous periods when it was pointed towards the sky but away from known bright sources. A serendipitous observation of the Diffuse X-ray Background (DXRB), which has a brightness of a few times 10^{-13} ergs cm^{-2} sec^{-1} (BBXRT field of view)$^{-1}$, resulted. Initial results from these observations are presented in this paper. As will be clear, this represents a progress report rather than finished work; it contains hints of the contributions that will be made to the study of the DXRB by BBXRT (when all the data are included in the analysis) and the next generation of spectroscopic experiments.

Among the well established results about the DXRB spectrum are that (a) above 3 keV the observed spectrum is well described by a 40 keV bremsstrahlung model (Marshall et al. 1980) which itself can be well approximated in the 3-10 keV range as a power law of photon index -1.4 and (b) below 1 keV the observed flux exceeds the extrapolation from higher energies. Despite careful work by several groups, the normalization of the 2-10 keV flux remains uncertain (to 30%, see figure 12 of McCammon and Sanders 1990). Proportional counter experiments have also been unable to determine at what energy the DXRB spectrum departs from the power-law approximation. The relatively high energy resolution of the BBXRT may be able to shed light on these questions.

The plan of this paper is as follows. We first review the relevant characteristics of the BBXRT instrument. The question which we hope to address with the BBXRT data is defined and discussed. We next review the available evidence about the behavior of the non-cosmic background in the BBXRT. This background is, for BBXRT as for other experiments, the limiting factor in determining the spectrum of the cosmic background, and we describe the data selection criteria which have been used. Finally we fit several models to the data. Although the BBXRT data alone cannot distinguish among the several models presented here, we can come to some general conclusions about where a break from the power-law description of the DXRB spectrum which is an adequate description of the data above 3 keV is required.

2 CHARACTERISTICS OF THE BBXRT

The BBXRT consists of two co-aligned telescopes with silicon solid state detectors in the focal planes. The in-flight performance of the entire system is summarized by Serlemitsos et al. (1991); the performance of the mirrors is summarized by Petre et al. (1991). The mirror system employs conical foil grazing incidence optics (Serlemitsos 1988). The use of many thin walled cones (118) allows a large fraction of the geometric area to be filled with X-ray mirrors although at some cost to the image quality. The individual cones are coated with lacquer to improve the smoothness of the surface and then gold to increase the X-ray reflectivity. The focal length of the telescopes is 3.8 m, a design chosen as the longest practical focal length which could fit inside the shuttle bay without requiring a mechanism to be deployed after the opening of the payload bay doors. The observed point source response is that 65% of the reflected photons are distributed as a gaussian with 1.3' HPD while the remaining 35% are distributed in a gaussian with 5.8' HPD. Observations in the vicinity of the Crab nebula demonstrate that there is also a ghost image from sources about 0.5-1.5 degrees off axis; this image comes from photons which are reflected from only one mirror. This response is unlikely to be important for any observations except of the diffuse background and amounts to ~2% of the peak. The ghost images probably have a larger effect on the absolute normalization than on the derived spectrum. This paper does not attempt any statements about the absolute normalization, and we have not yet included any corrections for the ghost images.

Each focal plane detector is segmented into 5 independent elements as shown in Figure 1, which shows an *Einstein* image of the Tycho supernova remnant projected onto the BBXRT focal plane. The central element is 4', in diameter. The outer elements have inner and outer diameters of 5' and 17'. There is an X-ray opaque mask covering the area near the pixel boundaries, which prevent events near pixel boundaries from being detected in

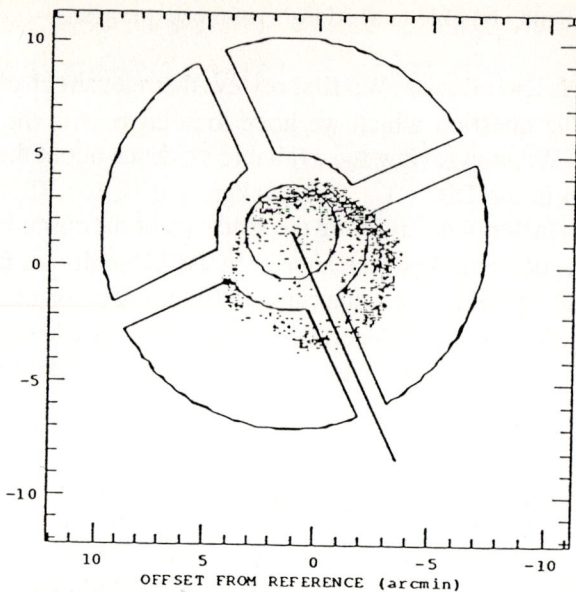

Figure 1: An Einstein image of the Tycho supernaova remnant projected onto the BBXRT focal plane. The outer diameter is 17'.

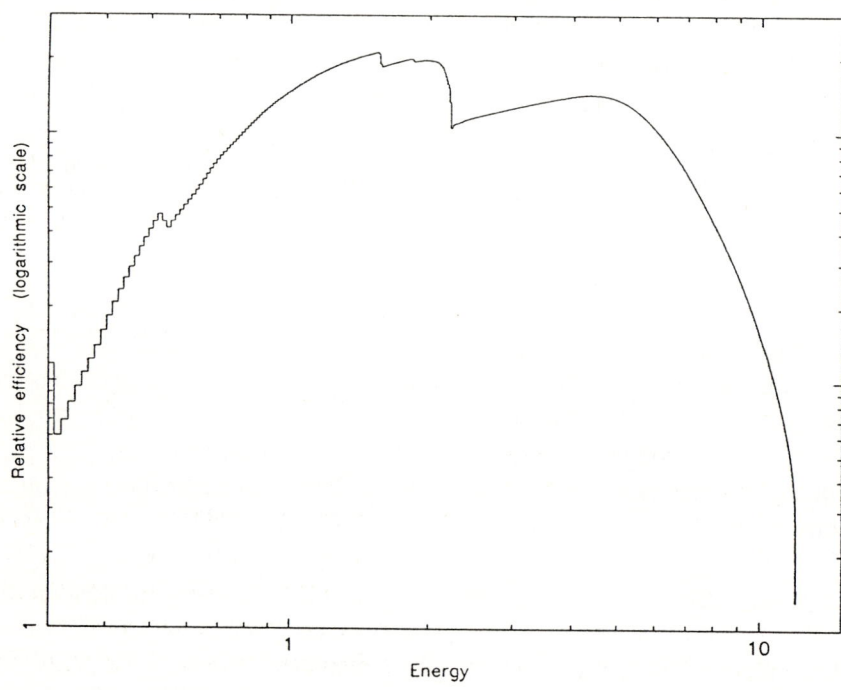

Figure 2: The relative effective area of the BBXRT for a source observed 8' off-axis

multiple pixels. The silicon is lithium doped and has noise limited resolution of 90 (central) and 110 (outer) eV at 1 keV and resolution of about 150 (central) and 170 (outer) eV at 6 keV. BBXRT therefore has better resolution than any other experiment which has contributed to the discussion at this meeting, although the resolution is not as high as that of the future experiments described elsewhere in this volume (CUBIC, Burrows; ASTRO-D, Inoue; DXS, Sanders; and UW/GSFC sounding rocket, McCammon).

The overall detector efficiency is limited by photo-electric absorption by filter elements in the detector system at low energies and by inefficiencies in the mirrors at high energies. Figure 2 shows the relative effective area for sources observed 8' off-axis; we use this approximation for the average response of the BBXRT telescope to a diffuse source (ray-tracing is in progress to determine the proper response for a diffuse source which fills the field of view). For sources observed on-axis, the relative efficiency is slightly higher at the larger energies. The substantial drop in effective area just above 2 keV is caused by the change in reflectivity just above and below the gold M edge. The change in reflectivity is primarily due to a change in the gold photoelectric cross section. As the cross-section increases above the edge, a smaller number of gold atoms participate in the reflection. The drop in efficiency at higher energies reflects the fact that increasingly steep grazing angles are required at higher energies and that fewer and fewer of the foils reflect X-rays at higher energies. The efficiency is low at small energies due to photoelectric absorption by an aluminum/parylene window, the gold electrode layer on top of the detector, and the silicon dead layer. The mirror response has been measured at several energies and intermediate values are predicted through ray-tracing. The ray tracing results have been verified at the measured energies.

At press time (October 1991) we are still making adjustments to our best estimate of the overall efficiency of the A telescope system (which collected all the data presented here). The remaining uncertainties are at levels < 5%, and should not have to great an influence on the conclusions presented here, which are limited by statistical precision. The B telescope is less efficient at lower energies; those data are not included here.

The essential features of the BBXRT response for the current analysis are (a) that there is enough area at high energies to fit the normalization of the -1.4 photon index power-law which gives a good description of the DXRB spectrum above 3 keV, and (b) sufficient energy resolution to determine where the observed DXRB spectrum deviates from this description.

3 SIGNIFICANCE OF THE QUESTION
McCammon (this volume) summarizes our knowledge about the intensity of the extragalactic X-ray flux between 0.2 and 10 keV in fig. 1. Note that below 2 keV we have

only upper limits to the extragalactic DXRB flux, and that the upper limits are well above extrapolations of the higher energy spectrum. Many models for producing the DXRB rely on the integrated contribution of numerous sources, each with an individual spectrum that turns up at lower energies (Schwartz, this volume; Rogers, this volume; Field, this volume). Knowing the extragalactic contribution at energies below 1 keV can constrain these models. Knowledge of where the -1.4 power-law ceases to fit the data can also constrain these models. On the other hand, there is excellent evidence that some of the observed flux at energies < 1 keV is thermal emission from hot gas in the disk or halo of the Milky Way (McCammon, this volume). Limiting the extragalactic component also places limits on the galactic component and constrains these models as well (Snowden, this volume). The unique feature of BBXRT for addressing these questions is the combination of bandwidth plus energy resolution.

4 QUALITY OF THE DATA

The chief complication in analyzing DXRB data is usually in separating the true cosmic signal from the instrumental background signal. This is certainly the case for BBXRT, where the instrumental background is about twice the cosmic signal. At the end of the mission several hours of data with the instrument covers closed were collected. Figure 3b shows the counting rate from 3 pixels of the A detector; all counts from channels 30-400 (~0.5-8 keV) are summed in 192 second bins during intervals where the covers were closed. (In this paper we choose to exclude the central pixel, which has a substantially smaller solid angle, and the fourth outer pixel which was noisier than the other pixels. The search for low energy upturns requires that these data be treated separately; we have not yet included them in our analysis)

These data appear to be more variable than can be explained by counting statistics alone. Figure 3a shows the guard rate for the A detector binned the same way. The guard is a plastic scintillator which surrounds > 3π sr around the detector. A phototube readout counts particle interactions in the guard. Typical rates vary from 800 to 2500 per 8 sec housekeeping frame outside the South Atlantic Anomaly; rates within the South Atlantic Anomaly peak above 20,000 per 8 second. When longer intervals are plotted, it is clear that the guard rate has two maxima per orbit, correlated with magnetic latitude. Figure 4 shows the data from figure 3 plotted against each other. Most of the data was taken at times when the guard rate was less than ~1200, and there is clear evidence that the overall event rate is correlated with the guard rate at higher rates.

The same effect can be seen in data taken with the covers open, i.e. sky looking data. This is shown in figure 5; the difference between this and the previous figure is just that the counting rates now include contributions from the cosmic as well as the instrumental background.

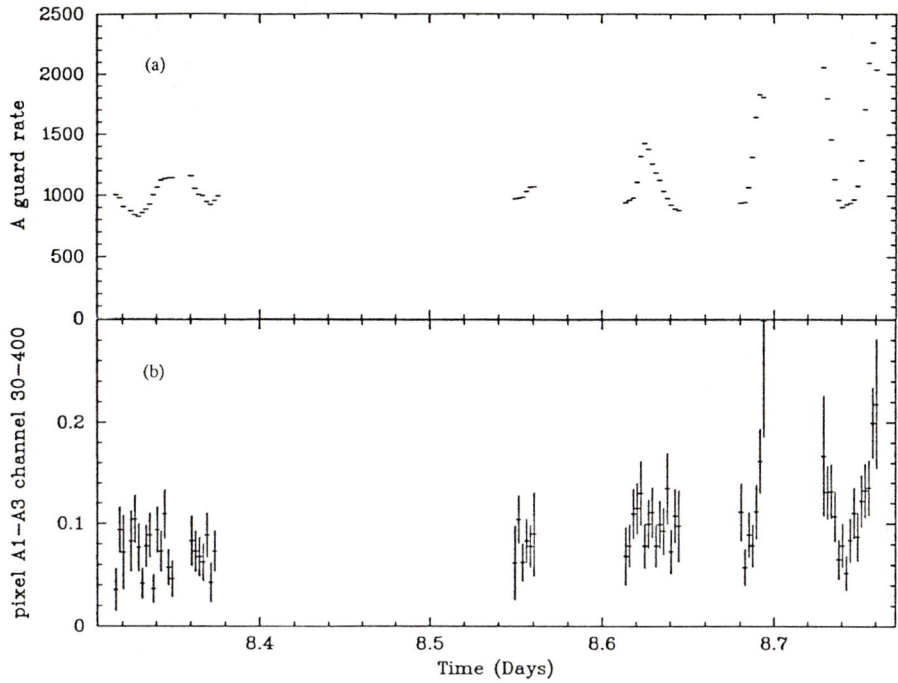

Figure 3: Guard rate (a) and event rate for three pixels, 1-8 keV (b) vs Mission Elapsed Time in days.

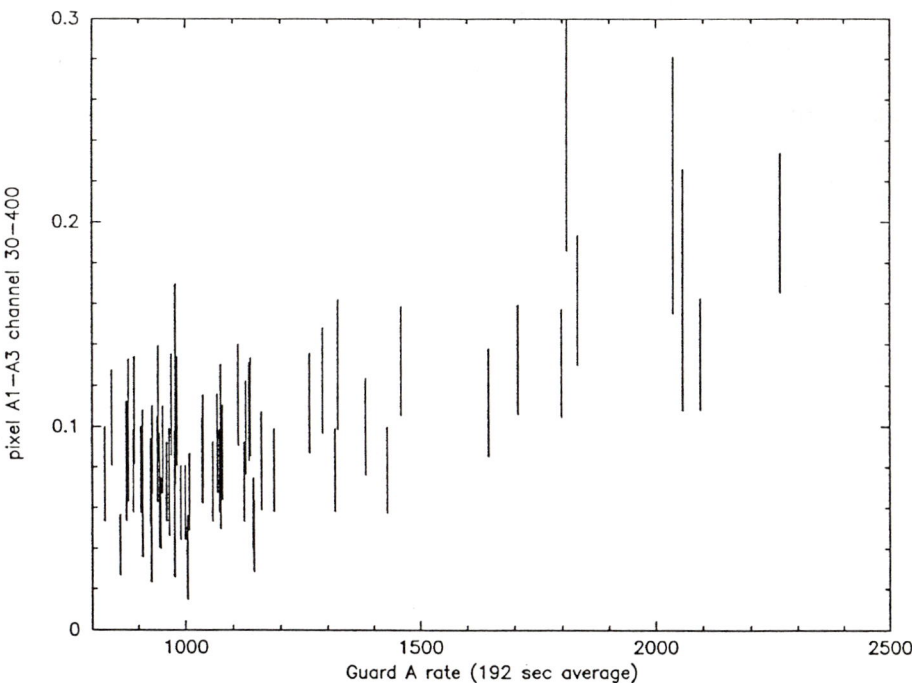

Figure 4: Event rate vs guard rate with covers closed. Same data as in fig. 3.

Inspection of figures 4 and 5 suggests that the instrumental background rate is not too variable so long as the guard rate is < 1200. To test the variability we compare the pulse height spectrum collected during periods with 800 < guard A rate < 1000 with that collected during periods where 1000 < guard A rate < 1200. Figure 6 shows the lower guard rate spectrum subtracted from the higher rate spectrum. The mean rate is 1.3 x 10^{-4} count sec^{-1} keV^{-1}, nearly consistent with 0, the reduced chi-square is 1.56. The eye suggests that there may be differences in the shape of the spectrum collected during these two periods (which could be due to instrumental background or due to looking at different directions in the sky). On the other hand, the differences are not nearly as large as when we compare subtract the spectrum taken with 800 < guard A rate < 1200 from that taken with 1200 < guard A rate < 2000, which is shown in figure 7. Here there is a clear difference from zero. The mean counting rate is 1.2 x 10^{-3} count sec^{-1} keV^{-1}. In this paper we choose to restrict ourselves to data obtained when the guard A rate was < 1200. Figure 6 may be used as an estimate of the variability of the internal background for this data set. We

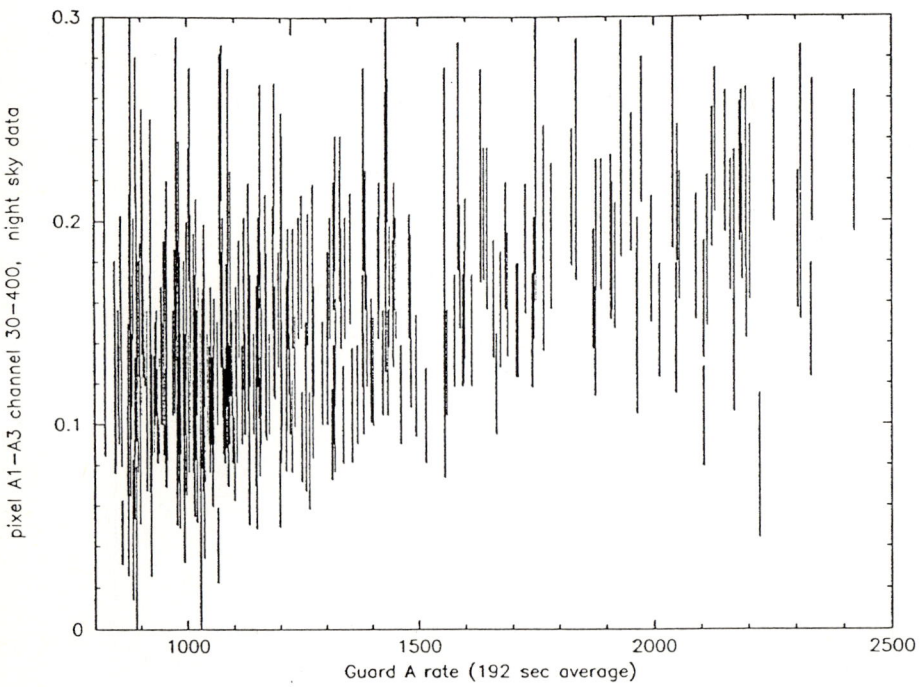

Figure 5: Event rate vs guard rate with covers open.

further restricted the data to night periods when the planned target was more than 0.5° off-axis and we required that the angle between line of sight and earth center was > 90° in order to avoid earth glow. (Day time data can be similarly included if the earth angle is > 120°;

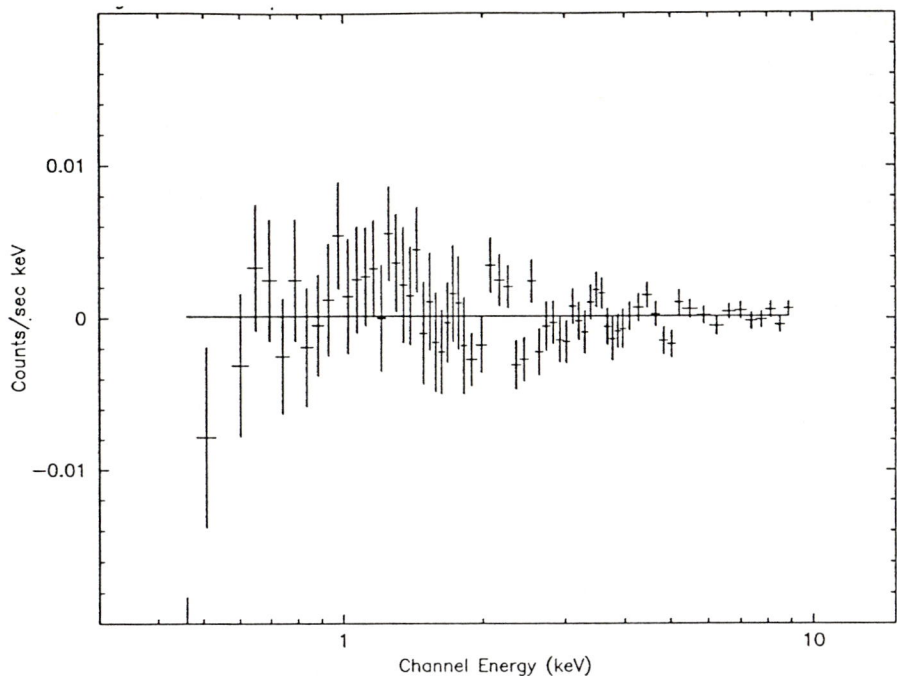

Figure 6: Data with 1000 < guard rate < 1200 minus 800 < guard rate < 1000. Solid line shows average.

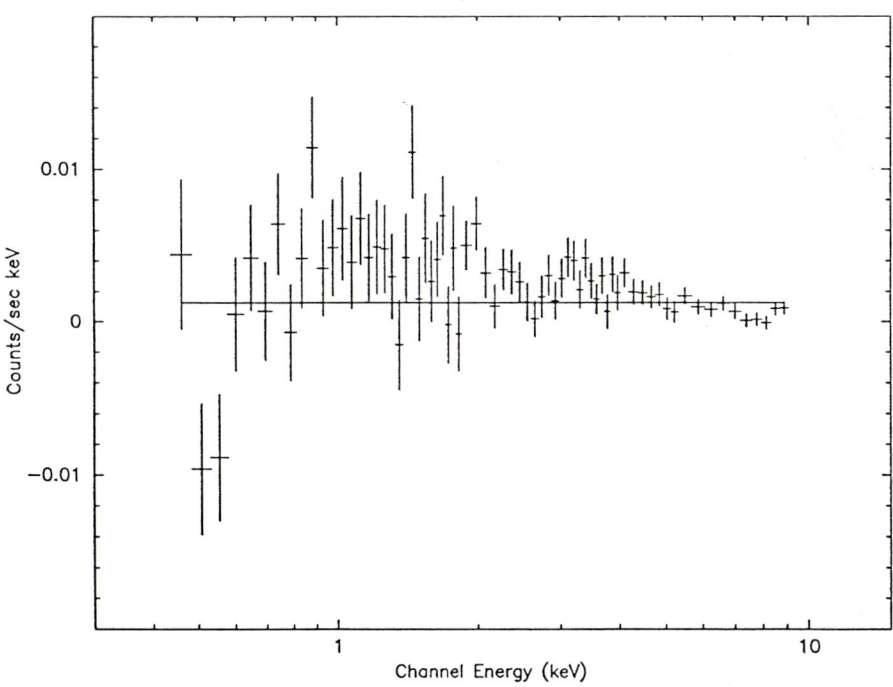

Figure 7: Data with 1200 < guard rate < 2000 minus 800 < guard rate < 1200. Solid line shows average.

because BBXRT generally pointed more than 80° from the sun there is relatively little such data and we have not yet included it.)

Figure 8 shows the background subtracted data for the three A detector pixels. The solid line represents a power-law of photon index -1.4, normalized to fit the data from 2.5 to 8 keV (bins 36-71), folded through the instrument response. All data in this paper have been binned as in figure 8; channel 1 (not displayed) contains unbinned channels below the instrumental discriminators. The total counting rate over the range displayed (0.4-9 keV) is 0.136 cts/s of which 0.080 ct/s have been subtracted as background. The inferred sky counting rate is 0.056 ± 0.003 ct/s; 32600 seconds of data are included so the spectrum includes about 1800 counts. It will become clear later that this data set is limited by statistical precision. Future work will concentrate on adding data at higher guard rate, data from the B telescope, and data obtained during the day. Inclusion of higher guard rate data and the B telescope data will each double the amount of data; including the day time data will increase the available data perhaps by 25%.

At energies below 2 keV there is a clear excess above the power-law extrapolation. There appears to be a line like feature at 1.78 keV. We believe that this is an instrumental, and not cosmic, feature. The best evidence that this feature is instrumental comes from comparing the data in figure 8a to the B detector data, shown in figure 8b. The B detector shows no feature at 1.78 keV; there is no reduction in chi-squared for a line fixed at 1.78 keV and only a change in chi-squared of 3 for a line allowed to choose its energy (1.89 keV) and normalization. In the A detector, on the other hand, the change in chi-squared is 10 for the addition of the two free parameter line (energy and normalization) and the line appears consistent with the energy of neutral Silicon fluorecence (1.78 +/- 0.31, 1 σ for one interesting parameter, compared with 1.74 keV). The origin of this line, and its non-appearance in the B detector, is not understood.

Another interesting feature of the B detector data is the reduced efficiency at lower energies. Although we intend in the future to include these data in studies of the XRB spectrum, we have not yet adequately calibrated the low energy response of the B detector. In the fits that follow, we always include a narrow line at 1.78 keV in order that these channels do not influence the fit of the other models too greatly. It is necessary to remember that we regard this as an artifact either of the detector or of the background subtraction.

5 MODEL FITTING

We have fit several models to the DXRB data. We cannot simultaneously fit for a slope and a normalization of the spectrum above 2 keV as very small changes in the background subtraction make large changes in the fit power-law index. In all cases we assume that the data may be described in this range by a power law of photon index -1.4. Our knowledge

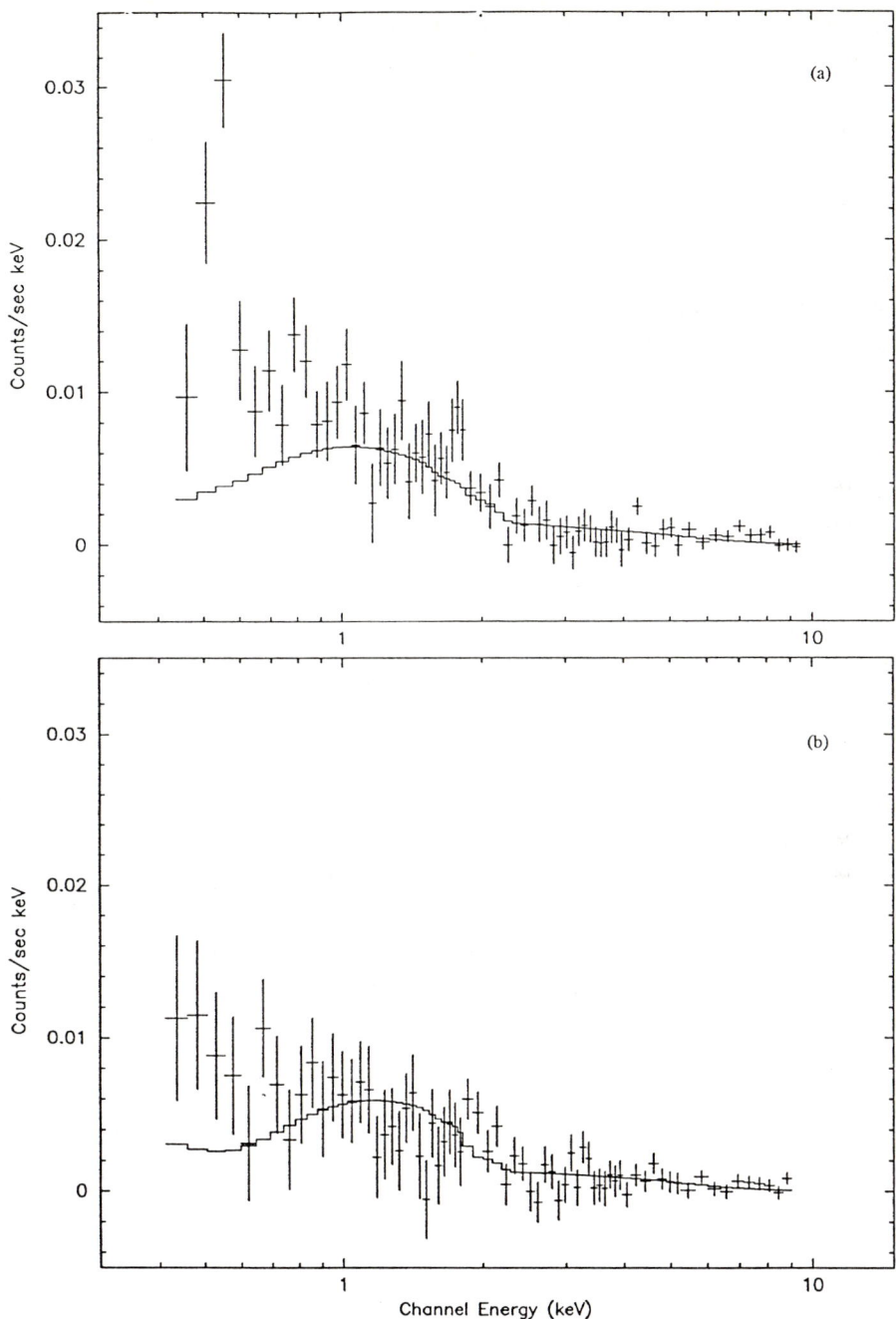

Figure 8: (a) Data from A detector and 1.4 photon index power-law folded through instrument response and normalized to channels above 2.5 keV and (b) data from B detector with response from (a).

of the absolute calibration is not sufficiently advanced, yet, to attach any significance to the actual value of the normalization, which is determined to a statistical precision of about 10%. We choose to focus instead on the energy at which the BBXRT data require an additional component or a steepening in the power law description. This representation will also allow a statement about the excess fraction required at 0.5 or 1.0 keV.

In all cases we use the data in figure 8a. We subtract a non X-ray background spectrum, keeping track of statistics, and evaluate our models with a chi-square statistic. Although we are developing tools to simultaneously fit the instrumental background and the cosmic background, we use the approximately correct method of subtracting a background spectrum here.

The models we will consider are (a) a broken power-law, (b) two power-laws, and (c) a power-law plus thermal component. All models include absorption by a galactic neutral hydrogen column of 2×10^{20} cm^{-2}. The fits are insensitive to the precise value of this column due to the limited efficiency below 0.5 keV. The first two models do not have a physical reality; the parameterization, however, is easily interpretable within our framework.

5.1   Broken power-law model
The model is given by
$$A(E) = K (E^{-\alpha 1}) \qquad \text{for } E < E_b$$
$$= K E_b^{(\alpha 2 - \alpha 1)} E^{-\alpha 2} \quad \text{for } E > E_b,$$
where α_1 is the photon index for E < break energy, Eb is the break point energy in keV, α_2 is the photon index for E > break energy (fixed in this case at 1.4), and K is the normalization in photons keV^{-1} cm^{-2} s^{-1} at 1 keV. The formal best fit parameters are summarized in figure 9. Our convention is to mark parameters that were not allowed to vary with asterisks. Note that galactic absorption and the background line at 1.78 keV are included in the fit. The chi-square is 63 for 65 degrees of freedom.

Figures 9a and 9b show the unfolded spectrum and data and model folded through the instrument response. Figure 10 shows the confidence contours for two interesting parameters, E_b and α_1. The contours are drawn at $\Delta\chi2 = 1, 2.71, 4.61$, and 9.21 which represent the 40, 68, 90, and 99% confidence contours. We see that the break energy is constrained to be < 1.2 keV.

We take the 90% confidence limits on photon index/Break Energy to be (-3.4, 1.24 keV) and (-5.4, 0.95 keV). In this representation, an extension of the -1.4 power-law contributes 3-16% of the flux at 0.5 keV and 100-65% at 1.0 keV where the first limit refers to the (-5.4, 0.95 keV) case.   Changing the normalization, by 12%, of the

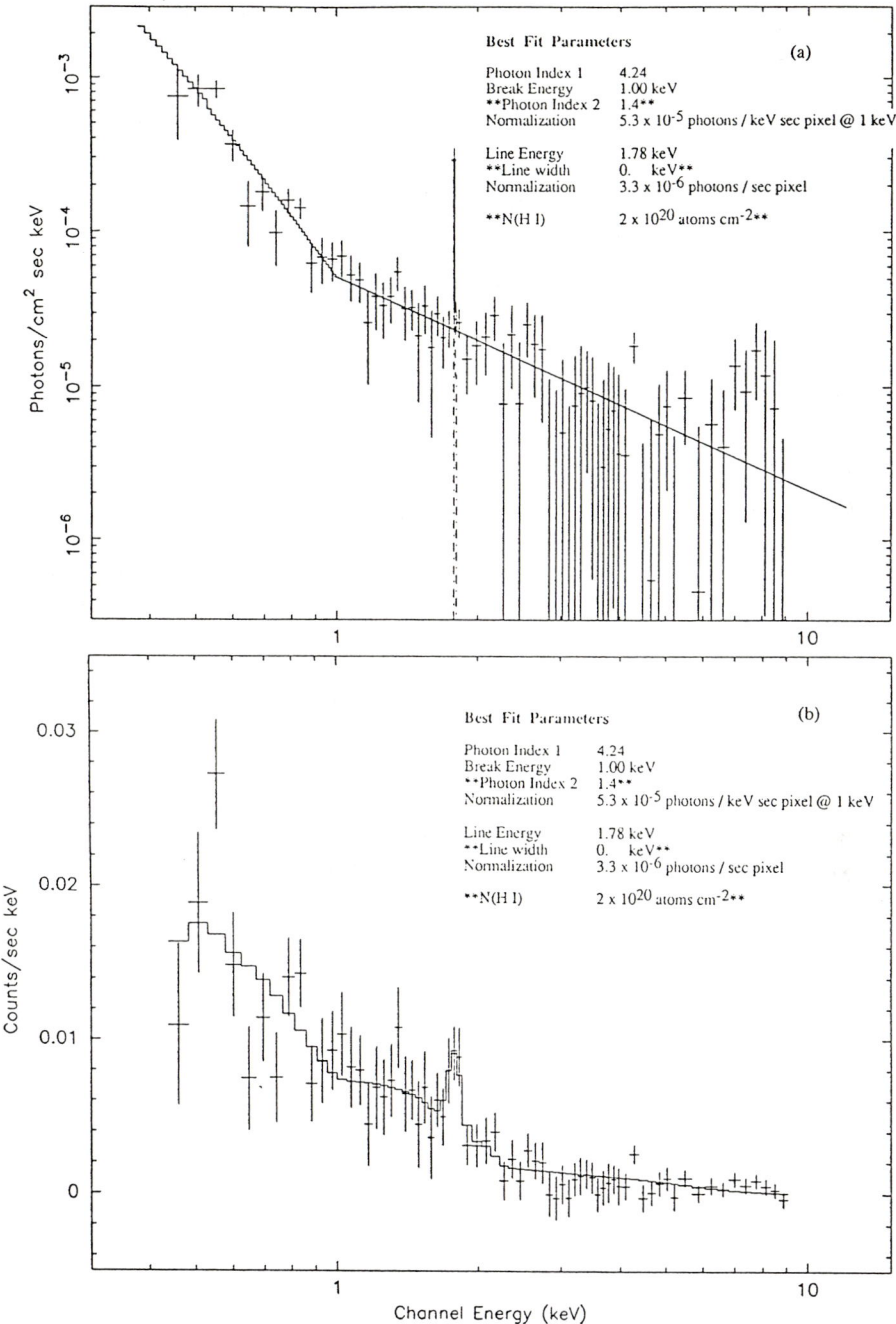

Figure 9: Best fit spectrum for broken power-law model: (a) in photon space and (b) count space. The line at 1.78 keV is believed to be an instrumental artifact.

background which is subtracted, using energy scales determined for each of the three different pixels included in the data, and using an effective area curve calculated for an on-axis source rather than a source 8' off axis each makes a very small difference in the derived results. The confidence contours in figure 10 each move by an amount small compared to the spacing between the indicated contours, so the conclusions about the location of the break are not affected.

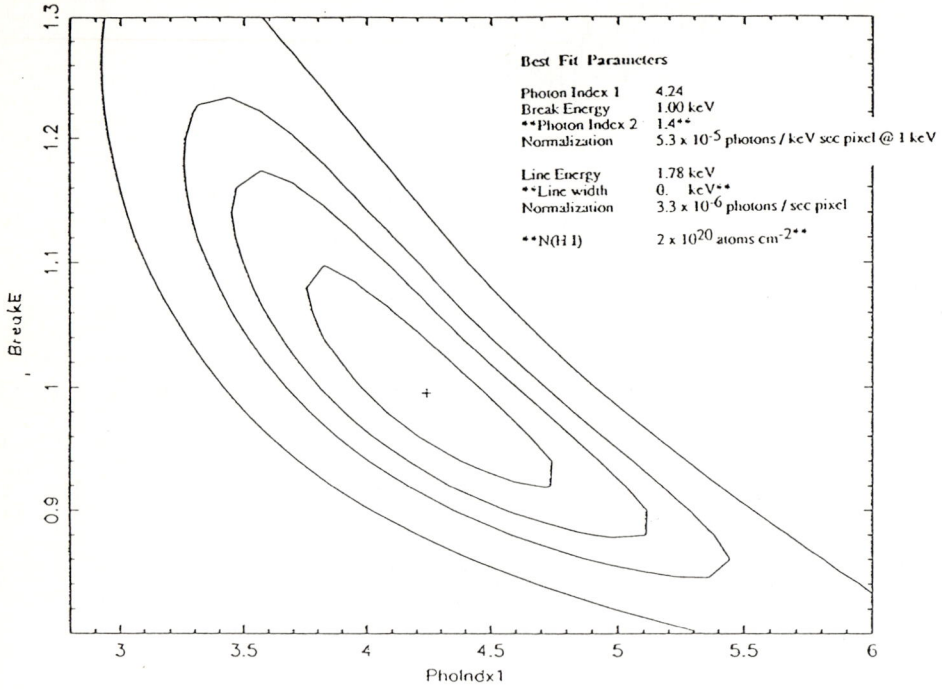

Figure 10: Confidence intervals for break energy and the photon index below the break. Contours represent $\Delta\chi2 = 1.0, 2.71, 4.61$, and 9.21 which give 40, 68, 90, and 99% confidence intervals for two interesting parameters.

5.2 Two power-law model

The model is given by $A(E) = K_1 E^{-\alpha 1} + K_2 E^{-\alpha 2}$, with the normalization constants defined as in the previous model. This model is similar to the broken power-law model although the break is more gentle. This has the effect that the soft power law can be even steeper. The best fit parameters are summarized in figure 11. We have examined the variation of α_1 with both normalization coefficients, as shown in figure 12. This leads us to conclude that the 90% extremum triples (k1, k2, a1) are (3.2 x 10^{-5}, 4 x 10^{-5}, 4.5) (6 x 10^{-5}, 5.8 x 10^{-5}, 7.7). For these two cases, the energy at which the two power-laws make equal contributions are 0.93 and 0.71 keV. The fraction of the total contributed by

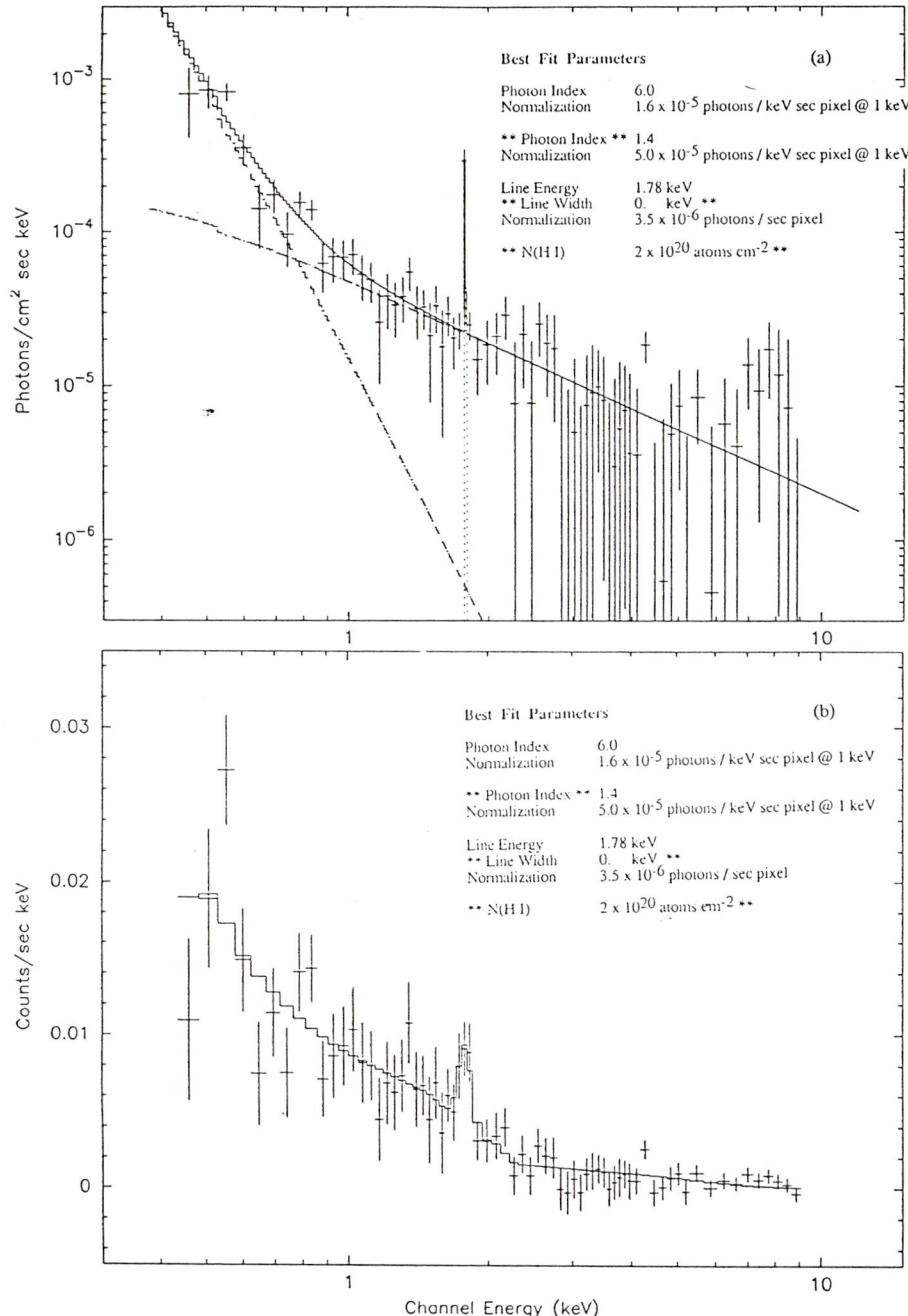

Figure 11: Best fit spectrum for two power-law model: (a) in photon space and (b) count space. The line at 1.78 keV is believed to be an instrumental artifact.

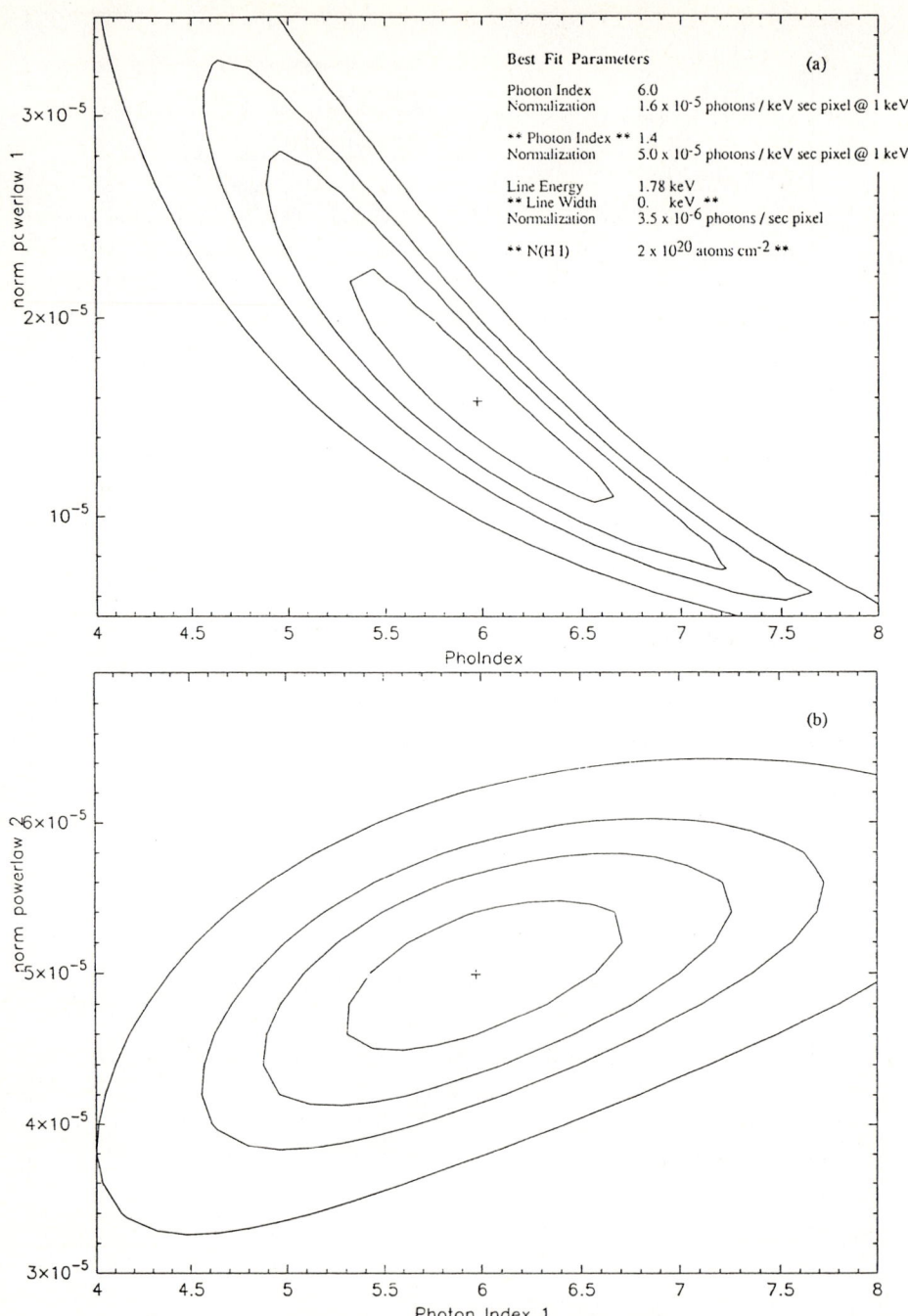

Best Fit Parameters (a)

Photon Index 6.0
Normalization 1.6 x 10⁻⁵ photons / keV sec pixel @ 1 keV

** Photon Index ** 1.4
Normalization 5.0 x 10⁻⁵ photons / keV sec pixel @ 1 keV

Line Energy 1.78 keV
** Line Width 0. keV **
Normalization 3.5 x 10⁻⁶ photons / sec pixel

** N(H I) 2 x 10²⁰ atoms cm⁻² **

Figure 12: Confidence intervals (a) for soft power-law photon index vs soft power-law normalization and (b) for soft power-law photon index vs hard power-law normalization. Contours are as in figure 9.

the high energy extrapolation is 13-10% at 0.5 keV and 56-90% at 1.0 keV for the 90% extremum parameter triplets.

5.3 Power-law plus thermal components

This model has more physical motivation behind it than the others discussed so far. McCammon and Sanders (1990) review the arguments for the existence of a galactic soft X-ray component and the arguments that this must be a thermal component rather than a continuum process. The model consists of a power law of photon index -1.4 plus a Raymond-Smith equilibrium model of a thermal plasma. Figures 13a and 13b show the best fit model in count space and photon space, and summarize the best fit parameters. It seems clear that the lowest channels are better fit by the thermal component than by either of the continuum models considered, although the overall chi-square is virtually identical.

The fraction of the model flux contributed at a particular energy is less meaningful here as the fraction is quite variable due to the dominance of line emission. However, it is clear from the figure that the thermal component contributes less than a few percent at all energies above 1 keV, but 80% or more at lower energies.

The best fit temperature, 0.14 keV or nearly 2 million degrees, is intermediate between the two temperatures typically required to explain the University of Wisconsin data (McCammon et al. 1983). The contour surface, figure 14a, which shows temperature vs normalization of the thermal component demonstrates that as temperature decreases, the emission measure must go up in order to match the observed flux in the Oxygen lines around 500-600 eV.

The contour surface, figure 14b, which shows temperature versus power-law normalization shows that the temperature must be less than 0.2 keV in the one thermal component model. Although the chi-square is formally acceptable for this model (63 for 65 degrees of freedom), as it was for the continuum models, we attempted to add another thermal component in order to compare our result with the Wisconsin paradigm and the ROSAT results (Hasinger, this volume). We did not obtain meaningful results, probably due to fact that our greatest remaining uncertainties in the response matrix are near the aluminum and silicon K edges and the gold M edge, i.e. just the band that would be sensitive to the temperature of the second component.

6 DISCUSSION

We have attempted to characterize the deviation from a power-law description of the DXRB spectrum. The preliminary results can be characterized as follows. There is no significant deviation from the power-law description at energies above 1.2 keV. The power law contributes over half of the observed flux in the 0.5 to 1.0 keV band. We have not had success in fitting the 3 component model of Hasinger, primarily due to limited statistics and

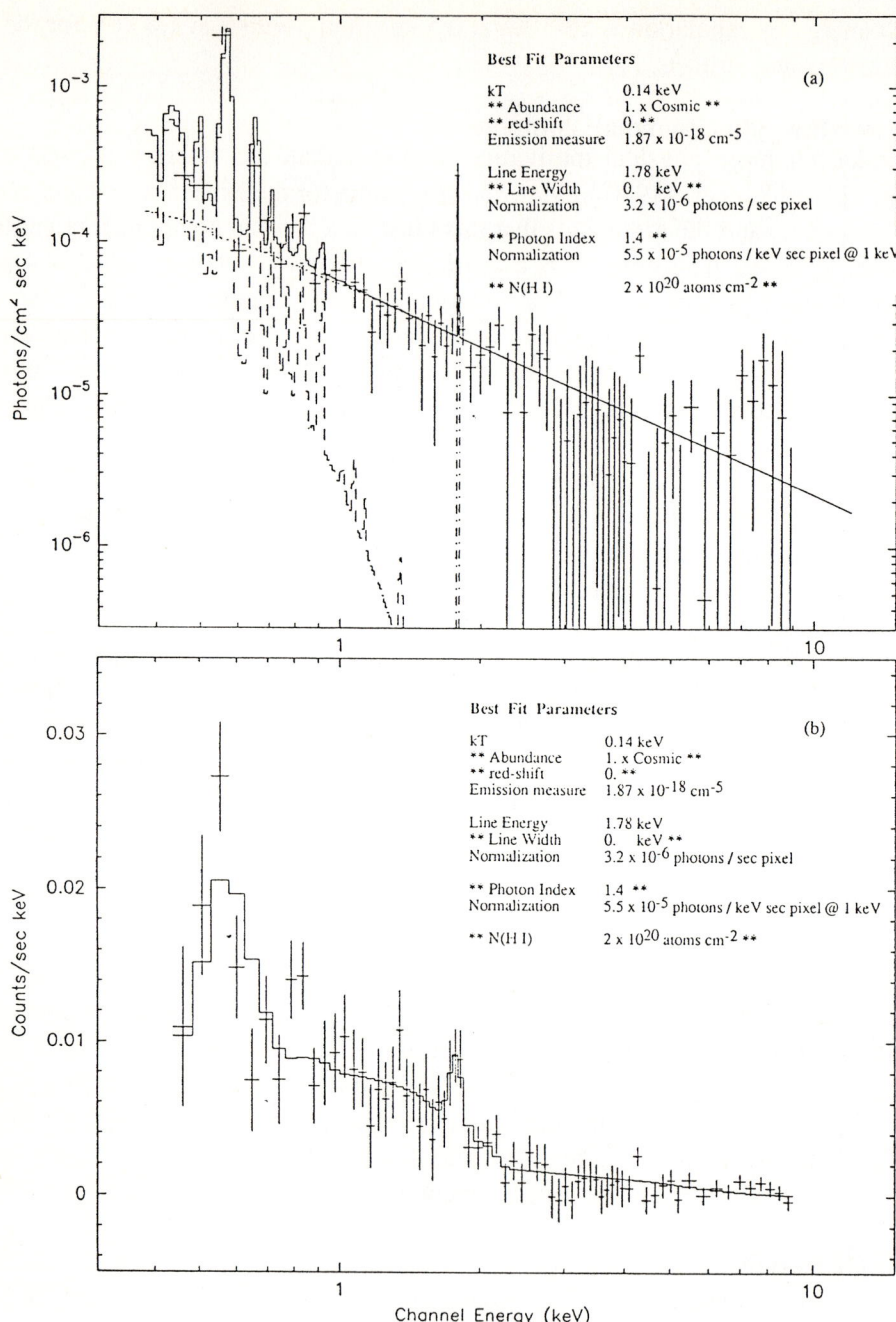

Figure 13: Best fit spectrum for power-law plus thermal component model: (a) in photon space and (b) count space. The line at 1.78 keV is believed to be an instrumental artifact.

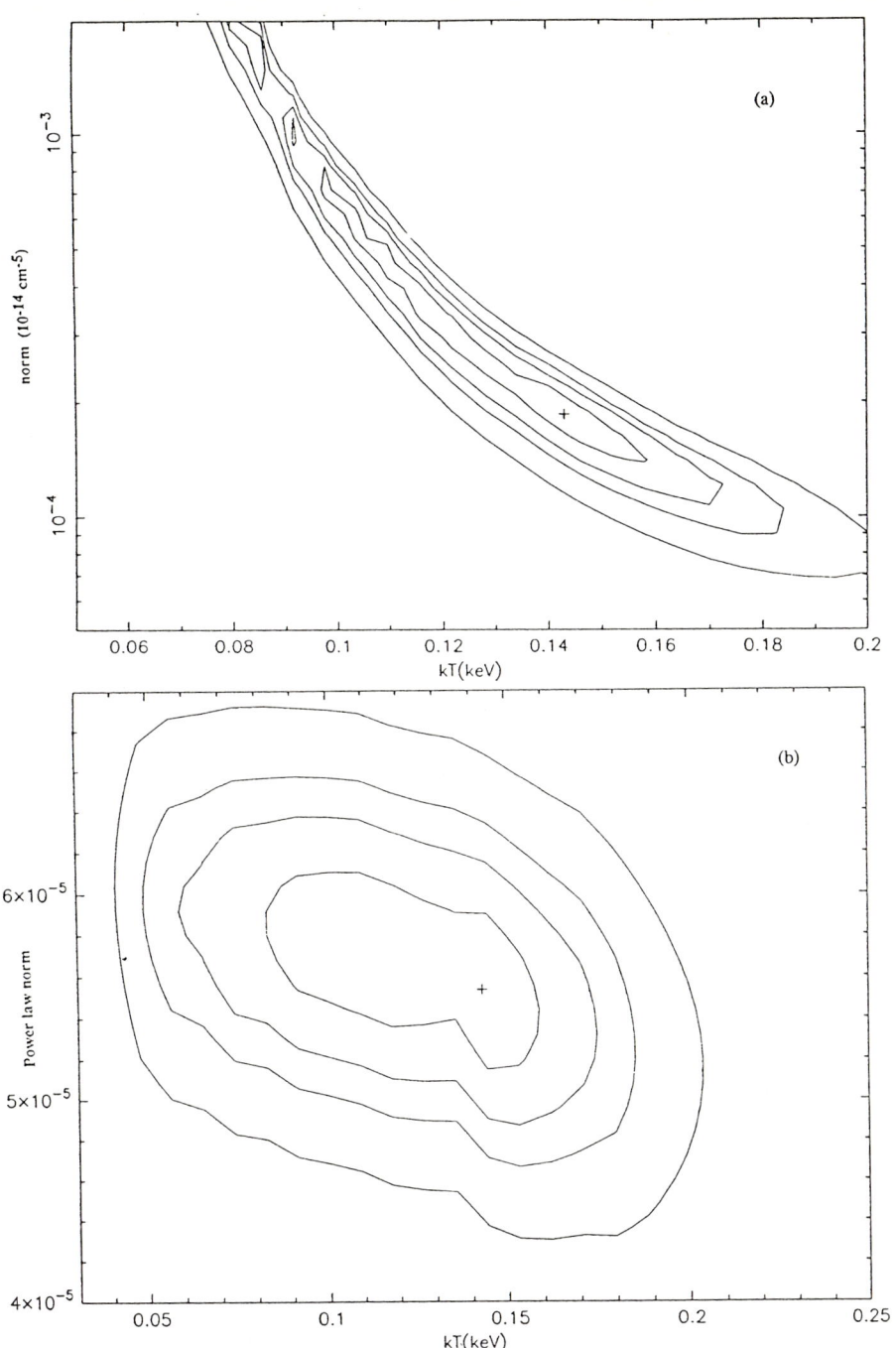

Figure 14: Confidence intervals (a) for thermal component temperature vs
its own normalization and (b) for thermal component temperature vs
power-law normalization. Contours are as in figure 9.

some residual uncertainties in the response. We expect progress in both of these areas. We have not attempted to fit a more complicated model such as the one proposed by Schwartz (this volume) which will be similarly statistics limited. A model like Schwartz proposes plus a thermal component would probably require relatively less thermal emission. We hope to quantify this statement when all of the BBXRT data are used.

ASTRO-D and CUBIC, which will observe the DXRB for months, serendipitously or by design, will have great statistical precision and provide insight on the nature of the DXRB spectrum in ways we have just begun to imagine.

REFERENCES

Blair, W. P. and Gull, T. R. 1990, Sky and Telescope, 79, 591.

Marshall, F. E. et al. 1980, Astrophysical Journal, 235, 4.

McCammon, D. and Sanders, W. T. 1990, Ann. Rev. Astr. Astrophys., 28, 657.

Petre, R., Serlemitsos, P. J., Marshall, F. E., Jahoda, K., and Kuneida, H. 1991,
        Proceedings of SPIE technical conference 1546, San Diego, July 1991, in press.

Serlemitsos, P. J. 1988, Applied Optics, 27, 1447.

Serlemitsos, P. J. et al. 1991, in Proceedings of 28th Yamada Conference "Frontiers of
        X-Ray Astronomy", Nagoya, Japan, April 1991, Y. Tanaka and K. Koyama,
        eds. (Tokyo: Universal Academy Press) in press.

ACKNOWLEDGEMENTS

Space flight experiments are never the results of individual effort. In addition to the science team, the BBXRT instrument and mission were made possible by the efforts of countless engineers and technicians at the Goddard Space Flight Center, Spacelab operations engineers at the Marshall Space Flight Center, and the NASA infrastructure which supports shuttle missions. We thank them all.

The Contribution of Ginga to Studies of the X-ray Background

G.C. STEWART

Department of Physics and Astronomy, Leicester University

1 INTRODUCTION

The properties of the LAC detectors (Turner *et al*, 1989) on board the *Ginga* satellite (Makino *et al.*,1989) are such that a significant improvement in the understanding of the X-ray spectra of a wide variety of X-ray sources has come from observations with this satellite. Here I will concentrate on the results obtained by *Ginga* on the spectra of extragalactic sources (in particular AGN) and on the sky density of these sources. I will discuss the implication of the *Ginga* results in these areas for the origin of the cosmic X-ray background in the 2-30 keV bandpass where, of course, most of the power emitted in the X-ray background is observed.

2 THE SPECTRA OF ACTIVE GALAXIES

Given the extra-galactic origin of the X-ray background the discovery that active galaxies were luminous X-ray sources gave rise to the suggestion that a substantial part (if not all) of the background arose from these sources. For this to be the case requires (at least) 2 criteria to be satisified:
• the source number counts and fluxes must allow the energy density in the X-ray background at some energy to be produced by AGN.
• the integrated (or average) AGN spectrum must be similar to that of the background in the observational frame.
Deep imaging observations with low energy telescopes such as Einstein and ROSAT have indeed shown that the number density of X-ray emitting active galaxies is large (of order 100 per square degree) down to X-ray flux levels of $\sim 10^{-14}$ergs cm^{-2} s^{-1}. Indeed at energies of $\sim$ 1keV approximately 40% of the background has been resolved and comes mainly from QSOs (Hasinger *et al.*, 1991, Shanks *et al.*, 1991). While it is still not clear how much more can be produced by similar sources at fainter flux levels the truly crucial question is what is the contribution of these AGN at higher energies? The answer to this question requires some knowledge of the spectra of the AGN at higher energies. This is where *Ginga* comes in.

2.1 Seyfert 1 Galaxies

It is in this area that the high quality of the *Ginga* data has produced the most dramatic developments. Prior to the launch of *Ginga* the canonical picture of the medium energy spectra of Seyfert galaxies was that they were featureless power-laws with an average energy index of $\sim$ 0.7

(Mushotzky, 1984, Turner and Pounds, 1989). Notable exceptions to this included NGC 4151 and Cen A whose spectra exhibited strong iron lines with equivalent widths of 200eV. The natural interpretation of these lines was fluorescence from the same (spherically distributed) material which gave rise to the line-of-sight absorption observed in these sources, which were equivalent to a neutral hydrogen column density of 10^{23} cm^{-2}.

Ginga observations, however, have revealed that iron lines are in fact ubiquitous and that equivalent widths of 150eV are typical (Nandra, 1991). As the typical line of sight columns to most of these sources are negligible these lines cannot arise from fluorescence in evenly spherically distributed cold material.

This result is now generally interpreted in terms of 'Compton reflection' from cold matter out of the line of sight, as postulated by Guilbert and Rees, (1988) and Lightman and White (1988) at around the time *Ginga* was launched. This interpretation is strengthened by the almost universal evidence for spectral flattening above 10 keV where the combined effects of the reduced ratio of photo-electric absorption to the Thompson cross-section and Compton down-scattering of higher energy photons cause photons to pile up to form a 'Compton hump'. Figure 1 shows an updated version of the 'Ginga 12' spectrum presented by Pounds *et al.*, 1990. This is the sum of 40 spectra of 25 Seyfert 1 Galaxies fitted with a simple power-law spectrum. The shape of the residuals is characteristic of the Compton reflection component (e.g George and Fabian 1991). Further evidence for the hump comes from the flattening of the observed spectrum at energies above 10keV as revealed by fitting simple powerlaw models to the 2-10keV and > 10keV data separately. The strength of both the iron emission feature and the 'reflection hump' are such that the reflector must subtend a solid angle of $\sim 2\pi$ steradians at the X-ray source. This leads to the supposition that an accretion disc is the reflector although currently alternative geometries such as a screen of optically thick clouds cannot be ruled out.

2.2 Seyfert 2 Galaxies

Here again the sensitivity of *Ginga* has shown for the first time the complexity of the spectra of Seyfert 2 galaxies. The best studied example, NGC 1068 (Koyama *et al.*, 1989), has a spectrum which is best interpreted as being a Seyfert 1 Galaxy whose nuclear spectrum is completely obscured. The hard X-rays we see are Compton scattered in the same manner as the broad-line region which is only seen in polarised light (Antonucci and Miller, 1985). The nuclear regions of the other Seyfert 2 galaxies detected with *Ginga*, however, are not totally obscured (e.g. Awaki 1991, Hayashida, these procedings). Instead they are seen through a large column density (typically of order 10^{23} cm^{-2}) of absorbing material. In all cases strong iron lines consistent with fluorescence in the absorbing material is seen.

These results are of particular importance with regard to the 'effective' sky density of such sources seen by the low energy telescopes. The reduction in the low energy flux (either by photoelectric absorption or by scattering) is such that sources of this type may be severely under-represented. Their contribution at higher energies could, however, be substantial.

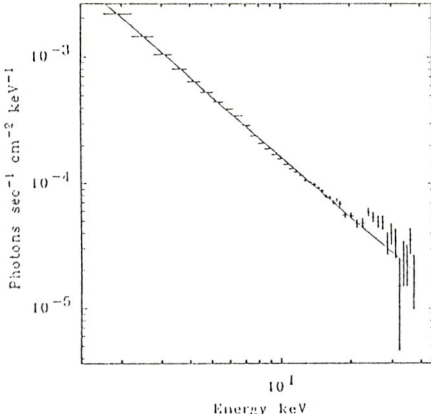

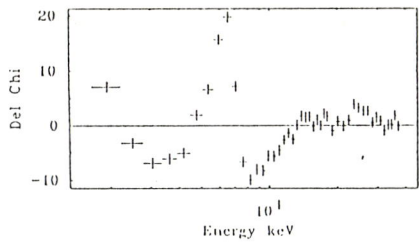

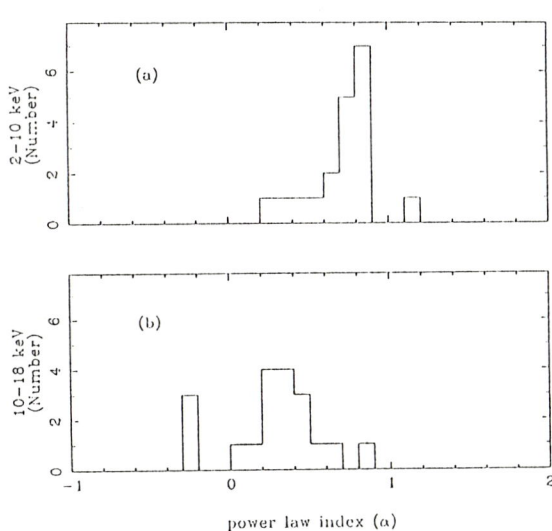

Fig. 1: Top) The result of a power-law fit to the sum of 40 Seyfert spectra showing the residuals characteristic of the Compton reflection model

Bottom) Power-law fits to (a) data below 10 keV and (b) data above 10 keV of the individual spectra showing the flattening at high energies.

2.3 Quasars

No universal picture similar to those found for Seyfert galaxies has appeared for the spectra of the high luminosity AGN. Two examples of iron emission lines have been discovered (3C273, Turner *et al.*, 1990 and 1821-178, Kii *et al.*, 1991). The emission line strengths of 50eV and 350eV are atypical of those found for the Seyfert galaxies but are still larger than could be produced by any absorbing matter in the line of sight as both objects have low column densities. Recent ROSAT results have also shown that a possible contribution to the iron emission line from a surrounding cluster of galaxies in the 1821 spectrum can be ruled out.

There is also little evidence for a flattening of the QSO spectra at energies above 10keV due to Compton reflection of the type found for Seyfert galaxies. Upper limits are in general poor, however, and the tightest limit is about 30% of the reflection expected from material subtending 2π steradians as might be expected for reflection from an accretion disc. Whether this could be due to an enhancement of the direct power-law flux due, perhaps, to beaming of the primary flux or to a reduction in the reflection efficiency (the effective solid angle of the reflector) is not clear.

One result is clear from the QSO spectra. That is that at moderate redshifts the high luminosity AGN do not have spectral indices similar to that of the X-ray background. While the spectral indices of radio-loud and radio-quiet QSOs are different with the means being $\Gamma \sim 1.9$ and $\Gamma \sim 1.7$ respectively (Williams *et al.*, 1992) neither class is as flat as the effective slope of the X-ray background in the 2-10 keV region ($\Gamma \sim 1.4$).

3. SOURCE NUMBER COUNTS

Of course, to understand the possible contribution of such sources to the X-ray background it is necessary to know the sky density of the sources as a function of intensity (their $\log N - \log S$ relationship) in addition to their spectra. Clearly the best method of doing this would be a faint imaging survey where the sources can be individually detected. Unfortunately, proportional counter detectors with collimators, which remain the only instruments available in the medium energy band, become confusion limited at moderately bright fluxes. Some information on the source number counts at fluxes fainter than the confusion limit can be retrieved using the P(D) technique (e.g. Condon 1974) on the fluctuations in the measured 'sky' intensities and further information on the source properties can be derived by using auto-correlation analyses.

3.1 Discrete Source Number counts.

All-sky survey instruments such as ARIEL-V and HEAO-1 measured the flux density of resolved sources down to fluxes of order 3×10^{-11}erg cm^2 s^{-1} in the 2-10keV energy range. The extragalactic sources showed a Euclidean $\log N - \log S$ (e.g. Piccinotti *et al.*, 1981). All these all-sky surveys were consistent with a normalisation of $\sim 2 \times 10^{-15}$(erg cm^{-2} s^{-1})$^{1.5}$per steradian. *Ginga* has been able to extend this to fainter flux levels, albeit for only a small area of sky. Kondo, 1991, finds a total of 11 new sources at fluxes above $\sim 7 \times 10^{-11}$ erg cm^{-2} s^{-1} within an area of a few hundred square degrees. This is consistent with a Euclidean extension of the HEAO-1 results with a normalisation of 1.6 ± 0.6(erg cm^{-2} s^{-1})$^{1.5}$per steradian.

3.2 Number counts from fluctuation analyses

The smaller field-of-view of the *Ginga* collimators, larger effective area of the LAC and longer exposures per pointing mean that *Ginga* both measures the variations in the sky background caused by P(D) noise with a higher signal to noise ratio than previous experiments and, because P(D) analyses are sensitive at roughly the 'one source per beam' level, also provides information on sources at a lower flux level. The data used for the analysis described here comes from a total of 132 separate pointings at galactic latitudes with $|b| > 20°$.

For this dataset we find that average value for the sky intensity, I, over the 4-10keV band is ~ 7ct s^{-1}. The typical statistical and systematic error associated with each of these measurements is of order 0.1ct s^{-1} while the rms dispersion of the sky measurements is $\sigma(I) \sim 0.4$ct s^{-1}. These values can be compared with the values obtained for the HEAO-1 A2 experiment data used by Shafer for which a mean sky intensity of ~ 20 ct s^{-1} in the 2-10keV band was found, with an associated statistical error of ~ 0.2 ct s^{-1} and a dispersion of ~ 0.3 ct s^{-1}. Thus while *Ginga* finds a value of $\sigma(I)/I \sim 6\%$ with a 'signal:to:noise ratio' of 4, HEAO-1 worked with a S:N of ~ 1 at a value of $\sigma(I)/I \sim 2\%$

Analysis of these data with standard P(D) fitting techniques reveals that they are consistent with a Euclidean slope for the $\log N - \log S$ relation (the best fit slope for the differential relation is 2.59 ± 0.25). Figure 2 shows the constraints placed on the source number count relationship by the Ginga P(D) measurements normalised to the fiducial Euclidean $\log N - \log S$ of the Piccinotti *et al.* sample. Clearly the normalisation of the fluctuation measurements is consistent with that of the discrete source number counts and with the HEAO-1 fluctuation results. The source counts are constrained down to a flux level of order 10^{-13} erg cm^{-2} s^{-1}

This level is similar to that achieved by the Einstein Observatory medium sensitivity survey (EMSS - Maccaccaro *et al.*, 1991) in the 0.5-3.5 keV energy band. who also find a number count relationship for extragalactic sources which is close to the Euclidean. To directly compare the number counts requires a conversion factor which depends on the spectra of the sources. As has been noted before, (e.g. Warwick and Stewart 1990), using a conversion factor appropriate for the $\Gamma \sim 2$ spectral index associated with AGN (the dominant component of the EMSS sources) results in a normalisation of the EMSS $\log N - \log S$ in the 2-10keV band which lies well below that found by the fluctuation results.

As analysis of the number counts from both the Einstein Observatory and ROSAT survey obser-

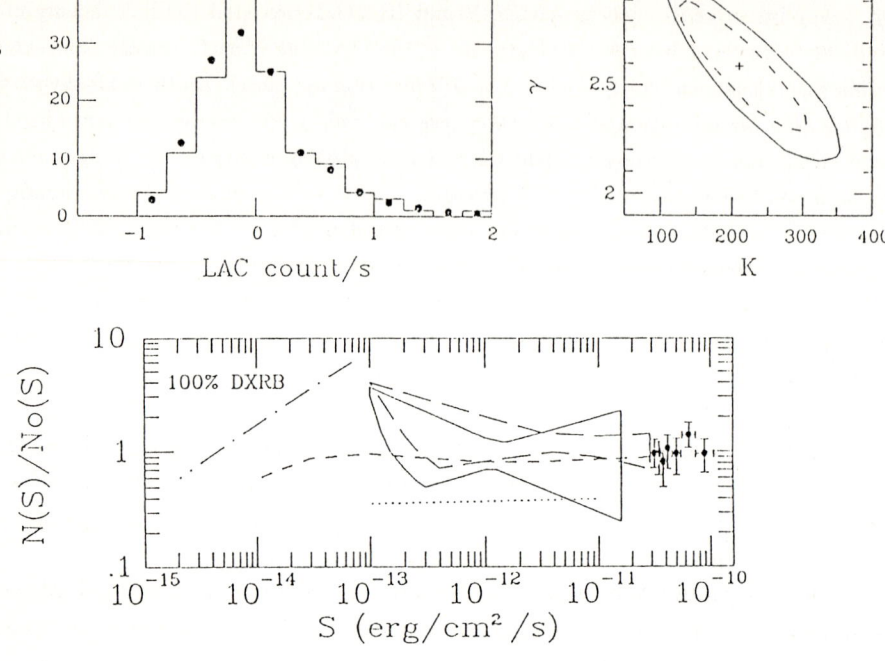

Fig. 2: a) The observed and best fit P(D) curves

b)The constraints on the differential $\log N - \log S$

c)The differential $\log N - \log S$ normalise to $N_0(S)dS = 2 \times 10^{-15}S^{-2.5}dS$. The data points are from Piccinnotti *et al.*, the solid 'fish' from Shafer, and the dashed fish the *Ginga* results. The dotted line represents the EMSS source counts transformed to the 2-10 keV band.

vations suggest that AGN of the type which are detected in soft X-ray observations can contribute at most $\sim$ 50% of the X-ray background at around 2 keV the origin of this discrepancy is clearly important for understanding the make up of the 2-10 keV background. Is the difference evidence for a population of sources not detected in the soft X-ray surveys? Is the spectral assumption used in the conversion correct? Is some other factor such as source clustering responsible?

3.3 The spectrum of the fluctuations

Clearly of importance in deciding between these options is a measurement of the spectrum of the fluctuations. The spectrum obtained by subtracting the spectra of the negative fluctuations from those of the positive fluctuations as first presented by Warwick and Butcher is shown in figure 3. As the fluctuations are dominated by sources at the one source per beam flux level, this should be representative of the spectrum of a 'typical' source at a level by which $\sim$ 15% of the X-ray background is produced.

Analysis of the 2-10 keV spectrum, using standard spectral fitting procedures, shows that for an assume power-law spectral form the derived values for the power-law photon index and low-energy absorption column density are $\Gamma = 0.84^{+.10}_{-.06}$ and $n_H < 3 \times 10^{21}$ cm^{-2}. This is very similar to the parameters derived for for the spectra of individual active galaxies described earlier.

As a corollary it is of interest to consider the extrapolation of this spectrum to higher energies and also to examine the residuals to the fit at energies around the 6-7 keV region. The spectral flattening at higher energies coupled with the positive residuals around 6 keV is reminiscent of the evidence for Compton reflection from cold matter seen in the spectra of individual AGN of moderate luminosity.

4. DISCUSSION

The *Ginga* fluctuation measurements effectively constrain the number density of sources in the 2-10 keV band down to a flux level of $\sim 10^{-13}$ erg cm^{-2} s^{-1}. The total flux contribution of the sources above this level is $\sim 25 \pm 5\%$ of the background flux in this energy range. The spectra of the fluctuations are remarkably similar to those of the bright nearby Seyfert spectra for which *Ginga* has achieved such remarkable results. It seems reasonable therefore to make the conclusion that at least approximately 25% of the medium energy X-ray background comes from these or similar objects.

This conclusion, however, must be treated with some caution. First, the fluctuations have not been identified and some contribution will therefore come from other sources such as clusters of galaxies. Estimates suggest that the contribution from clusters is small. The mean temperature of clusters is low (e.g. Edge and Stewart,1991) and they would be redshifted. Their contribution therefore will be mainly at low energies. Similar conclusions can be reached for possible contributions to the fluctuations from galactic sources. *Ginga* has also been able to obtain results on the spectra of some starburst galaxies (e.g. Makashima *et al.*, 1991). These indicate that the spectra are soft and that starburst galaxies therefore will also only make a substantial contribution at lower energies.

Secondly, I have ignored the possible influence of source clustering on the P(D) analysis - this is dealt with elsewhere in these procedings. This may introduce a reduction in the source counts of a

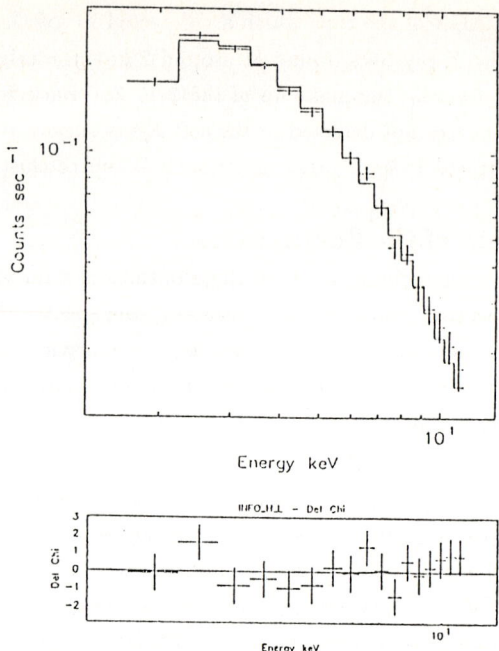

Figure 3) The best fit power-law spectrum to the *Ginga* fluctuations spectrum with, below, the result of extrapolating that spectrum to higher energies.

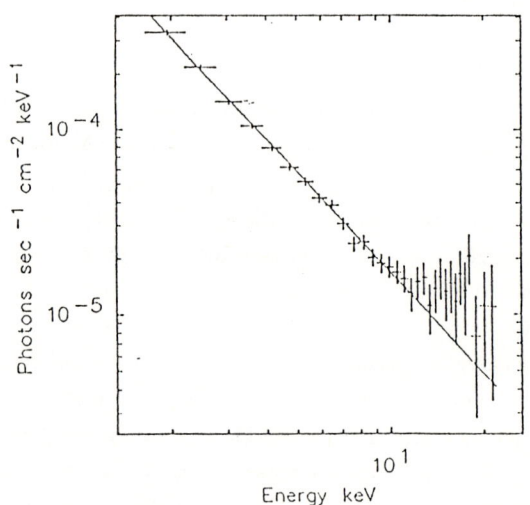

factor of $\sim 25 - 30\%$.

However, given these caveats, the question remains. Can the remainder of the X-ray background come from AGN similar to the Seyfert 1s we observe locally. The answer is probably not. The spectral paradox - that is the difference in slopes between the background and the AGN remain. In addition, results on the auto-correlation function of the *Ginga* data on scales up to a feew degrees (Carrera *et al.*, 1991, these procedings) suggest that at most $\sim 50\%$ of the X-ray backgrund can be produced by AGN with simialr clustering properties to those currently observed.

As other speakers have noted, however, this problem can be circumvented.

The first remedy is to adopt the approach first put forward by Fabian, 1989, on the basis of the *Ginga* results for Seyfert 1s. That is to postulate a population of AGN in which the direct power-law component is supressed and hence the Compton-reflection component dominates. The effect of red-shift, the spectral flattening at 10keV and the Compton roll-off then combine to mimic the spectrum of the background.

The second variation on this theme, again motivated by the *Ginga* data, is to incorporate a substantial contribution from Seyfert 2 galaxies (assuming essentially that they are highlty absorbed Seyfert 1s.) Such a model has been investigated by Awaki and does indeed show some promise.

This latter model would also explain why the number count relations found from the EMSS and the P(D) analyses differ. Highly cut off AGN would just not be detected in the low energy survey. Indeed Warwick and Butcher, 1991, have shown that with reasonable assumptions for the luminosity function and the fraction of absorbed sources at different luminosities the EMSS results and the *Ginga* P(D) results can be brought into consistency.

The *Ginga* fluctuations spectrum should be affected by the inclusion of a substantial number of absorbed sources. Why then is there no evidence for absorption? This may be explained by the neglect of the soft sources which will help to fill in any absorption. Further detailed studies will be required to fully evaluate this point.

5. CONCLUSIONS

Ginga results have indeed substantially improved our understanding of the spectra of AGN. These results have re-opened the possibility of AGN being the origin of the X-ray background and this is certainly consistent with the results on source number counts which come from the P(D) analysis of the fluctuation results. The question remains open, however, and will require input from the next generation of X-ray telescopes capable of imaging in the 2-10 keV band.

6. ACKNOWLEDGEMENTS

It is a pleasure to acknowledge the contribution of the many Japanese and british scientists and engineers responsible for the construction and and operation of the LAC detectors and the it Ginga satellite without which the results discussed here could not have been achieved. It is also a pleasure to acknowledge the contributions of J. Butcher, P. Nandra, R. Warwick, K. Pounds, A.C. Fabian, X. Barcons and F. Carrera to the results presented here. Finally I acknowledge support from the SERC through an advanced research fellowship and the British council.

7. REFERENCES

Antonucci, R.R.J. & Miller, J.S. 1985. *Astrophys. J.*, **297**, 621.

Awaki, H., Koyma, K., Kunieda, H., & Tawara, Y. 1990. *Nature*, **346**, 544.

Fabian, A.C., George, I.M., Miyoshi, S. & Rees, M.J. 1990. *Mon. Not. R. astr. Soc.*, **242**, 14P.

George, I.M. & Fabian, A.C. 1991. *Mon. Not. R. astr. Soc.*, **249**, 352.

Guilbert, P.W. & Rees, M.J. 1988. *Mon. Not. R. astr. Soc.*, **233**, 475.

Hasinger, G., Schmidt, M. & Trumper, J. 1991. *Astron. and Astrophys.*, **246**, L2.

Kii, T. *et al.*, 1991. *Astrophys. J.*, **367**, 455.

Kondo, H. 1991. Ph.D. Thesis, University of Tokyo.

Koyama, K. *et al.*, 1989. *Publ. Astr. Soc. Japan*, **41**, 731.

Lightman, A.P. & White, T.R. 1988. *Astrophys. J.*, **335**, 57.

Maccaccaro, T., Della Cecca, R., Gioia, I.M., Morris, S.L., Stocke, J.T. & Wolter, A. 1991. *Astrophys. J.***374**, 117.

Makino, F. *et al.*1987. *Astrophys. Lett. Comm.*, **25**, 223.

Mushotzky, R.F. 1984. *Advances in Space Research*, bf 3:10-12, 157.

Piccinotti *et al.*, 1981. *Astrophys. J.***253**, 485.

Pounds, K.A., Nandra, K., Stewart, G.C., George, I.M. & Fabian A.C. 1990. *Nature*, **344**, 132.

Shanks, T.S., Georgantopoulos, I., Stewart, G.C., Pounds, K.A., Boyle, B.J. & Griffiths, R.E. 1991. *Nature*, **353**, 315.

Turner M.J.L. *et al.*, 1989. *Pub. Astr. Soc. Japan*, **41**, 345.

Turner, M.J.L. *et al.*, 1990 *Mon. Not. R. astr. Soc.***244**, 310.

Turner, T.J. & Pounds, K.A. 1989. *Mon. Not. R. astr. Soc.***240**, 833.

Warwick, R.S. & Butcher, J. 1991. Frontiers of X-ray Astronomy, in press.

Williams, O.R. *et al.*, 1992. *Astrophys. J.*in press.

Deep ROSAT Observations in QSO Survey Fields

R. E. GRIFFITHS (1), T. SHANKS (2), B. BOYLE (3), G. C. STEWART (4) AND I. GEORGANTOPOULOS (4)

(1) Space Telescope Science Institute, (2) University of Durham, (3) University of Cambridge, (4) University of Leicester

SUMMARY

ROSAT pointed survey exposures of moderate depth ($\sim$ 30 ksec) are dominated by AGN, at median redshifts of $\sim$ 1.5, but with objects up to z $\sim$ 3. Such exposures help to determine directly the contribution of AGN and other discrete sources to the soft X-ray background (SXRB). The sky density of confirmed X-ray AGN is 70 ± 10 deg^{-2}, contributing a minimum of 30% of the SXRB at 1 keV. The best estimate of the total contribution of AGN is $\sim$ 50%, although fluctuation analysis shows that discrete sources contribute at least $\sim$ 75% of the total. The AGN detections are consistent with point sources, with no evidence of any extensions which might be taken as evidence of cluster environments. On the possible origin of the non-AGN component of the SXRB, we report the detection of a minority of starburst galaxies in the surveys, at z $\sim$ 0.3 $-$ 0.4. The latter objects may have relatively greater significance to the XRB above 2 keV.

1. Introduction

Previous attempts to resolve directly the discrete source component of the SXRB with Einstein resulted in the detection of, *e.g.* 14 QSOs in the Pavo field, with a surface density of 32 ± 9 deg^{-2} (Griffiths et al. 1992). These sources were detected largely with the HRI and had greater density than the 5σ sources detected with the IPC which accounted for 10% of the XRB at 2 keV (Hamilton, Helfand and Wu 1991). HRI fluxes could not be well determined, but the Einstein HRI sources probably accounted for about $\sim$ 20% of the XRB at 2 keV.

2. The ROSAT Observations

Two $\sim$ 30000s ROSAT PSPC exposures were made in July 1990, in fields QSF1 and QSF3 of Boyle *et al.* (1990), which had been systematically surveyed for QSOs using the UV-excess technique. Note that galaxies, including starburst galaxies, were rejected by this analysis, unless sufficiently compact to escape detection as extended sources. In QSF1 (3), these observations identified 17 (19) UVX, z < 2.2 QSOs as well as 5 (3) compact, narrow emission line galaxies in a 40' diameter field. At the

B=21.2 mag survey limit the average UVX QSO surface density is $\sim 46 \deg^{-2}$. A total of 46 (39) X-ray sources were detected in the central $0.35 \deg^2$ area of each field, to a detection limit of $S(0.1-2.4 \, \mathrm{keV}) \sim 9 \times 10^{-15} \, \mathrm{erg \, cm^{-2} \, s^{-1}}$, with an internal X-ray astrometric precision of $\sim 6''$. Our overall source surface density is $120 \pm 13 \deg^{-2}$, compared with $8 \deg^{-2}$ in the Extended Medium Sensitivity Survey (Gioia et al 1990) and $36 \pm 9 \deg^{-2}$ in the Einstein Pavo field.

3. The Optical Identifications: AGN

In fields QSF 1 (3), we immediately identified 10 (12) of the X-ray sources with UVX survey QSOs which were brighter than $B = 21.0$. Further spectroscopic identifications in QSF1 (3) resulted in 15 (13) new QSOs (emission line FWHM $> 1000 \mathrm{kms^{-1}}$), including 2 (3) probable QSOs, making a total of 25 (25) QSOs detected by ROSAT in the 40' diameter areas, and an X-ray QSO surface density of $70 \pm 10 \deg^{-2}$ at a detection flux density limit of $S(0.5-2.0 \, \mathrm{keV}) = 7 \times 10^{-15} \, \mathrm{erg \, cm^{-2} s^{-1}}$. The new AGN had B mags. between 21 and 22 in most of the cases, although a few were brighter than the UVX survey limit but had not been recognised as AGN.

Table 1 summarises the optical identification content of the two fields surveyed by ROSAT thus far. Thirteen (8) sources remain unidentified : these sources are either fainter optically than the others, or else lack strong emission lines, or have both attributes. Galaxies without strong emission lines are included here, as are BL Lac objects.

At the survey completeness limit of $S(0.5 - 2.0 \, \mathrm{keV}) = 1 \times 10^{-14} \, \mathrm{erg \, cm^{-2} s^{-1}}$ the equivalent QSO density is $63 \pm 10 \deg^{-2}$. This compares to the QSO surface density of $32 \pm 9 \deg^{-2}$ at $S(1-3 \, \mathrm{keV}) = 2 \times 10^{-14} \, \mathrm{erg \, cm^{-2} s^{-1}}$ found in the Pavo field (Griffiths et al 1992). The X-ray QSO number-redshift relation (Fig. 1) has a median $z = 1.52$, higher than the $z = 0.9$ value derived for the Pavo deep survey with Einstein. However, the redshift distribution is very similar to that for the UVX QSOs identified in faint optical surveys (Boyle et al. 1990). This emphasises the degree to which these two samples overlap at the flux levels reached here.

4. Star-forming Galaxies

Two X-ray sources in QSF1 have so far been classified as star-forming galaxies – further optical spectroscopy is required to establish the true number in the field, since the unidentified sources include faint galaxies which may not have strong emission lines and may be star-forming or post-starburst galaxies.

Griffiths and Padovani (1990) showed that star-forming galaxies may contribute at least $20 - 30\%$ of the SXRB at 2 keV, based on the correlation between X-ray and infrared luminosities and on the assumption of moderate evolution of IRAS galaxies. These galaxies were predicted to show up in the Einstein Medium-Sensitivity Survey in small numbers, as confirmed spectroscopically by Fruscione and Griffiths (1991) and in the Einstein deep surveys (Griffiths et al 1992; Danziger and Gilmozzi 1990), as well as the ROSAT surveys. The numbers predicted for a 30 Ks survey are small, only a few per sq. degree, i.e. about one within the 40 arcmin circle, but the predictions were based heavily on infrared selection rather than X-ray selection.

The identified star-forming galaxies are at moderate redshift (~ 0.3) with X-ray luminosities of $\sim 10^{42-43}$ ergs s^{-1}. The X-ray emission may be dominated by X-ray binaries, either massive objects in low-metallicity Population I regions or low-mass binaries in post-starburst regions (Griffiths and Padovani 1990). To make a substantial contribution to the XRB at 3 - 10 keV, these galaxies should have harder X-ray spectra than the AGN. The detected objects have insufficient counts for determination of their spectra, but one of them was preferentially detected in the harder channel (1-2 keV).

5. The Number Density of X-ray Sources

The derived $\log N : \log S$ diagram for the non-Galactic sub-sample is shown in Fig. 2, showing good agreement at the higher flux levels with the Einstein results. Analysis of the fluctuations in the PSPC background has further constrained the surface density of sources at fluxes below the individual discrete source detection limit. These are fit to the observed distribution and the 1σ limits on $\log N : \log S$ are also shown in Fig. 2 as dashed lines at top left. The best fit value for the slope is -1.2 ± 0.15, significantly flatter than the slope of the source counts at higher flux levels. An extension of the Euclidean slope below $\sim 3 \times 10^{-15}$ erg cm^{-2} s^{-1}, is ruled out at the 95% level even if there are *no* other discrete sources. This is a much more stringent limit than that determined from similar analyses of Einstein data. We also show the QSO $\log N : \log S$ predicted from the EMSS QSO/AGN luminosity function evolution results (Maccacaro et al 1992), with similar behaviour to that found from the P(D) analysis with *i.e.* the number count showing a gradual flattening from the Euclidean beyond our source detection limit. By a flux level of around $\sim 10^{-15}$ erg cm^{-2} s^{-1}, the EMSS prediction falls below the limits of the P(D) analysis. This could be evidence for a new population of faint sources. However, the QSO count could be higher than the EMSS prediction if a slightly higher rate of evolution were adopted or if the faint end slope of the QSO/AGN luminosity function is significantly steeper at $z > 0.5$, or if the low-redshift EMSS sources have steeper spectra than those at higher redshift observed in the ROSAT band. Alternatively, the lower bound from the P(D) analysis may not be a strict lower bound if some of the observed variance is due to instrumental effects.

We conclude that the upper limits from the P(D) analysis implies at least a gradual flattening in the observed source logN:logS at faint X-ray fluxes, similar to what is expected from predictions based on the evolution of the QSO X-ray luminosity function.

6. Contributions to the X-ray Background

The ratio of integrated source counts above 0.5 keV to the total number of counts detected in the inner 20' radius of the ROSAT field gives a value of $\sim 30\%$ for non-Galactic sources. The source density required to produce 90% of the background is $\sim 4000\,\mathrm{deg}^{-2}$ integrating to $5 \times 10^{-16}\,\mathrm{erg\,cm}^{-2}\,\mathrm{s}^{-1}$, a density enlarged as a result of the flattening in the $\log N : \log S$ slope from $S^{-1.5}$ to $S^{-1.2}$. Discrete sources may thus account for almost all of the 0.5-2keV X-ray background.

The EMSS QSO luminosity function model (Maccacaro et al 1992) predicts that 35-40% of the background is due to QSOs. However as noted above, uncertainties in this model could raise (or lower) this contribution. For example, the model currently somewhat underpredicts the total ROSAT source density at $1 \times 10^{-14}\,\mathrm{erg\,cm}^{-2}\,\mathrm{s}^{-1}$ (see Fig. 2). If details of the evolution model or the AGN spectra were adjusted to improve this agreement then this could increase the predicted QSO contribution to the background.

There is some evidence for an upturn in the X-ray background spectrum in the 0.5-2keV band. If the extragalactic spectrum measured at higher energies is extrapolated into the ROSAT passband then only $\sim 60\%$ of the ROSAT background counts are contributed by this component. The summed spectra of the QSO sources and of the background are shown in Fig.3. The detected QSO contribution to the background peaks at $\sim 1\,\mathrm{keV}$ at approximately 35%. In the range $1.0 - 2.0\,\mathrm{keV}$, power law fits to the source spectrum suggest an energy index, $a_x = 1.3 \pm 0.2$, while for the background we find $a_x = 0.7 \pm 0.2$. The observed steepening of the background spectral slope relative to energies above 3 keV suggests that the background in the ROSAT passband may be dominated by QSOs with a spectral index of around 1.

There may therefore be two dominant extragalactic background components at 1 keV, since QSOs have spectra too steep to explain the extragalactic background in the 2-10keV energy range. QSOs could account for $\sim$half the background at 1 keV, with the other source component having a flatter, $a_x = 0.4$, spectrum explaining the other half at 1keV and a larger fraction at harder energies. Suggested components for the second component include heavily absorbed AGN and starburst galaxies.

Further details of preliminary aspects of this work appear in Shanks et al 1991.

References

Boyle, B.J., Fong, R., Shanks, T. & Peterson, B.A., 1990, MNRAS **243**, 1-56.

Danziger, J., and Gilmozzi, R., 1990, ESO preprint.

Fruscione, A., and Griffiths, R. E., 1991, Ap. J., **380**, L13

Gioia, I.M., Maccacaro, T., Schild, R.E., Wolter, A., Stocke, J.T., Morris, S.L. & Henry, J.P., 1990, Ap. J. Supp. Ser. **72**, 567-619

Griffiths, R.E. & Padovani, P., Ap. J., 1990, **360**, 483-489

Griffiths, R.E., Tuohy, I.R., Brissenden, R.J.V., and Ward, M. J., 1992, MNRAS, in press.

Hamilton, T. T., Helfand, D. J., and Wu, X., 1991, Ap. J., **379**, 576.

Maccacaro, T., Della Ceca, R., Gioia, I.M., Morris, S.L., Stocke, J.T., Wolter, A., 1992, Ap. J., in press.

Shanks, T., Georgantopoulos, I., Stewart, G. C., Pounds, K. A., Boyle, B. J., and Griffiths, R. E., 1991, Nature, **353**, 315.

Table 1: Summary of Identification Content of QSF1 (3)

10 (12) Previously known UV-excess QSOs (B<21)
15 (13) Newly-identified QSOs (B<22)
2 (3) Galaxies
6 (3) Stars (B < 12)
13 (8) Unidentified

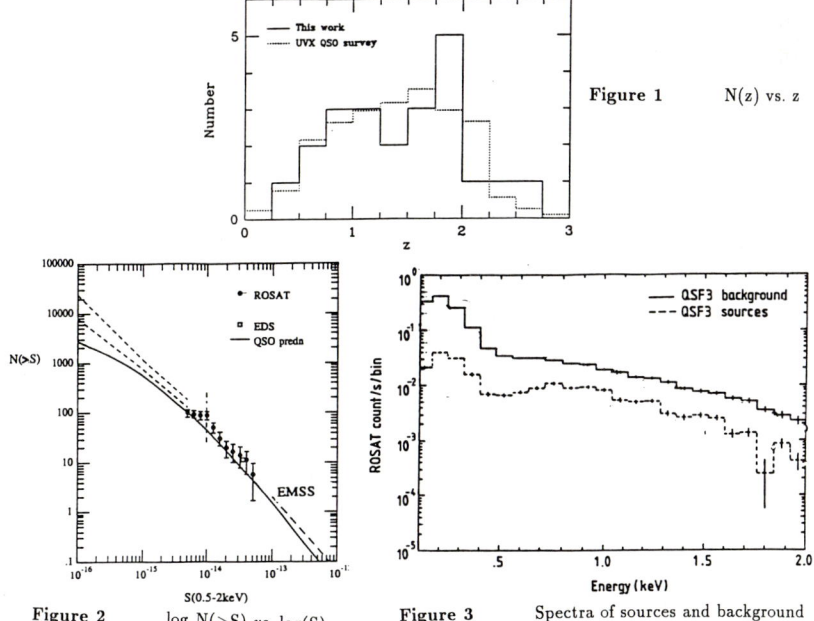

Figure 1   N(z) vs. z

Figure 2   log N(>S) vs. log(S)

Figure 3   Spectra of sources and background

Chapter VII

Future missions

The CUBIC Experiment

D.N. BURROWS, G.P. GARMIRE, D.H. LUMB, J.A. NOUSEK

Department of Astronomy and Astrophysics, Penn State University

1 ABSTRACT

The *Cosmic Unresolved X-ray Background Instrument using CCDs (CUBIC)* is designed to obtain spectral observations of the Diffuse X-ray Background (DXRB) with moderate spectral resolution ($E/\Delta E \sim 10-60$) over the energy range $0.2-10$ keV using mechanically collimated CCDs. This experiment will have an effective area-solid angle product exceeding 0.05 cm^2 sr from $0.4-10.0$ keV and substantial sensitivity down to ~ 0.2 keV, making it competitive with other planned X-ray experiments for study of the spectrum of the diffuse background. Observations will consist of 1–2 day exposures of each target direction, resulting in a series of high quality spectra. Over the anticipated 3 year lifetime of the satellite, *CUBIC* will be able to study up to 50% of the sky with $5° \times 5°$ spatial resolution.

2 EXPERIMENT DESCRIPTION

CUBIC is being built at Penn State for launch on the joint NASA/Argentine mission *SAC-B*, a small satellite being built by the Argentine space agency, CONAE. *SAC-B* is a three axis stabilized spacecraft designed to obtain solar and cosmic X-ray and gamma ray data. The satellite Z axis will be pointed at the Sun, allowing the *CUBIC* experiment, which looks out along the X axis, to observe any point in the sky in a six month period. *SAC-B* is scheduled for launch in July 1994 into a 550 km, 38° inclination, circular orbit.

The *CUBIC* experiment contains a CCD camera with two mechanically collimated CCDs having square fields of view of $5° \times 5°$ and $7° \times 7°$ (FWHM). The two CCDs are mounted on a single cold head, which will be cooled by a TEC to a temperature of approximately -100 C. Heat from the TEC will be radiated to space from the top plate of the experiment. Data from the CCDs will be processed on-board to identify valid X-ray events, which will be stored in 2 MBytes of static RAM.

Although an X-ray is absorbed in a single pixel in the CCD, the charge cloud produced by each X-ray interaction can spread into adjacent pixels as it drifts into the charge collection channel, or it can be spread into trailing pixels during readout. We have found that the highest energy resolution is obtained for X-ray events that produce charge in only one or two pixels; multiply-split events have inferior energy resolution due to charge loss and to additional readout noise. Fortunately, high resistivity CCDs designed for use as X-ray detectors collect most events in one or two pixels, so that little quantum efficiency is lost by restricting data collection to those events with the best energy resolution. Our on-board processing will store only single-pixel and double-pixel events for telemetry to the ground.

In addition to selecting the highest quality X-ray events, this processing simultaneously rejects charged particles with over 99% efficiency, since these interact throughout the volume of the CCD and produce large events encompassing many pixels. Thus, the CCD acts as a five-sided anti-coincidence detector, and the unrejected charged particle background is expected to be nearly two orders of magnitude below the sky flux for *CUBIC*.

CUBIC will use CCDs designed for this experiment by Jim Janesick of JPL and built by Loral Aerospace. The format is 1024 x 1024, with 24 micron pixels. A special gate structure called OPP (Open Pinned Phase) will be used to obtain good low energy sensitivity, and the devices will be built on high resistivity wafers to obtain good high energy quantum efficiency. A plot of the anticipated effective area-solid angle product of the CCD camera is shown in Figure 1. This plot includes the transmission of a 1000Å Al filter required to block optical and UV light.

Table 1: Sensitivity comparisons

	$A\Omega$ (cm² sr) (< 2 keV)	$A\Omega$ (cm² sr) (~ 6 keV)	Exposure (m² s sr)	ΔE (eV) (6 keV)
CUBIC	> 0.084	> 0.084	~ 100/year	~ 120
HEAO-1 A2 LED [1]	0.61		45	
MED/HED [2]		~ 2.2	80	~ 950
BBXRT [3]	~ 0.013	~ 0.0025	~ 0.05	~ 160[4]
Astro-D GIS [3]	~ 0.42	~ 0.19		~ 470
SIS [3]	~ 0.076	~ 0.035		~ 120
AXAF (ACIS) [3]	~ 0.01	~ 0.002		~ 120

[1] Garmire *et al.* 1991
[2] Marshall *et al.* 1980
[3] No significant response below 0.5 keV
[4] 155 eV in the central pixel, 165-170 eV in the outer pixels

---

The sensitivity of *CUBIC* to the DXRB compares favorably with other recent and planned experiments, as shown in Table 1. *CUBIC* is expected to obtain an exposure of about 100 m² s sr yr⁻¹ (assuming only 25% operating efficiency due to Earth occultations, solar restrictions, and SAA passages), compared to 80 m² s sr for the *HEAO-1 A-2 MED/HED* experiment (Marshall *et al.* 1980). The *Astro-D GIS* will be more sensitive to diffuse radiation than *CUBIC*, but with lower energy resolution and with no response below about 0.6 keV. In comparison with other experiments with similar energy resolution, only the *Astro-D SIS* can compete with *CUBIC* for diffuse background sensitivity. Neither *BBXRT*, *SIS*, nor *ACIS* has significant response below 0.5 keV where diffuse emission from hot galactic gas dominates. We expect *CUBIC* to make substantial contributions to the understanding of this component.

Radiation damage to CCDs has become a major concern in recent years with the realization that scientific grade CCDs are very susceptible to bulk displacement damage by protons, which adversely affects Charge Transfer Efficiency (CTE) and therefore degrades the X-ray resolution. The *CUBIC* CCDs will be hardened by employing channel notch architecture, low operating temperatures, and shielding. In spite of these measures, the energy resolution is expected to degrade with time. The severity of this degradation is under investigation.

3 *CUBIC* PERFORMANCE

The results of spectral simulations are shown in Figures 2-4. Figure 2 shows an incident spectrum, derived from a spectral fit to proportional counter data for a typical

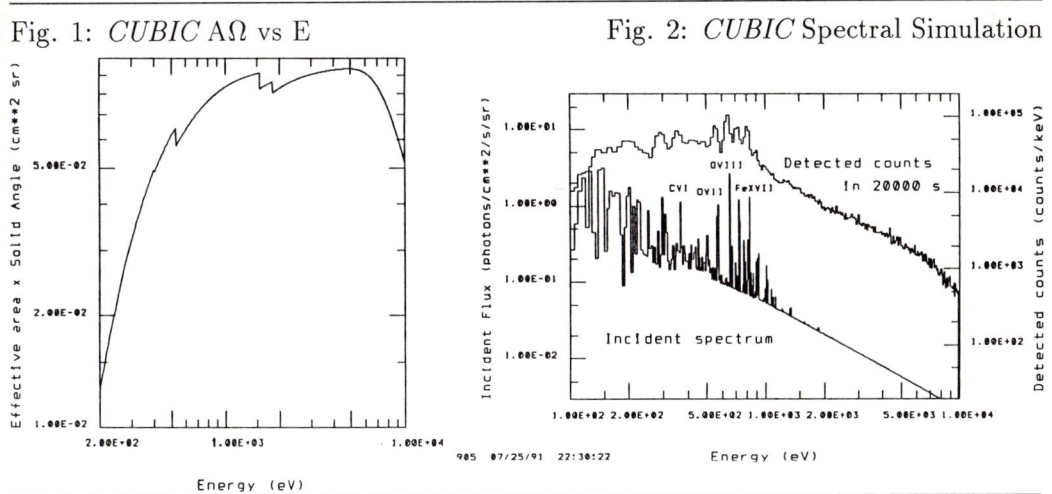

Fig. 1: *CUBIC* AΩ vs E   Fig. 2: *CUBIC* Spectral Simulation

diffuse background region, together with the resultant CCD pulse-height distribution. The incident spectrum contains three components: a $10^{6.0}$ K thermal component (Raymond and Smith 1990), a $10^{6.5}$ K thermal component, and an $11E^{-1.4}$ power law.

Prominent line groups are identified in the predicted CCD pulse-height distribution. The CCD resolution is sufficient to resolve line groups from ions of C, O, and Fe. Although this resolution is not sufficient to provide detailed plasma diagnostics, it will give us a wealth of information about the emitting gas. We will use these data to study variations in the temperature, ionization structure, and elemental abundances of the hot phases of the interstellar medium.

Above about 1.5 keV, *CUBIC* will provide a very sensitive measurement of the extra-galactic component of the DXRB, which we refer to as the cosmic X-ray background (CXRB). The long-standing discrepancy between the intensity of the CXRB measured by thin-window gas-flow proportional counter experiments ($\approx$ 11; McCammon *et al.* 1983, Garmire *et al.* 1991) and that measured by thick-window sealed detector experiments ($\approx$ 8; Schwartz 1979, Marshal *et al.* 1980) has led to speculation that there may be a spectral break in this component near 3 keV. *CUBIC* spans this energy region and will be able to accurately measure the intensity and spectrum of the CXRB with an independently calibrated instrument utilizing very different detector technology than previous experiments. Figure 3 compares the predicted pulse-height distributions resulting from a single power law and from a spectral break model. We will be able to clearly distinguish between these models in the first week of observations.

Fig. 3: Broken power law model   Fig. 4: Fe continuum model

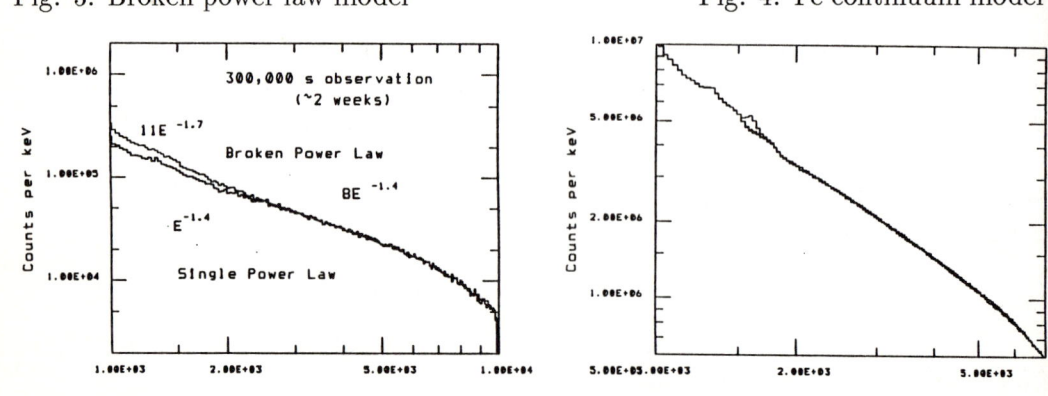

Finally, *CUBIC* will be able to make a very sensitive search for line emission in the CXRB. Figure 4 shows the incident spectrum and pulse-height distributions for two models including a red-shifted Fe line from AGNs (Schwartz 1991), overlaid on the pulse height distribution from the "standard" model shown in Fig. 2. The pulse height distributions of Fig. 4 are for 10^7 s of data, which represents nearly a year of observations. The Fe continuum models differ in their assumptions about the

AGN luminosity function and evolution; the top line in the figure is for a model with strong luminosity evolution (with power law index of 1.8), while the middle line is for a less strongly evolving model with an index of 1.5. The lowest line is for the "standard" model without any Fe continuum. The "turn-on" of the AGNs results in a feature at 1.6 keV (assuming $z_{max} = 3$) which is clearly observable for the strongly evolving model, but otherwise the deviations from a standard power law are only a few percent. Although the counting statistics available from *CUBIC* may permit such small deviations to be detected, the principal difficulty in making this measurement will be in understanding the instrumental characteristics to better than 1% over a long observing period. It remains to be seen whether this will be a practical measurement.

The *CUBIC* experiment will provide unprecedented spectral information on the galactic components of the diffuse X-ray background, including the first sensitive measurements in the 0.3–0.5 keV range, which is inaccessible to plastic window proportional counters. It will also provide very sensitive spectral measurements of the extragalactic component of the X-ray background. Above 0.5 keV, the sensitivity and energy resolution of *CUBIC* will complement the *Astro-D SIS*. Below 0.5 keV, *CUBIC* will provide unique high quality spectra of bright sources, SNRs and the diffuse background, and will complement the higher spectral resolution measurements of the Wisconsin *DXS* experiment.

ACKNOWLEDGEMENTS
Our CCD development work is being carried out by Andy Collins and Jim Janesick at JPL under contract with NASA. *CUBIC* is funded by NASA grant NAG 5-100.

REFERENCES
Garmire, G. P., *et al.* 1991, ApJ, submitted

Marshall, F. E., *et al.* 1980, ApJ, 235, 4

McCammon, D., *et al.* 1983, ApJ, 269, 107

Raymond, J. C., and Smith, B. W. 1990, private communication; update to Raymond and Smith 1977, ApJ Suppl, 35, 419

Schwartz, D. A. 1979, in X-ray Astronomy, ed. W. A. Baity and L. E. Peterson (Pergammon Press, Oxford), 453

Schwartz, D. A. 1991, ApJ, submitted

The Spectrum X–Γ Project

A. GIMÉNEZ

Laboratorio de Astrofísica Espacial y Física Fundamental Estación de Villafranca, Apartado 50727, 28080 Madrid, Spain

1 INTRODUCTION

Spectrum X–Γ is one of a series of astrophysical observatories to be launched by the Soviet Union. These satellites, including also Radioastron and Spectrum UV, use a common platform for astronomical missions adapted from previous soviet space programmes.

The objectives of the Spectrum X–Γ mission are the obtention of high–resolution spectroscopy having imaging capabilities with high angular resolution (around 20 arcsec), and to cover a broad wavelength range (from 0.02 to 10000 keV) with high sensitivity (up to around $3\,10^{-15}\,\mathrm{erg\,cm^{-2}\,s^{-1}}$). Moreover, the possibility to observe time variations and to monitor the whole sky to detect transient events is also considered.

In spite of the current difficult situation in the Soviet Union, the Spectrum X–Γ mission is proceeding on schedule and the launch date has been recently confirmed for the end of 1994. In order to assure state-of-the-art technology for the X-ray mirrors and focal plane instruments, a wide international cooperation has been established which included, apart from the Soviet Union, Denmark, Great Britain, Germany, Italy, Spain, Finland, the United States, etc.

Spectrum X–Γ will contribute, together with ASTRO-D and SAX to cover the X-ray gap of missions after ROSAT and before the two Great Observatories being developed by NASA (AXAF) and ESA (XMM). Overlap with AXAF is expected but the contribution to X-ray astronomy of Spectrum X–Γ will be made before and with unique characteristics. In fact, it will be faster because of its larger collecting area, and will be more efficient at higher energies than both AXAF and XMM.

2 THE MISSION

The observatory will be placed in a high-elliptical orbit with a period of around 4 days by means of a Proton launcher. With all the operational constraints taken into account, there will be access to about 80 per cent of the celestial sphere at any time. An average lifetime of 5 years is expected with a guaranteed minimum of 3 years.

The spacecraft consists of a service module, which uses the basic Spectrum astrophysical platform, and the scientific instruments which can weight up to 2.5 tons. The service module provides an absolute pointing accuracy of 2-5 arcmin and 3-axis stabilization with 30 arcsec at 10^{-4} degrees/sec. It also supplies the necessary power to the scientific instruments and provides communication links with the ground station. Data dumping will be performed at a high rate usually few times per day but real-time data transmission of some science will also be possible at a slow rate. An average of 15 pointings per day is foreseen, leading to the observation of around 5000 targets per year. The pointings will be defined by the primary instruments though all of them are co-aligned, exception made of the all-sky monitors.

3 THE INSTRUMENTS
The following instruments will be available:

3.1. XSPECT/SODART
This is the main instrument of the satellite and consists of two identical telescopes, with 60cm in diameter and a focal length of 8m each, using grazing incidence optics with thin-foil mirrors. Conical thin shells approximation to Wolter 1 geometry permits to have 154 elements and thus a large effective area and energy range (from 0.1 to 20 keV) with an angular resolution of 2 arcmin. Detectors at the focal plane include microstrip proportional counters for the high (HEPC) and low (LEPC) energy ranges for each of the telescopes, two imaging detectors, one silicon X-ray array (SIXA) optimized for the detection of the Fe line at 6.7 keV, and one stellar X-ray polarimeter (SXRP). On the other hand, an Objective Cristal Spectrometer (OXS) for high-resolution Bragg spectroscopy of bright sources will be available in one of the X-ray telescopes. The sensitivity of the instruments will permit the detection, with 5-sigma confidence, of $3\,10^{-14}$ erg cm^{-2} s^{-1} fluxes in 2000 sec. With respect to spectroscopy, as an example, the Fe line at 6.7 keV can be obtained for AGNs with a resolution of 150 eV using SIXA down to $2\,10^{-13}$ erg cm^{-2} s^{-1} or with a resolution of 5 eV using OXS down to fluxes of $5\,10^{-12}$ erg cm^{-2} s^{-1}.

3.2. JET-X
This instrument is formed by two co-aligned telescopes designed for accurate localization, spectrometry, timing and imaging of faint X-ray sources. The two telescopes consist of 12 nested replicated mirrors and passively cooled CCD detectors. With a focal length of 4m obtain an angular resoltion of 10-30 arcsec within an energy range from 0.2 to 10 keV. Energy resolution varies from 10 to 100 in terms of $E/\Delta E$ but is

enhanced to 350 at 7 keV. The maximum sensitivity is $3 \, 10^{-15} \mathrm{erg \, cm^{-2} \, s^{-1}}$ in 10^5 sec for a 3-sigma detection in the 0.5-10 keV range. A co-aligned optical monitor and an attitude reconstruction monitor will allow the localization of sources with 5 arcsec spatial resolution within a field of view of 5×5 degrees.

3.3. MART

This is a coded-mask telescope for imaging, timing and spectroscopy in the hard X-ray range from 4 to 150 keV, with a position sensitive detector which allows wide band spectroscopy at moderate resolution of around 10-20 per cent. Gas proportional counting provides sensitivities down to $3 \, 10^{-12} \mathrm{erg \, cm^{-2} \, s^{-1}}$ in 10^5 sec within the mentioned energy range.

3.4. EUVITA

This is a complex of 8 telescopes aligned along the main axis for the extreme ultraviolet region. Normal reflection from paraboloidal mirrors and multilayer coating is used together with microchannel plate detectors with position sensitive readout. Each of the telescopes has a central wavelength from 50 to 250 Å with a wavelength width of around 10 per cent of the central value.

3.5 All-Sky Monitors

Finally, two instruments are devoted to a continuous all-sky monitoring of transient events. MOXE is an array of 6 pinhole cameras with position sensitive proportional counters designed to monitor transient X-ray sources in the energy range from 2 to 12 keV with a sensitivity of 2 millicrabs in 1 day with a S/N of 5. On the other hand, SPIN is a monitor of gamma-ray bursts designed to work in the energy range from 10 keV to 10 MeV and a position resolution of 0.5° with a sensitivity of $10^{-7} \mathrm{erg \, cm^{-2}}$.

4 THE X-RAY BACKGROUND

There are three observational methods to identify the origin of the X-ray Background: the spectrum, the isotropy and the identified source content. For the specific purpose of investigating the XRB, Spectrum X–Γ will provide detailed energy distributions for the faintest sources, well below the Einstein Deep Survey limit, which may contribute to the observed spectrum of the XRB. Thus, discrimination between theories based on AGN evolved in their properties or a new type of objects as responsible for the extragalactic background will be possible.

On the other hand, the coverage of a broad energy range will permit the separation of galactic and extragalactic contributions as well as to have a deeper insight into possible different explanations for each energy range of the XRB. The spectrum above 3 keV will be determined and studies of the hard X-ray background and the spectra of active galactic nuclei may clarify the reality of the former as a truly diffuse phenomenon or establish the radiation production involved for the latter.

A combination of the high sensitivity of SODART and the high spatial resolution of JET-X, which observe the same regions simultaneously, will no doubt allow a break-through advance in understanding the nature of the XRB and prepare the coming of the great observatories AXAF and XMM at the end of the century. An average integration time with SODART will be around 10^5 s, which is more than enough to significantly detect the XRB at most energies. A large sky coverage and a detailed spectrum will therefore be obtained as a result of the pointed observations. The new generation of telescopes which are carried by the mission will permit counts of super-weak X-ray sources near the boundary of the observable Universe which may bring new light on the problem of the origin of the diffuse X-ray background.

The Astro-D Mission

H. INOUE

Institute of Space and Astronautical Science
3-1-1, Yoshinodai, Sagamihara, Kanagawa 229, Japan

1. INTRODUCTION

Astro-D is the fourth X-ray astronomy mission of ISAS, Japan, following Ginga (Makino et al. 1987), and is designed to be a high-capability X-ray observatory. It will be equipped with nested thin-foil telescopes which provide a large effective area over a wide energy range from below 1 keV up to 12 keV. It is capable of imaging the X-ray sky with a spatial resolution comparable to that of the IPC of the Einstein Observatory. Two different types of detectors, CCD cameras and imaging gas scintillation proportional counters (IGSPC), are employed for the focal plane instruments. The CCD's will provide a spectral resolution comparable or superior to that of SSS of the Einstein Observatory but cover a much wider energy range. The IGSPC's will provide a spatial resolution comparable to that of the IPC but have a much better background rejection rate. These features will characterize Astro-D to be the first high-throughput imaging and spectroscopic observatory covering a wide energy range over 10 keV.

Astro-D will be launched by the ISAS vehicle M-3SII into an approximately circular orbit of a 550–650 km altitude. The spacecraft will be three-axis stabilized. The absolute pointing accuracy will be approximately 1 arcminute. However, the spacecraft stability and the accuracy of post facto attitude reconstruction is expected to be better than 0.2 arcminutes.

2. HIGH-THROUGHPUT X-RAY OBSERVATORY

Astro-D will be a high-throughput X-ray observatory. Four identical mirrors will be mounted on the top plate of the extendible optical bench, while the focal plane instruments, 2 units of CCD cameras and 2 units of IGSPC's, will be placed on the base plate. The focal length will be 3.5m.

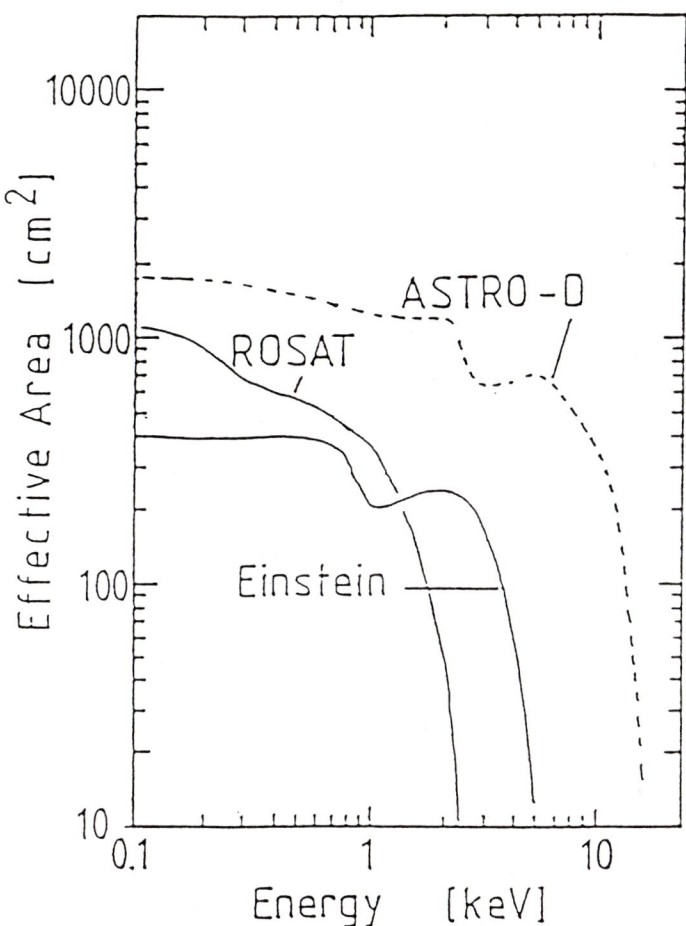

Fig.1 Total on-axis effective area for four sets of mirrors of Astro-D as a function of X-ray energy, in comparison with those of the Einstein observatory and the ROSAT.

2.1 X-ray telescopes

The X-ray telescope consists of four sets of nested thin-foil conical mirrors. This technology is developed by P.Serlemitsos (1981). The mirror system of Astro-D is prepared by a collaboration between P.Serlemitsos and his collaborators at NASA Goddard Space Flight Center and Nagoya University/Kyoto University/ISAS group.

The image size actually achieved is approximately 2.5 arcminutes half-power-diameter due to several error sources. Although the image resolution is moderate, this technology can provide a large effective area over a wide energy range with a very light weight. Therefore, it is particularly advantageous for Astro-D which is severely weight limited. The total on-axis effective area for four sets of mirrors of Astro-D as a function of X-ray energy is shown in Fig.1, in comparison with those of the mirrors on board the Einstein Observatory and ROSAT, respectively.

2.2 Extendible optical bench

The nose fairing of the present ISAS launch vehicle, M-3SII, cannot accomodate an optical bench of a 3.5m focal length. In order to achieve the required focal length, we employ an extensible optical bench which is folded short during the launch. It will be extended in orbit to a 3.5m focal length by means of a sliding mechanism, and latched precisely in position. The optical bench will be constructed by a truss structure composed of CFRP tubes in order to maintain the required accuracy against temperature changes.

2.3 Focal plane instruments

The focal plane instruments consist of two units of CCD cameras and two units of IGSPC's. These detectors will operate independently and simultaneously.

The CCD cameras are prepared jointly by G.Ricker and his collaborators at MIT and the Osaka University/ISAS team. The Penn. State University group is also involved in the software development. A CCD camera head consists of the Lincoln Laboratory 840×840 hybrid CCD of front-side illumination (four 420×420 CCD chips abuted side by side, each read out by a seperate preamplifier). They will be operated at a modest temperature of $-70°$C. The field of view coverage is 23×23 arcminutes^2. Since the CCD's provide a spectral resolution of about $100 - 150$ eV (FWHM) in the energy range from 0.5 keV to 10 keV, the CCD camera will be a powerful instrument for spectroscopic studies of lines from oxygen through nickel.

Each IGSPC consists of a gas chamber filled with Xe and sealed off with a thin berylium window (10 micron), and a position-sensitive photo-multiplier (HAMA-MATSU R2486QX). An energy resolution of 8% and a position resolution of 0.5 mm have been achieved for 5.9 keV (Mn-K) photons. The IGSPC's will cover the entire field of view of the telescope (larger than the coverage of the CCD camera) and will be characterized by their high background rejection rate. Non X-ray signals from IGSPC's are mostly rejected through the rise time discrimination and the light width discrimination. The background rejection rate in the 2-10 keV range is 96% and is comparable to that of the LAC on board Ginga (Turner et al. 1989).

3. SENSITIVITY

With the instruments described above, Astro-D will be the first X-ray astronomy satellite capable of imaging the X-ray sky in the 2-10 keV range. The sensitivity of Astro-D will be much better than that of non-imaging instruments such as the large area counters on board Ginga. In Fig.2, the expected detection limit of Astro-D is plotted on the log N − log S relation in the 2-10 keV range. The high sensitivity of Astro-D will enable us to observe much fainter sources than the limit of previous missions. As a result, a number of important astrophysical investigations will be carried out.

One of the most important scientific objectives is to resolve the origin of the cosmic X-ray background (CXB). The origin of the extragalactic isotropic component of the X-ray background is still not fully solved almost 30 years after its discovery (Giacconi et al. 1962).

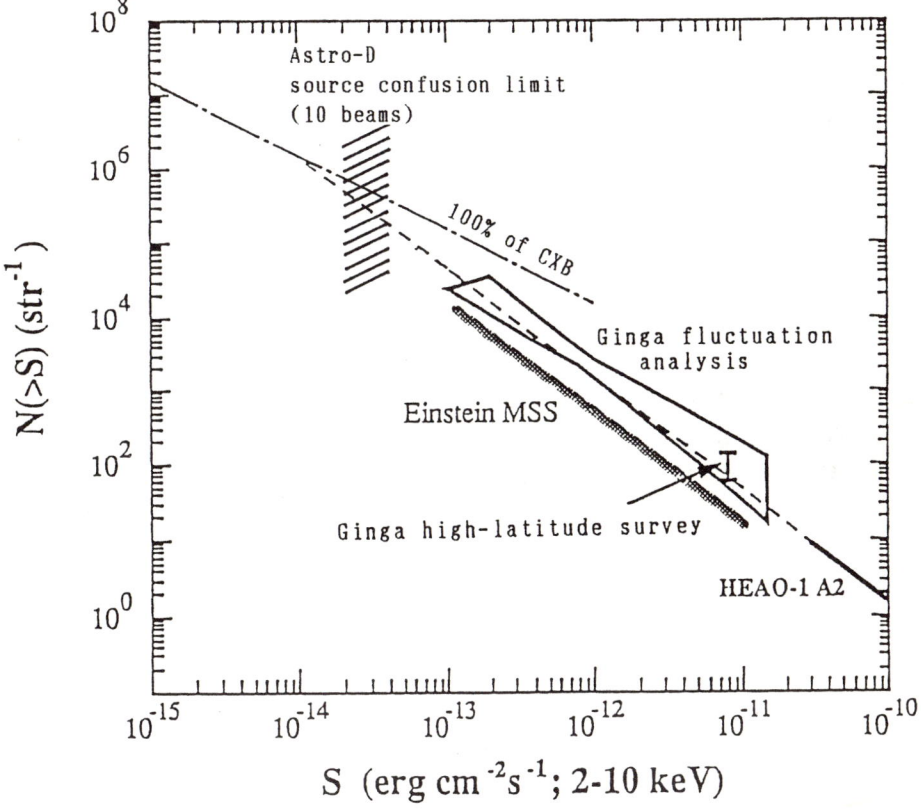

Fig.2 Expected sensitivity of Astro-D plotted on the log N-log S plane for X-ray sources in the 2-10 keV range. The log N-log S relations obtained by the Ginga high latitude survey (Kondo 1991), the Ginga fluctuation analysis (Hayashida 1990; Warwick and Stewart 1989) and the Einstein medium sensitivity survey (Gioia et al. 1984) are shown. The Ginga results are consistent with the $N \propto S^{-3/2}$ relation extraporated from the HEAO-1 A2 result (Piccinotti et al. 1982).

A simple solution is that the observed background flux is supplied by a collection of unresolved, weak point sources. Einstein Observatory obtained the number down to a level approximately 1000 times fainter than previously possible (Giacconi et al. 1979). However, Einstein observed X-ray sources in the energy range below 3 keV where the X-ray background is contaminated with the local Galactic component. The number–flux relation in the energy range above 3 keV was precisely obtained down to the Einstein medium sensitivity survey level by Ginga (Hayashida 1990; Warwick and Stewart 1989). As seen from Fig.2, the number of X-ray sources is yet insufficient to account for the CXB flux. If the number-flux relation observed by Ginga is extrapolated down to 10^{-14} erg cm^{-2} s^{-1}, the CXB flux can be explained as the collection of weak point sources. Astro-D will have the capability to resolve such sources down to about 10^{-14} erg cm^{-2} s^{-1}.

REFERENCES

Giacconi, R. et al., *Phys. Rev. Letters*, 9, 439 (1962).
Giacconi, R. et al., *Astrophys. J. (Letters)*, 234, L1 (1979).
Gioia, I.M. et al., *Astrophys. J.*, 283, 495 (1984).
Hayashida, K., *Ph.D.Thesis, University of Tokyo*, 1990.
Kondo, H., *Ph.D.Thesis, University of Tokyo*, 1991.
Makino, F. and the ASTRO-C team, *Astro. Lett. and Comm.*, 25, 223 (1987).
Piccinotti, G. et al., *Astrophys. J.*, 253, 485 (1982).
Serlemitsos, P.J., in: *X-Ray Astronomy in 1980's*, NASA TM 83848, 1981, p.441.
Turner, M.J.L. et al., *Publ. Astron. Soc. Japan*, 41, 345 (1989).
Warwick, R. and Stewart, G., in *Proceedings of the 23rd ESLAB Symposium*, ESTEC, Noordwijk 1989, p.727.

X-ray Background Investigations With The X-ray Timing Explorer

RICHARD E. ROTHSCHILD

University of California, San Diego

1 MISSION OVERVIEW

The X-ray Timing Explorer (XTE) is dedicated to the study of the spectral and temporal variability of astrophysical systems in the 2-250 keV x-ray range. These systems encompass phenomena at the extremes of physical conditions (magnetic fields, densities, temperatures), and are associated with several classes of compact objects, ranging from stellar mass white dwarfs and neutron stars to galactic mass black holes. X-rays are the paramount tool for studying these objects, since they provide information on the underlying processes occurring closest to the objects. Previous missions have discovered the existence of many phenomena which XTE will exploit to probe not only the systems involved, but also the structure of the individual collapsed objects themselves.

The XTE instrument complement features low background, continuous 2-250 keV response with large apertures — the 0.63 m^2 Proportional Counter Array (PCA) and the 0.16 m^2 High Energy X-ray Timing Experiment (HEXTE). The PCA and HEXTE form a set of 13 coaligned proportional counters and scintillators viewing the same 1° FWHM portion of the sky, and they are sensitive to the thousand brightest x-ray sources. The large collecting area of the PCA will generate counting rates in excess of 10^5 counts/second for bright sources, and the Experiment Data System (EDS) will not only be capable of processing such high input rates, but will also provide simultaneous scientific processing of all PCA data into multiple output formats. Rapid response (in hours) to temporal phenomena, e.g. transients, is available through constant surveillance of the sky with the All Sky Monitor (ASM), the capability of the dedicated XTE spacecraft to slew rapidly to any point on the sky, and the continual monitoring of XTE data in the XTE Science Operations Center (XSOC). The XTE mission is baselined for at least two years of operations, with launch projected to be at the end of the first quarter of 1996.

2 PROPORTIONAL COUNTER ARRAY (PCA)

The PCA consists of 5 large, multi-anode, multi-layer, collimated, xenon/methane gas proportional counters with net area of 1250 cm^2 each. These detectors are larger versions of the highly successful HEAO-1 Cosmic X-ray Experiment High Energy Detectors (Rothschild et al. 1979) that featured low background with efficient anti-coincidence processing, including side and rear xenon sections with a propane/neon top layer. The PCA is effective over the range 2-60 keV with 18% energy resolution at 6 keV. The 1° FWHM field of view yields a source confusion limit at $\sim$ 0.2 milliCrab.

The Crab nebula will yield 8700 counts/second (2-10 keV) and 1200 counts/second (10-30 keV) in the PCA. The background rates in the same energy ranges are expected to be 20 and 24 counts/second, respectively. Similarly, a 1 milliCrab active galaxy (2-10 keV) with spectral energy index of 0.7 will be detected at 2σ in 1 second (2-10 kev) and at 3 σ in 10 seconds (10-30 keV).

3 HIGH ENERGY X-RAY TIMING EXPERIMENT (HEXTE)

The HEXTE consists of 2 independent clusters of detectors, each cluster containing 4 NaI(Tl)/CsI(Na) scintillation counters in the phoswich configuration, sharing a common 1° FWHM field of view. These detectors are directly descended from the Low Energy Detectors of the HEAO-1 Hard X-ray and Low Energy Gamma Ray Experiment (Matteson 1978). The fields of view of the two clusters are modulated independently, and in orthogonal directions, to provide near real time x-ray background measurements from 4 independent regions surrounding the target source. Each detector has 200 cm^2 net open area, and covers the energy range 15-250 keV with better than 18% energy resolution at 60 keV. Automatic gain control, along with the aperture modulation, will be utilized to allow systematics free ($\leq$0.1% of detector background) observations over the entire energy range.

The Crab nebula will yield 170 counts/second (15-30 keV) and 130 counts/second (30-250 keV) in one cluster. The background in these energy bands will be 6 and 29 counts/second, respectively. A 1 milliCrab active galaxy with 0.7 spectral energy index will be detected at $\geq$3σ in each of a dozen energy resolution elements comprising the 15-120 keV band in 10^5 seconds.

4 ALL SKY MONITOR (ASM)

The ASM consists of three scanning shadow cameras on one rotating mount with a total net effective area of 90 cm^2. Each scanning shadow camera is a one dimensional Dicke camera (Dicke 1968) containing a one dimensional mask and a one dimensional, imaging, xenon/CO_2 proportional counter sensitive in the 2-10 keV range. The gross field of view of a single unit is 6° x 90° FWHM and the angular resolution in the narrow (imaging) direction is 0.°2. Two of the units view perpendicular to the rotation axis with a $\pm$5° inclination of the fan beams to the rotation axis. These crossed fields provide a positional error region of 0.°2 x 2° for a weak source and 3' x 30' for a brighter source. The third camera views along the axis of rotation. It serves in part as a rotation modulation collimator detector, and views the polar area of rotation not blocked by the spacecraft.

The ASM will make a complete revolution once every 90 minutes, and ~80% of the sky will be surveyed to a depth of ~3% of the Crab nebula. About 50 known x-ray sources are bright enough for detection each orbit. Frequent XTE maneuvers will insure that 100% of the sky is surveyed each day. In one day the limiting sensitivity is ~1% of the Crab nebula, or about 75 sources.

5 EXPERIMENT DATA SYSTEM (EDS)

The EDS comprises a high speed digital signal processor front end, microprocessor controlled, data compression and formatting system that will handle the high throughput generated by the PCA and will provide ~10 μs time resolution. The data stream can be binned and telemetered in several modes simultaneously. Each mode can be configured for optimum tradeoffs of timing and spectral resolution in accordance with the specific scientific objectives of each observation. Standard data modes, such as binned spectra, pulse folding, high time resolution event scaling, burst searches, and autocorrelations to 25 μs are planned.

A portion of the EDS is also dedicated to collecting and formatting the ASM data, as well as controlling the motion of the ASM. This data will be scrutinized each orbit in the XSOC to determine if a change has occurred in the x-ray sky and for long term monitoring of the brighter sources.

6 DEDICATED SPACECRAFT

In June of this year NASA decided to provide both XTE and FUSE with dedicated spacecraft and launches on Delta II expendable rockets. The spacecraft are in the process of phase B definition. The present XTE concept calls for twin steerable antennae for nearly constant real time communication between XTE and the ground through the multiple access channel of the TDRS system. For the ~20% of the orbits where TDRS is out of sight or unavailable, sufficient on board memory is provided to allow subsequent transmission to the ground. This large memory also allows PCA high rate telemetry (256 kb/s) to be taken based upon astronomical ephemerides. The dedicated spacecraft design will have the star trackers in close proximity to the detectors and aligned in the general direction of the PCA/HEXTE pointing axis. The spacecraft data system will be based upon that being developed for the NASA small explorers and it will employ the 1773 fiber optic network concept with a 3 Mips processor/coprocessor.

Studies are underway to determine the optimum orbit for XTE by taking into account the need for low and steady ambient particle fluxes (low inclination and low altitude), at least two year lifetime (not so low altitude as the solar maximum nears), Delta II performance, and XTE weight. Orbits outside of the magnetosphere were considered briefly, but the higher cosmic ray rate, solar variability, and, most importantly, the fact that the Delta II could not put the XTE weight in such orbits ruled against that option.

7 XTE SCIENCE OPERATIONS CENTER (XSOC)

XTE scientific operations will be run out of the XSOC located at GSFC on a 7 day per week, 24 hour per day basis. The XSOC will create the observing plan from successful observing proposals, create all scientific command sequences, and monitor all observations. The XSOC staff will also monitor the ASM data every orbit to determine if a dramatic change has occurred in any source, and, if so, will decide whether or not to break off from the present observing schedule to bring the full power of the pointed XTE instruments to bear on this source. Finally, the XSOC personnel will distribute data and analysis software to observers, and will assist them with basic analyses at the XSOC, if desired.

The XTE observing program will be devoted 100% to successful, peer-reviewed proposals. The instrument teams are not guaranteed any observing time and no scientific investigations are set aside for them ahead of time. Proposal evaluation will be conducted by a NASA appointed peer group, and NASA will fund domestic observers. Foreign observers are welcome to participate in the proposal process on a no-exchange-of-funds basis.

8 OBSERVATIONS RELATED TO THE X-RAY BACKGROUND

While XTE is principally designed for temporal studies, its excellent sensitivity and broad band coverage will allow it to pursue studies related to the question of the origin of the diffuse x-ray background.

8.1 Individual Active Galaxies

XTE sensitivity will allow for the study of hundreds of active galaxies and these studies will address essential questions about active galaxy physics, their evolution with redshift, and their contribution to the diffuse background. There are 32 high latitude Seyfert galaxies above $\sim$1 milliCrab in intensity in the HEAO-1 A2 survey (Piccinotti et al. 1982). HEAO-1 A1 data (Wood et al. 1984, Remillard, 1986), as well as scaling of the logN-logS for extragalactic objects, indicates 3-4 times as many objects are available at a flux greater than $\sim$0.6 milliCrab. These include 34 low (z<0.2) redshift quasars. At flux levels of $\sim$0.1 milliCrab diffuse background fluctuations will begin to be important, and at this level the logN-logS distribution of local extragalactic objects (active galaxies and clusters) implies about one source per 7 square degrees. Consequently, a range of redshifts and active galaxy types will be available for study by XTE.

With its extended energy coverage, XTE will be able to test the notion that active galaxy spectra must steepen around 100 keV or else they would overproduce the diffuse background at higher energies. Schwartz and Tucker (1988) have put forth a model that redshifts spectral breaks up to z=3 to produce the observed roll over in the diffuse flux at 40 keV. XTE will be able to observe individual galaxies and quasars to look for such steepenings in the spectra.

8.2 Reflection Spectra Of Active Galaxies

Guilbert and Rees (1988), Lightman and White (1988) and Rogers and Fields (1991) have pointed out the importance of reprocessing and absorption of x-rays in the vicinity of active galaxies. Their predictions call for a flattening of the "canonical" 0.7 power law spectra above 20 keV with a maximum in this additional component around 100 keV. Furthermore the presence of a 6.4 keV line from the fluorescence of cold iron and its attendant absorption edge at 7.1 keV are strong diagnostics of the covering factor in these sources. Ginga observations are just beginning to see such effects up to 20 keV (Pounds et al. 1990). XTE will be able to provide sensitive fits to reprocessing models for many active galaxies, and one will be able to begin to look for predicted differences between classes of active galaxies and for evolutionary effects.

8.3 Spectra Of Fluctuations Of The X-ray Background

A natural consequence of the HEXTE aperture modulation for background subtraction will be sets of off-source data surrounding most of the XTE targets. One will be able to difference these data sets, after eliminating those with known or discovered sources in the off-source fields, to produce the energy spectrum of the fluctuations in the diffuse flux for that position on the sky. The integral fluctuation spectrum for high latitudes, as well as individual spectra as a function of position can then be compared to that of the diffuse flux at other energies. Such analyses can then limit models of the origin of the diffuse flux, and aid in the refinement of successful models.

9 ACKNOWLEDGEMENTS The XTE mission is a collaboration of efforts by the instrument teams lead at MIT by Hale Bradt, at UCSD by Rick Rothschild, and at GSFC by Jean Swank, by the XTE project office at GSFC lead by Dale Schulz, and by NASA's Office of Space Science and Applications' Astrophysics Division lead by Charles Pellerin. The work at UCSD is performed under NASA contract NAS5-

30720.

10 REFERENCES

Dicke, R. 1968, *Ap. J. (Letters)*, **153**, L101.

Guilbert, P.W. and Rees, M.J. 1988, *MNRAS*, **233**, 475.

Lightman, A.P. and White, T.R. 1988, *Ap. J.*, **335**, 57.

Matteson, J.L. 1978, AIAA 16th Aerospace Sciences Meeting, paper 78-35 (unpublished).

Piccinotti, G. et al. 1982, *Ap. J.*, **263**, 485.

Pounds, K.A. et al. 1990, *Nature*, **344**, 132.

Remillard, R. 1986, *Ap. J.*, **301**, 742.

Rogers, R.D. and Field, G.B. 1991, *Ap J. (Letters)*, **378**, L17.

Rothschild, R.E. et al. 1979, *Space Sci. Inst.*, **4**, 269.

Schwartz, D.A. and Tucker, W.H. 1988, *Ap. J.*, **332**, 157.

Wood, K.S. et al. 1984, *Ap. J. Suppl.*, **56**, 507.

The Diffuse X-Ray Spectrometer Experiment

W. T. SANDERS and R. J. EDGAR

University of Wisconsin - Madison

1 INTRODUCTION

The Diffuse X-ray Spectrometer (DXS) experiment is designed to obtain medium resolution spectra ($\Delta E \approx 0.01$ keV) of the soft x-ray diffuse background in the energy range 0.15 - 0.28 keV. The soft x-ray diffuse background is discussed by D. McCammon (this volume) and was reviewed by McCammon and Sanders (1990). The DXS field of view is 15° x 15° (FWHM) and it will measure the spectrum of ten independent regions of the sky. DXS is to be flown as an attached payload aboard NASA's Space Shuttle in 1992.

2 EXPERIMENT DESCRIPTION

The DXS experiment consists of two Bragg crystal spectrometers, one mounted on each side of the Shuttle cargo bay. Each spectrometer has a detector assembly that uses a large area ($\approx$ 30 cm x 60 cm) array of lead stearate (PbSt) Bragg "crystals" to disperse incident x-rays across the face of a position-sensitive proportional counter. The spectrometers communicate with the Shuttle and the ground through the SPOC (Shuttle Payload of Opportunity Carrier) Avionics electronics package mounted on the starboard side of the Orbiter, next to the starboard DXS instrument. Figure 1 shows a DXS detector assembly mounted on a SPOC plate, which is then attached to the Shuttle. Other major assemblies are: the Control Electronics Assembly, the Power Converter Subsystem, the high pressure P-10 gas bottle, the detector latch, the radiation monitor (turns the proportional counter high voltage off and on when the count rate exceeds or falls below an adjustable threshold), the sun sensor (keeps the detector stowed when the experiment is in sunlight), the aperture seal (prevents sunlight from shining on the crystal panel when the detector is stowed), and the rotation motor (to scan the spectrometer field of view across the sky).

A cross sectional view of the spectrometer's detector assembly is given in Figure 2. The assembly extends more than 60 cm perpendicular to the plane of the page. X-rays enter the detector assembly through the aperture, are Bragg-reflected off the crystal panel, and pass through the collimator and thin window into the proportional counter, which is position-sensitive in one dimension. The magnets are to reject low energy (E < 20 keV) electrons. The Bragg crystal panel is a thin plastic sheet whose inner surface is coated with 200 layers of PbSt, producing a one-dimensional pseudo-crystal. The reflecting layers of lead have spacing 2d=101 Å. The PbSt may be destroyed by solar ultraviolet radiation and by atmospheric oxygen atoms hitting it at the Shuttle orbital velocity.

The DXS collimator is of a simple crossed slat design and restricts the field of view to 15° x 15° (FWHM), although at any one time x-rays are typically received from 4 or 5 different resolution elements. It supports a 100 line-per-inch nickel mesh, which supports the proportional counter thin window against the P-10 counting gas at a pressure of one atmosphere. Each counter has a large-area entrance window made of Formvar and a UV-absorbing agent, UV24, with a total mass thickness of about 90 $\mu g\ cm^{-2}$. The proportional counter has a front layer where the x-rays of interest are absorbed, and a back-plus-side veto layer, operated in anticoincidence with the front layer. Each layer contains an anode whose high voltage ($\approx$ 1700 V) is turned off whenever the orbiter is in the South Atlantic Anomaly (SAA). Between the two gas volumes is a plane of wires running perpendicular to the plane of the page and maintained at ground potential. The distribution of induced electrical charges among these ground plane wires allows the determination of the position of the incident x-ray across the counter, which is the dispersion direction of the Bragg crystals. The spatial resolution of the detector is about 0.5 mm, but the energy resolution of the spectrometer is determined by the 15° acceptance angle of the collimator.

3 EXPERIMENT OPERATION

The operation of the detector is as follows. Any one ground plane segment is constrained by the collimator to receive x-rays that are reflected from a particular segment of the crystal panel. The x-rays that are Bragg-reflected from that segment of the crystal panel have a particular angle of reflection, equal to their angle of incidence, θ, and a particular wave-

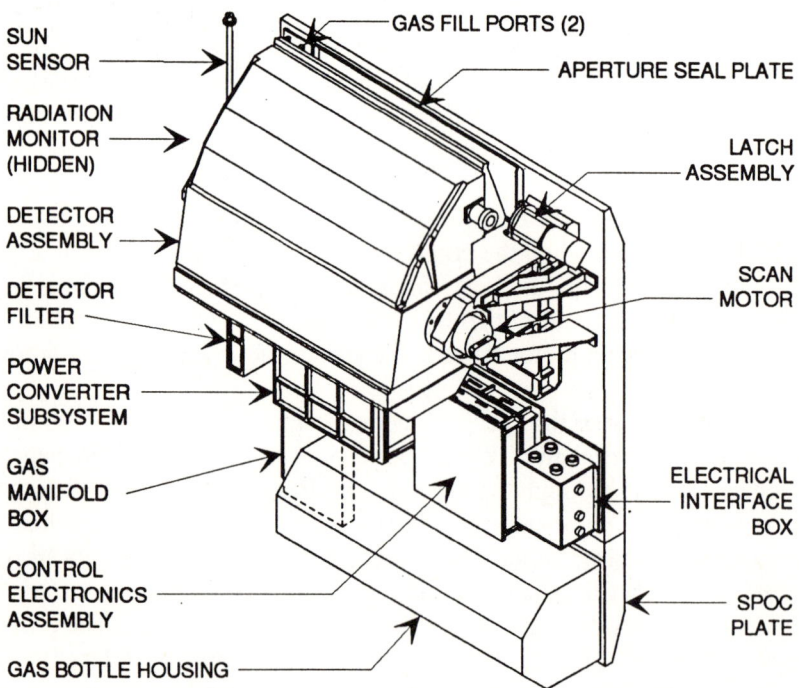

Figure 1 -- Isometric drawing of one of two DXS Bragg crystal spectrometers

length that satisfies the Bragg condition, $\lambda = 2d \sin\theta$. Given the position across the counter of a detected x-ray, both its wavelength and direction of origin on the sky are determined. The spectrum of the diffuse background is dispersed across the proportional counter at all times, but x-rays of different wavelengths come from different regions of the sky.

To get a complete spectrum from a single region of the sky, the detector assembly must be rotated through an angle $\approx 60°$. As the detector assembly is rotated, additional partial spectra are obtained from the adjacent sky regions, and as the rotation is extended beyond 60°, complete spectra from additional regions are obtained. Ideally, this instrument would be flown on a spinning satellite, continuously collecting data while rotating through 360°. Since DXS is Shuttle-mounted, the best it can do is rotate back and forth with a 157° view of the sky.

From our computer modeling, we find that the *minimum* useful exposure per 15° x 15° resolution element is 5000 seconds (and more than that is desirable). For ten resolution elements, DXS requires 50,000 seconds of good observing time during which the Shuttle maintains the same inertial attitude. Furthermore, these 50,000 seconds must be obtained

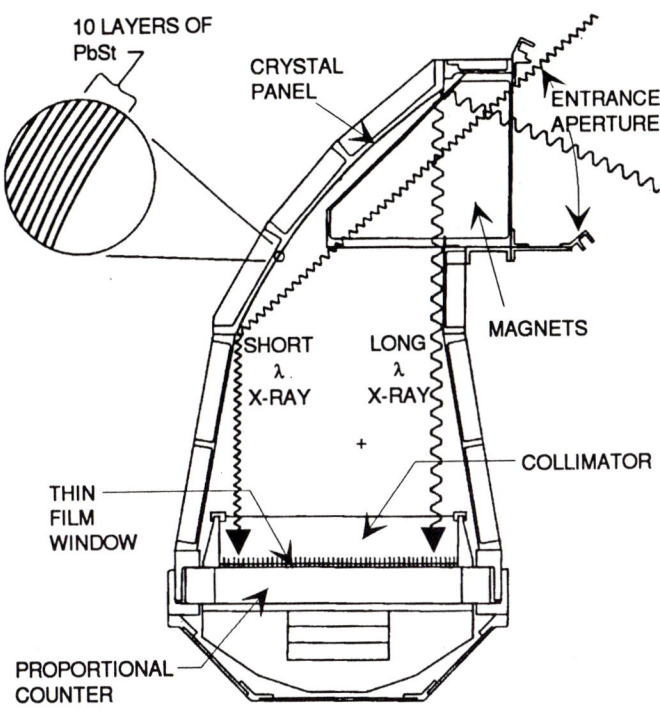

Figure 2 -- Cross sectional view of a DXS detector assembly

(a) at night so that the sunlight does not destroy the crystals, (b) when atmospheric oxygen is not directed into the detector aperture so the atomic oxygen won't erode away the crystals, and (c) when the Shuttle is not in the SAA, which is much larger for low energy x-ray instruments than it is for higher energy ones. Approximately 5 days on orbit are required to obtain the DXS data.

During the time that DXS is on-orbit, the science team operates out of a Payload Operations Control Center (POCC) at Goddard Space Flight Center, uploading time-tagged commands to the instruments and monitoring the data, which is telemetered to the ground in near real time. The proportional counters collect data for 10-20 minutes each orbit and the proportional counter gain is periodically checked by an on-board Al Kα X-ray tube.

The detector response function, including PbSt crystal geometry and efficiency, collimator effects, and mesh and window transmission is shown in Figure 3. When a typical diffuse background spectrum, a 10^6 K thermal plasma (Raymond & Smith 1977, 1987; hereafter RS) plus an absorbed ($N_H = 10^{20}$ cm^{-2}) 3×10^6 K (RS) plasma and an absorbed ($N_H = 5 \times 10^{20}$ cm^{-2}) 11 E$^{-1.4}$ ph cm^{-2} s^{-1} sr^{-1} keV^{-1} spectrum, is modeled with this response function, integrated for 10^4 seconds, and randomized assuming Poisson statistics, the result is as shown in Figure 4. The wavelength resolution is 2-3 Å, corresponding to about 0.01 keV, with a variation of a factor of three across the DXS bandpass. The

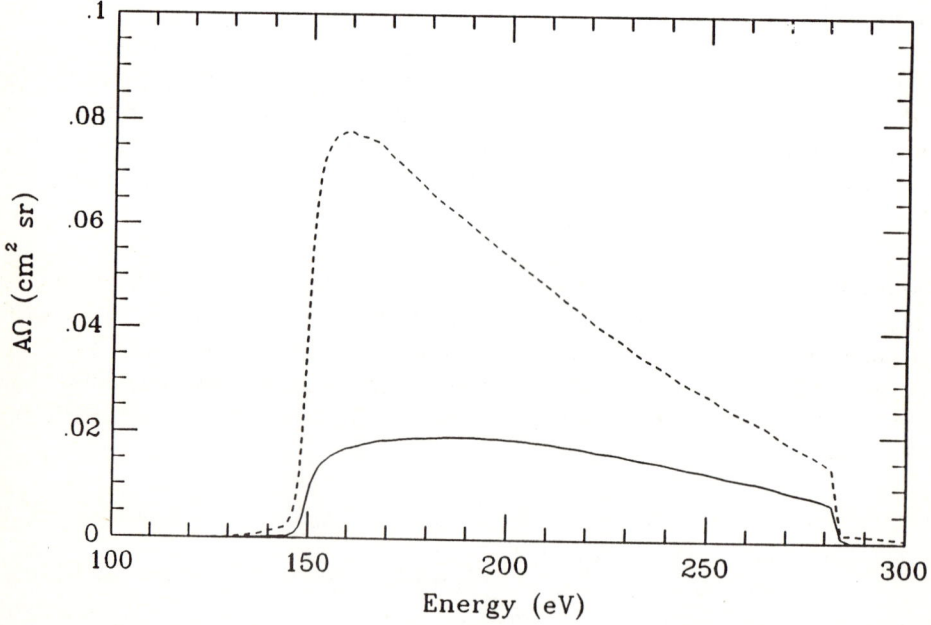

Figure 3 -- Response function for one DXS as a function of photon energy
(solid line); not including window transmission (dashed line)

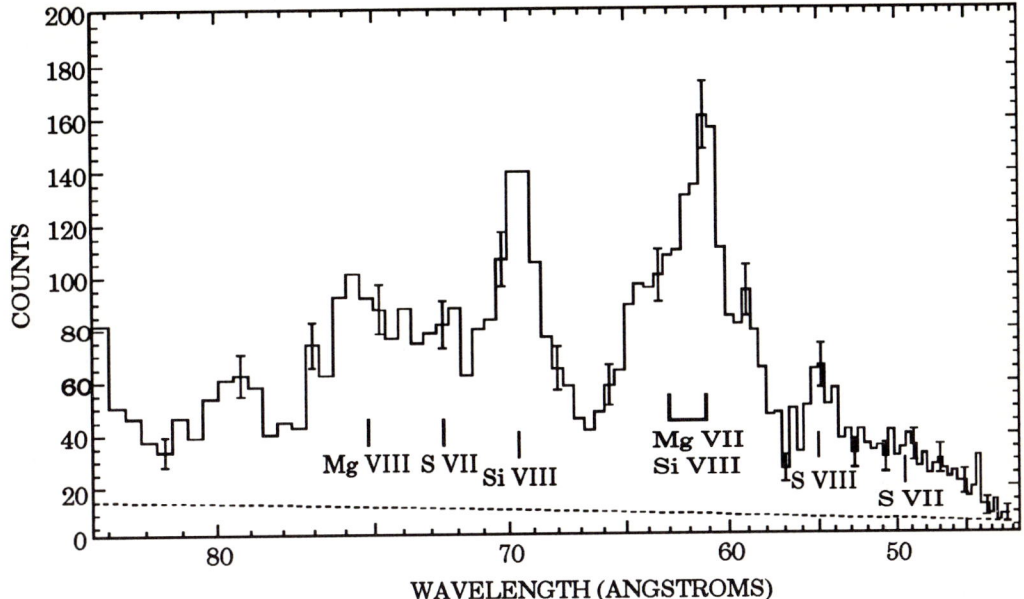

Figure 4 -- Diffuse background model spectrum folded through the DXS
response function

minimum detectable line in this spectrum has about 100 counts, corresponding to a
sensitivity of 0.5 ph cm^{-2} s^{-1} sr^{-1}.

Until recently, NASA also planned to fly a set of instruments similar to DXS on the Space
Station. This mission was called the X-ray Background Survey Spectrometer (XBSS)
and was designed to obtain spectra of resolution similar to that of DXS, but from the
whole sky and over a broader bandpass, 0.15 - 1.1 keV. However, in August 1991,
XBSS was "deselected" as a Space Station payload.

DXS was proposed by W. L. Kraushaar in 1978 and he has guided it since its inception.
The design and construction of the proportional counters were the work of Dan
McCammon and Kurt Jaehnig. Steve Snowden, Jeff Bloch and the staff of the University
of Wisconsin-Madison's Space Science and Engineering Center have contributed greatly
to the progress of DXS. This work was supported by NASA contracts NAS 5-26078 and
NAS 8-38664.

REFERENCES

Raymond, J. C., and Smith, B. W. 1977, ApJ Suppl, 35, 419
------, 1987, private communication (update of Raymond and Smith 1977)
McCammon, D., and Sanders, W. T. 1990, Annu. Rev. Astron. Astrophys., 28, 657

Chapter VIII

Summary

THE X-RAY BACKGROUND: SUMMARY

A.C. FABIAN

Institute of Astronomy, Madingley Road, Cambridge CB3 0HA, U.K.

1 INTRODUCTION

We have three observational handles on the X-ray Background; its spectrum, isotropy and the 'resolved component'. New results were presented at the Workshop on all of these topics. In summarizing the results, I will first review what we mean by the 'X-ray Background', for it is emerging that there may be more than one such distinct background. Different sources may dominate in different bands.

Three energy bands that now appear to divide it in this way are the soft band (photon energies $\epsilon \lesssim 2\,\mathrm{keV}$), the hard band ($2 \lesssim \epsilon \lesssim 100\,\mathrm{keV}$) and the soft gamma-ray band ($100\,\mathrm{keV} \lesssim \epsilon \lesssim 3\,\mathrm{MeV}$). *Einstein Observatory* and ROSAT data show that the soft band has a strong contribution from quasars. Simple arguments made over ten years ago for the soft gamma-ray band show that it is plausibly due to the hard emission from Active Galactic Nuclei (AGN) at low redshift (*e.g.* Bignami *et al.* 1979). It is the hard band, which dominates the energy density of the X-ray Background, that is largely still unexplained. Several arguments reviewed later in this Summary suggest that it cannot have either quasars or low redshift AGN as major contributors.

It is important to consider the whole X-ray energy band when reviewing the X-ray Background and its origin. It covers at least 4 decades of photon energy, equivalent to the range from the IRAS 100μ band to IUE/HST measurements at 1200 Å. No one would attempt to make estimates of the UV background based upon IRAS measurements without a very good model for the spectral energy distribution over those 4 decades. No one should make estimates of the hard and soft gamma-ray backgrounds without similarly detailed knowledge of the source spectra.

1.1 The Spectrum

The overall spectrum of the XRB has been well-reviewed by McCammon, Boldt, Gruber and others here. We still do not know exactly how and where the spectrum steepens below $\epsilon \sim 5\,\mathrm{keV}$. An important point to remember, however, is that the bulk of the energy density is found around 30 keV (Fig. 1).

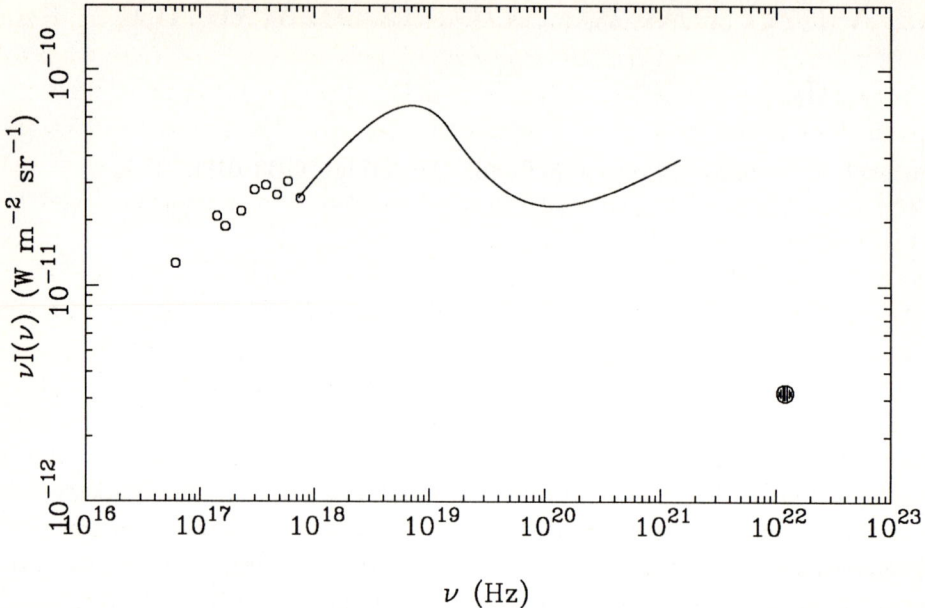

Figure 1. $\nu I(\nu)$ spectrum of the whole XRB, using the soft X-ray data points of Wu *et al.* (1991), Gruber's formula for the 3 keV to 10 MeV band and the SAS II point at 100 MeV.

Two assumptions are commonly made about the origin of the spectrum; there is either only one dominant population of source (*e.g.* the model of Schwarz & Tucker 1988), or there are several populations. In the first case, the sources observed with ROSAT are the source of the 30 keV XRB; in the second, it originates from sources which are unresolved (and perhaps even unresolvable) by ROSAT.

1.2 Isotropy

The XRB is smooth. It is not as smooth as the Microwave Background, but all observed graininess can be ascribed to known features of the nearby Universe. On scales of 180°, through $2 - 10°$ to about 10 arcmin, the mean upper limit to $\delta I/I$ varies from less than 1 per cent to about 3 per cent and back to a few per cent, respectively.

1.3 The Resolved Component

At $\epsilon \sim 1\,\mathrm{keV}$, between $30 - 40$ per cent of the observed intensity of the XRB is resolved by ROSAT observations (see the contributions by Burg, Griffiths, Hasinger and Shanks *et al.* 1991). In the $2 - 10$ keV band, only a few per cent are yet resolved and about 20 per cent is probed with fluctuation analyses. Of course this band has not yet been exposed to detailed imaging studies. We know that less than about 15 per cent of that XRB can be due to non-evolving local sources from the cross-correlation analyses of Lahav, Boldt *et al.* Of the 30 keV XRB, where most of the energy density resides, we know very little.

It remains possible that the soft extragalactic XRB resolvable with ROSAT is just an extra background due to quasars. This is a likely conclusion in my view. The direct counts of sources made in the ROSAT images indicate a flattening of the counts around $S \sim 10^{-14}$ erg cm^{-2} s^{-1}, as predicted from fluctuation $P(D)$ analyses and also from optical quasar counts (*e.g.* Schmidt & Green 1986). This means that the resolvable ROSAT XRB cannot be simply extrapolated to account all the intensity, even in that band. The confirmation of the fluctuation work (Hamilton & Helfand 1987; Barcons & Fabian 1990) by direct ROSAT counts also shows the power of the $P(D)$ approach for statistical studies. Indeed most of the *statistical* results obtained so far about the soft XRB from ROSAT are consistent with the results obtained from the *Einstein Observatory*.

If we assume that quasars do not contribute more than 50 per cent of the 1 keV XRB, then the residual background probably has an energy spectral index $\alpha \sim 0.2 - 0.3$. It is steeper than the spectrum obtained by cooling electrons ($\alpha > 0.5$). *It implies that we are dealing with sources which are optically–thick or in which photons are reprocessed.* We could manage with non-cooling electrons, but the best case for that, the hot IGM model, is ruled out by the COBE data (Mather *et al.* 1990).

2 NEW RESULTS PRESENTED AT THE WORKSHOP
I begin this section by listing what I recall as the new results reported at the Workshop;

2.1 The Soft XRB
• About 35% of the XRB intensity at 1 keV is resolved, mostly into quasars. The counts of the resolved sources flatten at low fluxes ($\sim 10^{14}$ erg cm^{-2} s^{-1}).
• The remaining XRB is so smooth that $\lesssim 10\%$ can be due to any source with a correlation length $r_0 > 25h^{-1}$ Mpc and $< 15\%$ with $r_0 > 8h^{-1}$ Mpc.
• The energy index of the observed sources $\alpha_{source} \sim 1.2$, whereas the index of the XRB $\alpha_{softXB} \sim 0.6$.
• Shadows are seen in the 0.25 keV XRB due to Galactic clouds, meaning that $\lesssim 10\%$ of the intensity at that energy can be extragalactic.
• Blotches have been seen in some ROSAT fields (*e.g.* the NEP ones; Hasinger, Burg *et al.*) which are probably due to low luminosity clusters. Perhaps the cluster

contribution to the residual soft XRB is high, but the AutoCorrelation Function (ACF) is low enough ($\lesssim 10^{-3}$) to argue against clusters.

• Einstein Observatory and ROSAT counts and spectra agree? Starbursts are common? (These points are debatable.)

• The evolution of quasars and AGN is confirmed to be slower in X-rays than in other bands.

• The Lockman hole is not observed to be bright in diffuse soft X-rays. However, the carbon-band soft XRB is sufficiently intense across the Sky, and the spectra of quasars not so steep, that an extremely low hydrogen column density does not particularly help in maximizing the number of detected sources in a deep field.

2.2 The Hard XRB

• New spectral fits have been made to the total intensity (Boldt, Gruber) and to the fluctuation spectrum (Stewart, Hayashida). The fluctuation spectral index $\alpha_{fln} \sim 0.7$ at $\epsilon < 10\,\mathrm{keV}$ is steeper than the XRB spectrum, but it appears to flatten to $\alpha_{fln} \sim 0.4$ above 10 keV. There may be an iron line in the fluctuation spectrum. The spectrum of the fluctuations is consistent with their origin in unresolved AGN of spectra comparable to those which are resolved in this band.

• There appears to be a real discrepancy between extrapolations of the EMSS and ROSAT source counts to the GINGA source counts, using reasonable, unabsorbed spectral shapes. Intrinsic absorption in AGN is considered by many to be the solution.

• The final mass of the black holes accreting with an efficiency $10\eta_{0.1}\%$, assuming that they make the XRB, $M_{BH} \sim 10^8 \eta_{0.1}(H_0/50\,\mathrm{km\,s^{-1}\,Mpc^{-1}})^{-3}\,\mathrm{M_\odot}$ (Daly).

• The cross-correlation amplitude between the intensity of the XRB and bright galaxies, $W_{Xg} = (3 \pm 1) \times 10^{-3}$ (Boldt, Lahav). A positive detection of the ACF has been reported (Mushotzky, Jahoda) at $W \sim 3 \times 10^{-5}$ on a scale of abut $10°$. The low limits and detection of W mean that $\lesssim 30\%$ of the total intensity can originate from a population with $r_0 \gtrsim 8h^{-1}\,\mathrm{Mpc}$.

• The spectra of (the few measured) quasars in the $2 - 10$ keV band has a spectral index $\alpha_{quasar} \sim 0.8$, significantly steeper than the observed spectrum. Starbursts also have steep spectra.

• Voids are detected in maps of the XRB, at a level of about 2%.

2.3 Comments

The residual XRB is very smooth in all wavebands. This may imply an origin in low luminosity galaxies at high redshift, which appear to cluster less than higher luminosity ones. It is interesting to note here that optical studies of faint galaxies show little clustering (Efstathiou *et al.* 1991). Certainly, the low cross-correlation signal between the XRB amd bright galaxies shows that strong evolution is required for any source of the XRB associated with galaxies.

We see evidence for nearby voids and superclusters, but not for distant ones. This may be weak evidence for recent formation of those features.

What is the ACF of the ROSAT–detected quasars and sources? If they show any significant clustering signal, then the fluctuation results on yet fainter sources probably *under*estimate the flattening of the source counts. This is also true if the residual background is clustered.

All direct evidence for the contribution of quasars to the hard XRB indicates that it is small. The spectra of quasars and their apparent optical clustering both argue against them being the source of a hard, smooth background.

Any intergalactic medium with a temperature of $T \gtrsim 3 \times 10^6$ K cannot be clumpy on scales greater than a few Mpc, without making the ROSAT images blotchier than they are. This is contrary to recent suggestions of Cen *et al.* (1990).

Can limits be set on dissipation in galaxy and large-scale structure formation from the soft XRB below 1/4 kev? This could potentially be a strong limit if absorption can be ignored.

3 DOES REFLECTION EXPLAIN THE RESIDUAL XRB?

The shape of the hard XRB, with a break at about 30 keV, requires that most of the constituent sources share a common spectrum. Otherwise the resultant spectrum could not break as sharply as observed. This suggests that some simple physical parameter controls that spectrum. We have suggested that it is the electron rest mass, expressed through the electron rest mass in sources where the reflected component dominates (Fabian *et al.* 1990). This model has been further elaborated by Rogers & Field (1991) and by Terasawa (1991).

A power-law, hard X-ray spectrum (the direct component) incident onto 'cold' gas will reflect a characteristic spectrum (the reflected component) with a break at about 150 keV. This is the basis of the model. What is puzzling is why the reflected fraction dominates over the direct fraction of the luminosity. It can be overcome by arguing that the direct fraction is anisotropic, so that little flux is emitted in our direction, which is away from the cold gas. Anisotropy can be induced in a relatively simple manner by the inverse Compton process (Ghisellini *et al.* 1991; Rogers) or by invoking beaming of the direct emission (*e.g.* Field). There is a problem in some of the modified models in that they are beginning to appear contrived. What we need is to observe a source with the required spectrum.

At present there may be strong selection against detecting such a source, since it may emit little flux below 20 keV. AGN studies with the Compton Observatory, Galactic Black Hole sources and of bright UV or IR sources might help to select candidate objects. Much of the emission in the reflection model is absorbed by the cold gas and must be re-emitted in the UV, or the IR if there is much dust in the source. In this sense the XRB will be strongly linked to the UV or IR backgrounds. The inefficiency with which the sources produce X-rays also means that black holes so produced will be more massive than assumed if $\eta \sim 0.1$.

4 THE FUTURE

In some ways, the problem of the origin of the XRB remains as intractable as ever. An excess component in the soft band has been identified and mostly resolved into quasars. All but 10 – 20% of the remaining spectrum is unexplained. It may also be due to quasars and luminous AGN, but the crucial linking evidence is missing and may not exist.

The XRB has strong potential for determining the integral properties of all X-ray emitting objects in the Universe. It constrains the lumpiness, $\delta\rho/\rho$ as a function of comoving scale and will eventually be used to measure its evolution with redshift. More new results can still be expected from HEAO-1 as well as GINGA, BBXRT and of course ROSAT. The next generation of orbiting telescopes will give us data on the hard XRB to rival what we now have in the soft band. ASTRO-D, Spectrum–X and later missions will make the next Workshop on the XRB as interesting as this one.

REFERENCES

I have not given explicit references where they are to elsewhere in this Volume.

Barcons X. & Fabian, A.C., 1990. *Mon. Not. R. astr. Soc.*, 243, 366.

Bignami, G.F., Fichtel, C.E., Hartman, R.C. & Thompson, D.J., 1979. *Astrophys. J.*, 232, 649.

Cen, R.Y., Jameson, A., Liu, F. & Ostriker, J.P., 1990. *Astrophys. J.*, 362, L41.

Efstathiou, G., Bernstein, G., Katz, N., Tyson, J.A. & Guathakurta, P., 1991. *Astrophys. J.*, 380, L47.

Fabian, A.C., George, I.M., Miyoshi, S. & Rees, M.J., 1990. *Mon. Not. R. astr. Soc.*, 242, 14P.

Ghisellini, G. George, I.M. & Fabian, A.C., 1990. *Mon. Not. R. astr. Soc.*, 248, 14.

Hamilton, T.T. & Helfand, D.J., 1987. *Astrophys. J.*, 318, 93.

Mather, J.C. *et al.* 1990. *Astrophys. J.*, 354, L37.

Rogers, R.D. & Field, G.B., 1991. *Astrophys. J.*, 370, L57.

Schmidt, M. & Green, R.F., 1986. *Astrophys. J.*, 305, 68.

Schwartz, D.A. & Tucker, W.H., 1988. *Astrophys. J.*, 332, 157.

Shanks, T., Georgantopoulos, I., Stewart, G.C., Pounds, K.A., Boyle, B.J. & Griffiths, R.E., 1991. *Nature*, 353, 315.

Terasawa, N., 1991. *Astrophys. J.*, 378, L11.

Wu, X. Hamilton, T., Helfand, D.J. & Wang, Q., 1990. *Astrophys. J.*, 379, 564.